Studies in Energy, Resource and Environmental Economics

Series Editors

Georg Erdmann, Former Chair of Energy Systems, Technical University of Berlin, Oberrieden, Zürich, Switzerland

Anne Neumann , Norwegian University of Science and Technology, Trondheim, Norway

Andreas Loeschel , Ruhr-Universität Bochum, Bochum, Nordrhein-Westfalen, Germany

This book series offers an outlet for cutting-edge research on all areas of energy, environmental and resource economics. The series welcomes theoretically sound and empirically robust monographs and edited volumes, as well as textbooks and handbooks from various disciplines and approaches on topics such as energy and resource markets, the economics of climate change, environmental evaluation, policy issues, and related fields. All titles in the series are peer-reviewed.

Rainer Quitzow · Yana Zabanova
Editors

The Geopolitics of Hydrogen

Volume 1: European Strategies in Global
Perspective

 Springer

Editors
Rainer Quitzow
Research Institute for Sustainability (RIFS)
Helmholtz Centre Potsdam
Potsdam, Germany

Yana Zabanova
Research Institute for Sustainability (RIFS)
Helmholtz Centre Potsdam
Potsdam, Germany

Technische Universität Berlin
Berlin, Germany

ISSN 2731-3409   ISSN 2731-3417 (electronic)
Studies in Energy, Resource and Environmental Economics
ISBN 978-3-031-59514-1   ISBN 978-3-031-59515-8 (eBook)
https://doi.org/10.1007/978-3-031-59515-8

This Springer imprint is published by the registered company Springer Nature Switzerland AG
The registered company address is: Gewerbestrasse 11, 6330 Cham, Switzerland

If disposing of this product, please recycle the paper.

Acknowledgments

The publication of this edited volume was financially supported by the German Federal Foreign Office within the framework of the project "Geopolitics of the Energy Transformation—Implications of an International Hydrogen Economy" (GET Hydrogen), funding reference number AA4521G125.

Contents

Contributors

Ines Bouacida Institute for Sustainable Development and International Relations (IDDRI), Sciences Po, Paris, France

Stefan Ćetković Institute of Political Science, Leiden University, The Hague, Netherlands

Per Ove Eikeland The Fridtjof Nansen Institute, Lysaker, Norway

Gonzalo Escribano Energy and Climate Program, Elcano Royal Institute, Madrid, Spain;
Department of Applied Economics, Universidad Nacional de Educación a Distancia-UNED, Madrid, Spain

Tor Håkon Jackson Inderberg The Fridtjof Nansen Institute, Lysaker, Norway

Mari Lie Larsen The Fridtjof Nansen Institute, Lysaker, Norway

Almudena Nunez Research Institute for Sustainability (RIFS), Helmholtz Centre Potsdam, Potsdam, Germany

Andrea Prontera University of Macerata, Macerata, Italy

Rainer Quitzow Research Institute for Sustainability (RIFS), Helmholtz Centre Potsdam, Potsdam, Germany;
Technische Universität Berlin, Berlin, Germany

Jon Birger Skjærseth The Fridtjof Nansen Institute, Lysaker, Norway

Michał Smoleń Instrat Foundation, Warsaw, Poland

Roelof Stam Centre for International Energy Policy (CIEP), The Hague, The Netherlands

Pier Stapersma Centre for International Energy Policy (CIEP), The Hague, The Netherlands

Janek Stockburger School of Politics and Public Policy, Technical University of Munich, Munich, Germany

John Szabo Institute of World Economics, Centre for Economic and Regional Studies, HUN-REN, Budapest, Hungary;
Brunel University London, London, UK

Ignacio Urbasos Energy and Climate Program, Elcano Royal Institute, Madrid, Spain

Coby van der Linde Centre for International Energy Policy (CIEP), The Hague, The Netherlands

Yana Zabanova Research Institute for Sustainability (RIFS), Helmholtz Centre Potsdam, Potsdam, Germany

Wojciech Żelisko Instrat Foundation, Warsaw, Poland

Introduction

Rainer Quitzow and **Yana Zabanova**

Abstract This introductory chapter places Europe's hydrogen ambition in the broader context of the evolving geopolitics of the transition to net zero. It highlights the growing geoeconomic rivalry among the world's leading economies, such as the European Union, the United States, and China. This process is marked by the resurgence of state intervention in markets and industries as well as by the increasing attention paid by governments to supply chain resilience and asymmetric dependencies. Clean hydrogen has been part and parcel of these developments. The EU and many of its Member States view hydrogen as essential to their climate goals, industrial competitiveness, and energy security. As a result, Europe has played an active role in promoting a European hydrogen economy and an international hydrogen market. The chapter frames EU hydrogen policy as the interplay of interests at EU- and Member State level, which can be a source of both tensions and synergies. It also discusses how this is manifested in the EU's external climate and energy policy and international partnerships. Finally, the chapter presents the structure of the edited volume, introduces the case studies and summarises the key analytical dimensions applied in individual chapters to examine the domestic and international components of European hydrogen policy.

1 Geopolitics of the Energy Transition

The transition to net zero involves a profound structural transformation of all economic sectors, going far beyond its initial focus on decarbonising the power sector. This transformation goes hand-in-hand with major changes in how access and control over energy resources influence the distribution of power in the global economy. While the spatial distribution of fossil fuels played a pivotal role in shaping

R. Quitzow (✉) · Y. Zabanova
Research Institute for Sustainability (RIFS), Helmholtz Centre Potsdam, Berliner Str. 130, 14467 Potsdam, Germany
e-mail: rainer.quitzow@rifs-potsdam.de

R. Quitzow
Technische Universität Berlin, Berlin, Germany

© Helmholtz-Zentrum Potsdam, Deutsches GeoForschungsZentrum GFZ 2024
R. Quitzow and Y. Zabanova (eds.), *The Geopolitics of Hydrogen*, Studies in Energy,
Resource and Environmental Economics, https://doi.org/10.1007/978-3-031-59515-8_1

power dynamics and inter-state relations during the twentieth century, the decarbonisation of the global economy is unleashing a shift to renewable energy resources as the new backbone of the energy system, with significant geopolitical and geoeconomic implications (Siddi, 2023; Vakulchuk et al., 2020). For countries and regions dependent on hydrocarbon rents, these developments represent a direct challenge not only to their economic model but also to the social and political model underpinning the distribution of power and influence in their societies (Goldthau & Westphal, 2019; Tagliapietra, 2019; Van de Graaf, 2023). Hence, this transition is also hotly contested with incumbent actors resisting change at different levels of governance (Geels, 2014), ranging from local protests against coal mine closures to intense international negotiations over the future role of fossil fuels in a carbon–neutral world economy.

But this transition not only implies a shift in the resource base that powers the global economy. Rather, it is a far-reaching transformation of socio-technical systems as climate-friendly technologies begin to replace carbon-intensive technologies and supply chains (Rohracher, 2018). This transition unlocks new opportunities for industrial development and value creation with countries and economic blocs vying for technological leadership (Goldthau, 2021). It opens up the potential for capturing early mover advantages in an emerging carbon–neutral economy and for positioning local industries in related value chains (Lema et al., 2020). In particular, China and other leading emerging economies have seen the rise of green industries as a window of opportunity for pursuing strategies to secure industrial leadership. Indeed, the rise of renewable energy technologies, the most prominent success story to date, has been accompanied by the rapid growth of Chinese renewable energy industries, challenging Western dominance of the global economy (Quitzow, 2015).

China's increasing success in dominating a growing number of emerging green industries is closely related to a deterioration of the relationship between China and Western market economies, with geopolitical considerations progressively overshadowing the principles of economic efficiency and market liberalisation. The US increasingly views China's growing industrial capabilities as a challenge to the existing economic order and US hegemony (Lippert & Perthes, 2020). China and the US have engaged in ever more hostile trade-related disputes, and multilateral fora for engagement are under strain (Kwan, 2019). More generally, the rise of Chinese state-centred capitalism and its growing influence around the world has raised concerns among OECD countries of asymmetric dependencies and waning industrial competitiveness (Gertz & Evers, 2020). Specifically, frontrunners in the development of renewable energy technologies, including the US and Europe, are increasingly disillusioned by the shrinking market shares of Western firms in the sector.

Major crises and shocks like the Covid-19 pandemic and the war in Ukraine have further reinforced these concerns. They have led to severe disruptions in critical supply chains, with the result that leading economies around the world are now paying far greater attention to geopolitical risks (Cui et al., 2023; Pujawan & Bah, 2022). Raising the resilience of critical supply chains—through reshoring, friend-shoring and the diversification of suppliers—is becoming a priority for many governments

(Maihold, 2022). This in turn is beginning to erode the principles of free trade and open markets that have dominated global discourse over the past decades.

More generally, this rise of geoeconomic rivalry coupled with the need to rapidly decarbonise the economy is leading to a re-emergence of state intervention in markets and industries (Bulfone, 2022). Governments are setting decarbonisation targets and designing support schemes to promote new, climate-friendly technologies. They are initiating policies to phase-out carbon-intensive industries, while cushioning the related socio-economic impacts. Moreover, these policies are increasingly combined with techno-nationalist discourses in the West, concerned with securing industrial value creation and maintaining control over key technologies and value chains to confront the rise of China (Van Manen et al., 2021). Western economic governance is partially adopting elements that have long characterised China's industrial policy regime, such as the strategic development of emerging industries and supply chains with large-scale investment subsidies and domestic content requirements (Lewis, 2015).

However, there are also differences in the Western response to the perceived threat of China. The US has taken a very vocal and at times belligerent stance towards its rival. It has framed the relationship as one dominated by systemic confrontation and aims at decoupling global supply chains and curtailing China's access to critical technologies (Mearsheimer, 2021). The European Union has taken a more measured approach, formulating a vision of strategic open autonomy (Miró, 2023) and de-risking—rather than decoupling—of economic interdependencies. Similarly, with the adoption of the Inflation Reduction Act (IRA) in 2022, the US government has launched an extensive subsidy scheme to counterbalance China's state-driven economic model and unleash large-scale investments across clean energy value chains. The EU, though formulating similar ambitions, remains constrained not only by the rules of its single market but also by diverging interests across different Member States and their constituencies (Prontera & Quitzow, 2021, 2022). Correspondingly, the EU's proposed Net-Zero Industry Act, widely considered a response to the US IRA, has largely fallen short of the European Commission's initial ambition to match US investment incentives.

2   The Emerging Role of Hydrogen in the Geopolitics of the Global Energy Transition

It is against the background of the global trends outlined above that hydrogen has emerged as a new arena of competition in a global energy transition, increasingly characterised by geopolitical contestation. Following the formulation of carbon-neutrality targets following the Paris Agreement in 2015, governments around the world have acknowledged the crucial importance of this energy carrier for addressing a range of decarbonisation challenges that renewable electricity alone cannot solve. Most prominently, hydrogen can help decarbonise hard-to-abate industrial sectors, like steel and cement, as well as parts of heavy transport, most notably aviation and long-distance shipping. It also has a role to play as a storage medium in a carbon–neutral energy system.

Indeed, the properties of hydrogen and derivatives like ammonia and methanol as a medium for storing and transporting energy may give it a role in a future energy system that exhibits similarities to that currently held by fossil fuels (Van de Graaf et al., 2020). Thus, it raises similar energy security questions as traditional energy commodities, pertaining to the supply and trade within a global trading system (Dejonghe et al., 2023). The development of infrastructure for the production and transport of hydrogen comes with important implications for the control over future energy flows. That said, the relative difficulty of storing and transporting hydrogen compared to fossil fuels make it unlikely that it will rise to the same prominence as today's hydrocarbons (Ansari et al., 2022).

Moreover, as an energy carrier—rather than a source of energy—hydrogen is not intrinsically linked to a natural geography of supply. There are a number of different production pathways that could play a role, at least temporarily, in the transition to carbon neutrality. These differing options represent another point of contention in the emerging geopolitics of hydrogen. Most prominently, hydrogen can be produced from natural gas in a carbon-intensive process known as steam methane reforming. If combined with carbon capture and storage technologies, the carbon footprint of this process can be reduced though not eliminated entirely. Upstream emissions of methane during extraction and transport of the natural gas feedstock as well as residual emissions from the capture and storage of CO_2 cannot be avoided entirely (Howarth & Jacobson, 2021). It also remains uncertain how cost-effective this is at levels of carbon capture approaching 100 percent. Nevertheless, incumbents from the natural gas industry are promoting this as an important option for accelerating hydrogen use on the way to a carbon–neutral future (Szabo, 2020). Opponents are warning of further exacerbating fossil-fuel lock-ins in the energy system.

The main alternative to this is the production of renewable hydrogen. Produced from water molecules via a process known as electrolysis, this can reduce emissions to zero. However, in a power system that still includes other forms of power generation, this raises points of contention regarding alternative uses of renewable energy as well as the role that nuclear power should play as a low-carbon option for powering hydrogen production. The latter has sparked intense debates in the European context, while the former raises important concerns regarding equity and justice within an emerging global hydrogen economy. Among other things, it has triggered discussions regarding possible trade-offs between the export of renewable hydrogen to industrial centres in the Global North, on the one hand, and the expansion of energy access in the Global South, on the other (Müller et al., 2022).

More broadly, it raises questions regarding the distribution of costs and benefits from the production of hydrogen across actors along the value chain, pointing to another important spatial dimension in the development of a global hydrogen economy. While renewable hydrogen can be produced anywhere using local renewable energy resources, the relative cost of renewables constitutes a critical factor in determining the final production cost. However, as alluded to above, relative to the fossil-fuel economy, the geography of hydrogen production is less determined by natural resource endowments. While it is likely that the costs of hydrogen will play a key role in the formation of regional centres of hydrogen supply, the production of hydrogen close to the sites of its use also comes with important benefits. Potential buyers have to consider not only the additional cost of hydrogen transport but also questions of energy security and other geopolitical considerations when making sourcing decisions (Quitzow, Mewes et al., 2023; Quitzow, Triki et al., 2023).

Moreover, the question of spatial distribution extends beyond the narrow sphere of hydrogen *production*. As producers of industrial goods consider investments in new, climate-friendly production facilities, they may also consider co-locating industrial production at sites where they have ready access to a low-cost renewable hydrogen supply. In this vein, scholars have entertained the idea of a "renewables pull effect" where the availability of low-cost renewables drives locational investment decisions in future decarbonised industrial production (Samadi et al., 2023). This raises the prospect that regions that are well-endowed with renewable energy resources might out-compete existing centres of industrial production with a less favourable resource endowment (Eicke & De Blasio, 2022). However, again, this is far more than just a question of economic efficiency. Rather, it is a subject of political negotiation as stakeholders from the private and public sector explore the development of international hydrogen value chains. Countries with favourable resource endowments have an incentive to leverage their assets to ensure industrial value creation and local socio-economic benefits, while existing centres of industrial production are keen to secure a reliable and low-cost supply of hydrogen to retain current levels of value creation.

Finally, hydrogen—like other clean energy technologies—represents an important sphere of technological competition, both between rival technologies and between the major economic blocs vying for technological leadership (Van de Graaf et al., 2020). While Europe has traditionally held a leading position in the field of hydrogen technologies, competition has increased markedly in the past years, with China rapidly emerging as a low-cost competitor in electrolyser technologies. Linked to this are considerations of critical minerals, like nickel and platinum, that play an important role in the production of hydrogen-related technologies (Ansari et al., 2022; Pepe et al., 2023). Concerns over asymmetric dependencies are driving efforts to secure technological sovereignty and to diversify the emerging supply chain relationships.

3 The Case of Europe in the Emerging Geopolitics of Hydrogen

The European Union as a self-identified global climate power has been at the forefront of the global energy transition and the development of a climate-friendly hydrogen economy. With its landmark European Green Deal (2019), the EU became the first major economic player to set a carbon neutrality target, which has been legally enshrined in the European Climate Law adopted in 2021 (Knodt, 2023). In addition to relying on carbon pricing as a key mechanism to promote emissions reductions, the EU has also been bolstering its green industrial policy. It has long promoted innovation in climate-friendly technologies by funding R&D and their early industrial deployment. In this context, it has strengthened its support for public–private partnerships and industrial alliances to bolster the competitiveness of European producers (Veugelers & Tagliapietra, 2023). More recently, the bloc has tried to broaden the scope of its industrial policy to include measures to retain and expand its manufacturing capacity for net-zero technologies deemed strategically important. These include electrolysers and fuel cells, wind and solar power components, batteries, heat pumps, carbon capture and storage (CCS), as well as biogas/biomethane and grid technologies (European Commission, 2023).

With the launch of its hydrogen strategy in 2020 (European Commission, 2020), the EU also positioned itself early on in the global hydrogen race. It has identified hydrogen as a strategic field of technology and a strategic commodity for the decarbonisation of European industrial production. It has formulated ambitious plans to rapidly boost the supply of renewable hydrogen for this purpose, aiming to develop both domestic hydrogen production and large volumes of hydrogen imports. Following the invasion of Ukraine by Russia and the ensuing gas crisis, it has further increased its ambitions, identifying the transition from natural gas to hydrogen as an avenue for eliminating its reliance on Russian natural gas. To this end, the EU has drastically increased its clean hydrogen targets, aiming for 10 million tons produced domestically and the same amount in imports by 2030 (European Commission, 2022).

In this way, the EU has framed the hydrogen economy as a crucial pillar of its future energy and economic security (Pepe et al., 2023; Quitzow, Mewes et al., 2023; Quitzow, Triki et al., 2023). However, as a continent scarce in renewable energy resources (Eicke & De Blasio, 2022), the EU faces challenges in developing reliable access to renewable hydrogen. In this, the EU differs from the US and China who have the potential to produce sufficient domestic renewable energy to decarbonise their economies (Eicke, 2023; Gong et al., 2023; Quitzow & Gong, 2023).

Moreover, due to its limited financial resources and its unique nature as a supranational entity and a union of 27 Member States, the EU faces significant constraints in investing into its domestic hydrogen economy. While the US has rapidly expanded its fiscal incentives for hydrogen investments (Eicke, 2023), EU efforts remain limited by the willingness of Member States to endow the European Commission with the required mandate and the corresponding financial resources. These challenges are particularly pronounced in the field of energy policy, where the EU lacks the competence to interfere with national decision-making (Prontera & Quitzow, 2021). In this context, the interests and positions among the EU Member States are crucial to the development of the EU hydrogen policy, emerging at the interface between national and EU-level decision-making.

4 Europe's Domestic Hydrogen Politics

This mismatch between European hydrogen ambition and the EU's capacities to act is met by a heterogeneous mix of national hydrogen strategies among its Member States. They reveal not only differing levels of ambition (Quitzow, Mewes et al., 2023; Quitzow, Triki et al., 2023), but also important disagreements on some of the critical questions highlighted above. Points of contention include the shape of future hydrogen infrastructure, acceptable hydrogen production pathways, the future role of imports as well as the design of European-wide support schemes and regulations for hydrogen. Coordinating and aligning these divergent interests is a major challenge as the EU seeks to position itself as a key player in the global hydrogen economy.

These diverging approaches are a reflection of pre-existing political and economic legacies and the structural features of their national energy systems. In many cases, they replicate long-standing tensions between Member States, such as the continued contention between France and Germany over nuclear power (Szulecki et al., 2016). In other cases, differing assets and starting points within the emerging hydrogen economy, such as pre-existing industrial legacies and pipeline infrastructure or the relative availability of renewable energy, translate into differing interests across countries. Moreover, broader political agendas and priorities, such as differing positions on Europe's stance vis-à-vis Russia (Varga & Buzogány, 2020), play a role in shaping diverging visions for a hydrogen economy. Domestic climate politics play a role, too: in several countries, there is growing unease with European climate policy, which is frequently viewed as overly stringent and costly and which has been subject to a populist backlash in some cases (Huber et al., 2021).

5 External Dimensions of the European Green Deal and the Role of Hydrogen

The manifold geopolitical considerations outlined above also give the EU's hydrogen policy a prominent international dimension. This aligns with the development of an increasingly active climate and energy diplomacy, which the EU recently declared a "core component" of its foreign policy (Council of the EU, 2023). Europe's limited renewable energy endowment and the importance of hydrogen imports to reach EU climate targets provide it with a further impetus for integrating hydrogen-related concerns in climate and energy diplomacy. As alluded to above, this sets it apart from other large players like China and the United States whose domestic renewable energy resources provide them with the potential for developing sufficient amounts of domestic hydrogen production. For the EU, establishing a functioning international hydrogen market and developing a diversified supplier base at acceptable costs is an important component of its strategy for decarbonising its economy while maintaining its industrial competitiveness and ensuring energy security (see chapter on EU hydrogen policy in this volume).

The development of hydrogen partnerships with potential exporters in the European Neighbourhood and the Global South also has a broader geopolitical dimension. In a context where Western-dominated multilateral institutions are under strain, it offers an opportunity to strengthen economic ties, promote European values and standards and bolster the EU's leadership role on the international stage (Quitzow, Mewes et al., 2023; Quitzow, Triki et al., 2023). Another important concern is the promotion of sustainable production practices in the hydrogen sector. This is not only demanded by European consumers and civil society, but it is also seen as playing a role in establishing mutually beneficial and stable economic partnerships with the Global South. Last but not least, the economic prospects of a hydrogen economy and climate-friendly industrial production for the Global South are critical for ensuring continued support for international climate policy goals (Newell et al., 2021).

Finally, developing a coordinated European approach for its international hydrogen policy promises important benefits. A Europeanised purchase mechanism for imported hydrogen, for instance, could help pool demand and lower costs, compared to Member States securing their supplies individually. Indeed, this is one of the lessons from the European scramble for gas and ballooning prices after Russia's invasion of Ukraine (Kuzemko et al., 2022). Similarly, speaking with one European voice would play an important role in garnering the support of potential partner countries and catalysing investment in this new economic sector where manifold uncertainties continue to hamper decision-making. In this context, the EU is beginning to integrate hydrogen-related projects into its existing connectivity initiative, the Global Gateway (European Commission, 2021). In doing so, it seeks to promote broad-based hydrogen cooperation with partners from emerging economies and the Global South. One of the key elements of the Global Gateway is the "Team Europe" approach, which is meant to pool the resources of the European Commission, EU Member States as well as European financial institutions to promote infrastructural links and investment in hydrogen production in non-EU countries. However, despite these efforts, concerted European action remains a challenge, mirroring the tensions alluded to above (see chapter on EU hydrogen policy in this volume). Proactive Member States like Germany have moved ahead rapidly in launching their own bilateral hydrogen diplomacy efforts (see chapter on Germany in this volume). Other Member States, which do not see the immediate need for importing hydrogen, lack incentives to be involved or may even oppose such efforts.

6 Understanding Europe's Role in the Emerging Geopolitics of Hydrogen: The Contribution of This Volume

This edited volume seeks to shed light on how the interplay between hydrogen policy and politics at the EU-level and in important Member States is shaping European hydrogen development as well as the role of the EU as an actor in the emerging global hydrogen economy. While scholarly reflection on the EU and its role in global politics is frequently limited to a consideration of EU-level structures and initiatives (see for example Keukeleire & Delreux, 2022), this volume places particular emphasis on how the political economy at both the national and supranational level is shaping the EU's role as a geopolitical actor. In this vein, this book conceptualises EU policy as the combination of policies at both levels of governance. This implies that the EU as a geopolitical actor is inherently fragmented in nature, exhibiting important contradictions and tensions. The aim of the book is to provide a better understanding of what those contradictions and tensions are and how they affect EU hydrogen policy both within the EU and externally.

The volume includes ten case studies examining hydrogen policy development at the EU-level as well as in nine members of the EU and the European Economic Area: Germany, France, Poland, Hungary, Italy, Spain, the Netherlands, Sweden and Norway. The selected countries represent different geographies, and they reflect different levels of economic and political status within the EU and differing assets and starting points within the hydrogen sector. Germany and France are included as the two largest Member States, both with highly active but strongly diverging hydrogen strategies. With Poland and Hungary, the volume discusses two important Central and Eastern European countries. Despite sharing a number of similarities in their endowments and capabilities, their diverging stances on Russia have translated into notable differences in their hydrogen strategies. Italy and Spain, the largest Southern European Member States, also exhibit significant differences in their ambitions as exporters and importers in both a European and a global hydrogen economy. The Netherlands occupies a key position in Europe's emerging hydrogen economy as both an important destination and a potential trading hub for ship-based hydrogen imports. Finally, Sweden and Norway represent two key Nordic countries, both well-endowed with renewable energy resources. Yet, both have very different stances on hydrogen. Sweden is keen to pursue domestic "fossil-free" hydrogen production from both renewables and nuclear energy, mainly for the decarbonisation of local industry. Norway, in turn, is seeking to build on its natural gas resources and the prospects of carbon, capture and storage technologies to pursue an export-oriented hydrogen strategy. Though not an EU Member State, Norway's endowments in both renewable and fossil energy make it a critical player in EU hydrogen developments.

Each of these case studies provides an overview of the domestic and international hydrogen policy in the EU and the selected European countries. Moreover, policy developments are linked to key domestic political economy drivers and their role in shaping the respective policy stance. To do so, the various chapters begin with an overview of national climate and energy policy, related political priorities and key points of contention. Against this background, they provide a review of national hydrogen policy, including a discussion of how structural features and stakeholder interests in energy and industry have influenced policy-making.

The review of domestic hydrogen policy and politics provides the basis for a subsequent examination of external hydrogen policy. This part provides a systematic review of international hydrogen policy action, while placing it against the background of each country's broader climate and energy foreign policy. Each chapter provides an overview of key dimensions of international cooperation in the hydrogen sector, covering both bilateral and multilateral engagement. The dimensions that are considered include diplomacy and political dialogue, efforts around international market and supply chain development, support for cooperation in research and development, the promotion of hydrogen-related regulations and standards, as well as capacity and skill development. Given the significant variance in scope and ambition of international hydrogen policy, each chapter provides a structure for presenting the review of activities that is tailored to the unique features of the respective case.

Each study concludes with a reflection of how national interests and priorities shape the respective country's stance vis-à-vis the EU and other key Member States, highlighting both points of tension and synergies. This includes a discussion of the evolving policy framework governing the EU hydrogen sector as well as an emerging EU hydrogen diplomacy. The chapter on the EU in turn discusses the role and constraints of EU-level policy making, both domestically and internationally.

The edited volume concludes with a chapter that synthesises the findings and key themes emerging from the individual case studies and situates them in the broader context of European and international hydrogen politics. It provides a review of how the interplay of national and EU-level politics and policies is influencing the EU's domestic and international hydrogen policy. It then highlights how the positions and interests of key Member States are shaping the EU's ambitions to become a major geopolitical player in a global hydrogen economy.

Acknowledgements Research for this chapter was financially supported by the German Federal Foreign Office within the framework of the project "Geopolitics of the Energy Transformation—Implications of an International Hydrogen Economy" (GET Hydrogen), funding reference number AA4521G125.

References

Ansari, D., Grinschgl, J., & Pepe, J. M. (2022). Electrolysers for the hydrogen revolution: Challenges, dependencies, and solutions. *Stiftung Wissenschaft und Politik (SWP)*. https://doi.org/10.18449/2022C57

Bulfone, F. (2022). Industrial policy and comparative political economy: A literature review and research agenda. *Competition and Change, 27*(1), 22–43. https://doi.org/10.1177/10245294221076225

Council of the EU. (2023). Council conclusions on climate and energy diplomacy. Bolstering EU Climate and Energy Diplomacy in a Critical Decade, 7248/23. https://www.consilium.europa.eu/media/62942/st07248-en23.pdf

Cui, L., Yue, S., Nghiem, X.-H., & Duan, M. (2023). Exploring the risk and economic vulnerability of global energy supply chain interruption in the context of Russo-Ukrainian War. *Resources Policy, 81*, 103373. https://doi.org/10.1016/j.resourpol.2023.103373

Dejonghe, M., Van de Graaf, T., & Belmans, R. (2023). From natural gas to hydrogen: Navigating import risks and dependencies in Northwest Europe. *Energy Research and Social Science, 106*, 103301. https://doi.org/10.1016/j.erss.2023.103301

European Commission. (2020). *Key actions of the EU hydrogen strategy*. https://energy.ec.europa.eu/topics/energy-systems-integration/hydrogen_en

European Commission. (2021). Joint communication to the European parliament, the council, the European economic and social committee, the committee of the regions and the European investment bank (JOIN (2021) 30). https://eur-lex.europa.eu/legal-content/EN/TXT/HTML/?uri=CELEX:52021JC0030

European Commission. (2022). *REPowerEU: Affordable, secure and sustainable energy for Europe*. https://commission.europa.eu/strategy-and-policy/priorities-2019-2024/european-green-deal/repowereu-affordable-secure-and-sustainable-energy-europe_en

European Commission. (2023). *The Net-Zero Industry Act: Accelerating the Transition to Climate Neutrality*. https://single-market-economy.ec.europa.eu/industry/sustainability/net-zero-industry-act_en

Eicke, L., & De Blasio, N. (2022). Green hydrogen value chains in the industrial sector: Geopolitical and market implications. *Energy Research and Social Science, 93*, 102847. https://doi.org/10.1016/j.erss.2022.102847

Eicke, L. (2023). US hydrogen policy: Paving the way for energy independence, technology leadership and decarbonization. *Research Institute for Sustainability Discussion Paper, 26*.

Geels, F. W. (2014). Regime resistance against low-carbon transitions: Introducing politics and power into the multi-level perspective. *Theory, Culture and Society, 31*(5), 21–40. https://doi.org/10.1177/0263276414531627

Gertz, G., & Evers, M. M. (2020). Geoeconomic competition: Will state capitalism win? *The Washington Quarterly, 43*(2), 117–136. https://doi.org/10.1080/0163660X.2020.1770962

Goldthau, A. C. (2021). The tricky geoeconomics of going low carbon. *Joule, 5*(12), 3078–3079. https://doi.org/10.1016/j.joule.2021.11.012

Goldthau, A., & Westphal, K. (2019). Why the global energy transition does not mean the end of the Petrostate. *Global Policy, 10*(2), 279–283. https://doi.org/10.1111/1758-5899.12649

Gong, X., Quitzow, R., & Boute, A. (2023). China's Emerging hydrogen economy: Policies, institutions, actors. *Research Institute for Sustainability Stusy*.

Howarth, R. W., & Jacobson, M. Z. (2021). How green is blue hydrogen? *Energy Science and Engineering, 9*(10), 1676–1687. https://doi.org/10.1002/ese3.956

Huber, R. A., Maltby, T., Szulecki, K., & Ćetković, S. (2021). Is populism a challenge to European energy and climate policy? Empirical evidence across varieties of populism. *Journal of European Public Policy, 28*(7), 998–1017. https://doi.org/10.1080/13501763.2021.1918214

Keukeleire, S., & Delreux, T. (2022). *The foreign policy of the European Union* (3rd ed.). Bloomsbury Publishing. https://www.bloomsbury.com/uk/foreign-policy-of-the-european-union-978 1350930483/

Knodt, M. (2023). Instruments and modes of governance in EU climate and energy policy: From energy union to the European Green Deal. In *Handbook on European Union Climate Change Policy and Politics* (pp. 202–215). Edward Elgar Publishing. https://www.elgaronline.com/edc ollchap-oa/book/9781789906981/book-part-9781789906981-27.xml

Kuzemko, C., Blondeel, M., Dupont, C., & Brisbois, M. C. (2022). Russia's War on Ukraine, European energy policy responses and implications for sustainable transformations. *Energy Research and Social Science, 93*, 102842. https://doi.org/10.1016/j.erss.2022.102842

Kwan, C. H. (2019). The China–US trade war: Deep-rooted causes, shifting focus and uncertain prospects. *Asian Economic Policy Review, 15*(1), 55–72. https://doi.org/10.1111/aepr.12284

Lema, R., Fu, X., & Rabellotti, R. (2020). Green windows of opportunity: Latecomer development in the age of transformation toward sustainability. *Industrial and Corporate Change, 29*(5), 1193–1209. https://doi.org/10.1093/icc/dtaa044

Lewis, J. I. (2015). *Green innovation in China: China's wind power industry and the global transition to a low-carbon economy* (p. 304). Columbia University Press.

Lippert, B., & Perthes, V. (2020). Strategic Rivalry between United States and China: Causes, Trajectories, and Implications for Europe. *Stiftung Wissenschaft und Politik. German Institute for International and Security Affairs, SWP Research Paper 4*. https://doi.org/10.18449/202 0RP04

Maihold, G. (2022). *A new geopolitics of supply chains: The rise of friend-shoring.* https://doi.org/10.18449/2022C45

Mearsheimer, J. J. (2021, October 19). The Inevitable Rivalry: America, China, and the Tragedy of Great-Power Politics. *Foreign Affairs, 100*(6). https://www.foreignaffairs.com/articles/china/2021-10-19/inevitable-rivalry-cold-war

Miró, J. (2023). Responding to the global disorder: The EU's quest for open strategic autonomy. *Global Society, 37*(3), 315–335. https://doi.org/10.1080/13600826.2022.2110042

Müller, F., Tunn, J., & Kalt, T. (2022). Hydrogen justice. *Environmental Research Letters, 17*(11), 115006. https://doi.org/10.1088/1748-9326/ac991a

Newell, P., Srivastava, S., Naess, L. O., Torres Contreras, G. A., & Price, R. (2021). Toward transformative climate justice: An emerging research agenda. *Wires Climate Change, 12*(6), 733. https://doi.org/10.1002/wcc.733

Pepe, J. M., Ansari, D., & Gehrung, R. M. (2023). The geopolitics of hydrogen. *Stiftung Wissenschaft und Politik (SWP).* https://doi.org/10.18449/2023RP13v02

Prontera, A., & Quitzow, R. (2022). Catalytic power Europe: blended finance in European external action. *JCMS: Journal of Common Market Studies, 61*(4), 988–1006. https://doi.org/10.1111/jcms.13442

Prontera, A., & Quitzow, R. (2021). The EU as catalytic state? Rethinking European climate and energy governance. *New Political Economy, 27*(3), 517–531. https://doi.org/10.1080/13563467.2021.1994539

Pujawan, I. N., & Bah, A. U. (2022). Supply chains under COVID-19 disruptions: Literature review and research agenda. *Supply Chain Forum: An International Journal, 23*(1), 81–95. https://doi.org/10.1080/16258312.2021.1932568

Quitzow, R. (2015). Dynamics of a policy-driven market: The co-evolution of technological innovation systems for solar photovoltaics in China and Germany. *Environmental Innovation and Societal Transitions, 17*, 126–148. https://doi.org/10.1016/j.eist.2014.12.002

Quitzow, R., & Gong, X. (2023). The geopolitics of hydrogen: The emerging role of China. *The Oxford Energy Forum, 137.* https://www.oxfordenergy.org/wpcms/wp-content/uploads/2023/08/OEF-137.pdf

Quitzow, R., Mewes, C., Thielges, S., Tsoumpa, M., & Zabanova, Y. (2023). Building partnerships for an international hydrogen economy: Entry-points for European policy action. *FES Diskurs.* https://library.fes.de/pdf-files/a-p-b/19921-20230215.pdf

Quitzow, R., Triki, A., Wachsmuth, J., Garcia, J. F., Kramer, N., Lux, B., & Nunez, A. (2023). HYPAT discussion paper 05/2023. Mobilizing Europe's full hydrogen potential: Entry-points for action by the EU and its Member States. *H2 Potential, HYPAT Discussion Paper.*

Rohracher, H. (2018). Analyzing the socio-technical transformation of energy systems: The concept of "sustainability transitions." In *Oxford Handbook of Energy and Society* (pp. 45–62). https://doi.org/10.1093/oxfordhb/9780190633851.013.3

Samadi, S., Fischer, A., & Lechtenböhmer, S. (2023). The renewables pull effect: How regional differences in renewable energy costs could influence where industrial production is located in the future. *Energy Research and Social Science, 104*, 103257. https://doi.org/10.1016/j.erss.2023.103257

Siddi, M. (2023). The geopolitics of energy transition: New Resources and Technologies. In J. Berghofer, A. Futter, C. Häusler, M. Hoell, & J. Nosál (Eds.), *The Implications of Emerging Technologies in the Euro-Atlantic Space* (pp. 73–85). Palgrave Macmillan, Cham. https://doi.org/10.1007/978-3-031-24673-9_5

Szabo, J. (2020). Fossil capitalism's lock-ins: The natural gas-hydrogen nexus. *Capitalism Nature Socialism, 32*(4), 91–110. https://doi.org/10.1080/10455752.2020.1843186

Szulecki, K., Fischer, S., Gullberg, A. T., & Sartor, O. (2016). Shaping the 'Energy Union': Between national positions and governance innovation in EU energy and climate policy. *Climate Policy, 16*(5), 548–567. https://doi.org/10.1080/14693062.2015.1135100

Tagliapietra, S. (2019). The impact of the global energy transition on MENA oil and gas producers. *Energy Strategy Reviews, 26*, 100397. https://doi.org/10.1016/j.esr.2019.100397

Vakulchuk, R., Overland, I., & Scholten, D. (2020). Renewable energy and geopolitics: A review. *Renewable and Sustainable Energy Reviews, 122*, 109547. https://doi.org/10.1016/j.rser.2019.109547

Van de Graaf, T., Overland, I., Scholten, D., & Westphal, K. (2020). The new oil? The geopolitics and international governance of hydrogen. *Energy Research and Social Science, 70*, 101667. https://doi.org/10.1016/j.erss.2020.101667

Van de Graaf, T. (2023). Barrels, booms, and busts: The future of petrostates in a decarbonizing world. In *Handbook on the geopolitics of the energy transition* (pp. 183–196). Edward Elgar. http://hdl.handle.net/1854/LU-01HD3CTVSS491FDM0QSQM4C4V5

Van Manen, H., Gehrke, T., Thompson, J., & Sweijs, T. (2021). Taming Techno-Nationalism. *The Hague Centre for Strategic Studies.*

Varga, M., & Buzogány, A. (2020). The foreign policy of populists in power: Contesting liberalism in Poland and Hungary. *Geopolitics, 26*(5), 1442–1463. https://doi.org/10.1080/14650045.2020.1734564

Veugelers, R., & Tagliapietra, S. (2023). *A green industrial policy for Europe.* https://www.bruegel.org/book/green-industrial-policy-europe

The EU in the Global Hydrogen Race: Bringing Together Climate Action, Energy Security, and Industrial Policy

Yana Zabanova

Abstract The European Union has identified clean hydrogen as essential to its climate targets, technology leadership and energy security in the decarbonizing world. The bloc is developing a comprehensive regulatory framework for a hydrogen economy, complete with supply-side policies and binding demand-side targets. In addition to boosting domestic production, the EU is planning to import large volumes of hydrogen and derivatives from third countries. Hydrogen is thus beginning to play a more prominent role in the EU's bilateral partnerships. The EU is also actively participating in multilateral hydrogen governance with the goal of creating a functioning international hydrogen market featuring strong sustainability standards. At the same time, aligning the diverging interests of Member States and various hydrogen stakeholders has been a challenge. As the global hydrogen race accelerates, the bloc has struggled to keep up with powerful players like the United States, which are offering massive subsidies to the hydrogen industry. This chapter examines the domestic and external dimensions of the EU's hydrogen vision, situating it within the bloc's wider climate and energy policy and recent geopolitical developments. It discusses key policies, regulations, and funding schemes for hydrogen in the EU, highlighting existing points of contention and the interplay between the EU and Member State level. It then goes on to analyze the EU's evolving international engagement on hydrogen and the challenges of fostering mutually beneficial green industrial partnerships that go beyond securing hydrogen supplies. It remains to be seen whether the EU succeeds in drawing on its early mover advantage and potential synergies to remain an attractive investment destination and build resilient clean hydrogen supply chains.

Y. Zabanova (✉)
Research Institute for Sustainability (RIFS), Helmholtz Centre Potsdam, Berliner Str. 130, 14467 Potsdam, Germany
e-mail: yana.zabanova@rifs-potsdam.de

© Helmholtz-Zentrum Potsdam, Deutsches GeoForschungsZentrum GFZ 2024   15
R. Quitzow and Y. Zabanova (eds.), *The Geopolitics of Hydrogen*, Studies in Energy, Resource and Environmental Economics, https://doi.org/10.1007/978-3-031-59515-8_2

1 Introduction

With its pioneering hydrogen strategy adopted in July 2020, the European Union has sent a clear message that it views hydrogen as essential to reaching its climate goals and maintaining technology leadership. Following Russia's 2022 invasion of Ukraine, hydrogen development has acquired an additional energy security dimension: the hope is that hydrogen can, in the future, make the EU less dependent on natural gas imports, in particular from Russia. This is expected to happen through both massively ramping up domestic production and sourcing large amounts of clean hydrogen from abroad. This, in turn, requires a solid regulatory framework, including internationally harmonized standards and certification schemes, a cross-border infrastructure within Europe and beyond, large amounts of public and private investment, and strong relations with prospective hydrogen exporters.

The EU has many advantages to draw on in the emerging international hydrogen economy. With a combined population of 447 million, it boasts a large and attractive single market and highly qualified human resources. Europe is home to the world's most innovative electrolyzer manufacturers and has a long history of funding public–private partnerships to promote R&D on hydrogen. It has the world's most developed natural gas pipeline infrastructure that could be repurposed to carry hydrogen, and its key ports such as Rotterdam and Hamburg are actively seeking to position themselves as hubs for handling clean fuels. The EU was the first global player to set ambitious—and legally binding—climate targets. As a result, Europe has emerged as an attractive location for hydrogen projects: as of October 2023, it was the world's leader in announced hydrogen projects, with total investment equaling 193billion USD, or 34% of the world's total (Hydrogen Council & McKinsey, 2023).

Yet Europe's lead on hydrogen is not uncontested. Other major players—notably the United States—are raising the stakes by offering massive public subsidies to hydrogen producers. However, this is precisely the area where the EU finds itself at a strategic disadvantage. With its origins steeped in economic liberalism and free trade, the EU has traditionally been uneasy with the idea of allowing state subsidies due to their potential to distort the single market. Given that the EU lacks the powers to directly collect taxes and has been, until recently, reluctant to borrow externally, it does not have significant resources of its own to undertake large strategic investments. This role falls mainly to the Member States, whose economic and geopolitical interests, as well as fiscal resources, vary significantly. In this regard, the EU functions rather as a forum for Member States to hash out their positions and disagreements on just how important hydrogen should be, how much and what kind of support is required and whether subsidies should come from national budgets or from a European-level facility—all of which slows down the legislative process. In addition, if hydrogen is to play a major role in the EU's economy, large-scale imports are unavoidable, continuing Europe's energy dependence on third countries, whereas China and the US can be largely self-sufficient.

The challenges the EU is facing in navigating its place in the emerging international hydrogen economy are formidable, and time is running short. This paper will examine the domestic and external dimensions of the EU's hydrogen vision, situating it within the bloc's wider climate and energy policy and recent geopolitical developments. It will discuss key domestic policies, regulations, and funding schemes for hydrogen in Europe, highlighting existing points of contention and the interplay between the EU and Member States. Then, the paper will turn to the EU's international engagement on hydrogen, including participation in multilateral hydrogen governance institutions, bilateral cooperation initiatives, EU efforts to promote cross-border trade in clean hydrogen and derivatives, and European investment into a hydrogen economy in third countries. The external repercussions of EU hydrogen policy will be addressed as a separate point. Finally, the conclusion will summarize the main themes emerging from the paper, as well as the issues the EU will need to tackle moving forward, both domestically and internationally.

2 The EU's Strategic Vision for a Hydrogen Economy

The EU has garnered a reputation as a global regulatory power (Bradford, 2019) and has been a pioneer in creating a regulatory framework for the transition to carbon neutrality. Its vision of using hydrogen to decarbonize hard-to-abate sectors is rooted in the steadily expanding scope and ambition of its climate policy. Among EU institutions, the European Commission in particular has emerged as the strongest advocate of the transition to net zero, including a clean hydrogen economy. In December 2019, the Commission proposed the European Green Deal, a strategy aimed at making the EU climate-neutral by 2050 while promoting sustainable economic growth. This was followed by the "Fit for 55" package in 2021, which contains legislative proposals and amendments to achieve a 55% reduction in greenhouse gas emissions in the EU by 2030 (from the 1990 baseline) and touches on a wide range of fields, including renewable energy deployment, energy efficiency, emissions trading, use of alternative fuels in the maritime industry and aviation, a Social Climate Fund, and many others. The European Climate Law adopted in 2021 includes a legally binding 2050 net-zero target and requires all Union policies and legislation to be consistent with this goal. The EU's approach to climate policy features a strong emphasis on timelines and numerical targets, and the bloc has also consciously positioned itself as a "policy laboratory" of sorts, developing and testing innovative policies and instruments that are often adopted by other jurisdictions as well (Rayner et al., 2023, pp. 1–3). The EU has also worked towards tackling emissions beyond its borders, such as from international aviation and shipping, where the key sectoral organizations—the International Civil Aviation Association and the International Maritime Organization—have been criticized for dragging their feet (Vogler, 2023). Geopolitics, too, is beginning to play a more prominent role in shaping the EU's green transition. Faced with geopolitical tensions, supply chain risks, and the intensifying global competition for clean technology leadership, the EU is now seeking to reduce its asymmetric dependencies on other actors, including in the area of key clean technologies, hydrogen and critical raw materials.

In July 2020, the European Union adopted a "Hydrogen Strategy for a Climate-Neutral Europe" as part of the European Green Deal package. The Strategy contains a vision of clean hydrogen as an energy carrier that would play a key role in decarbonizing Europe's economy and helping maintain European technology leadership. The document also lists ambitious targets, such as installing 40 GW of electrolyzer capacity in Europe to produce 5.6 million tons of green hydrogen by 2030. The Covid-19 pandemic played a central role in further galvanizing policy support for hydrogen development in Europe, as the European Commission successfully pushed for hydrogen to be integrated into "green recovery" efforts. In July 2021, the EU introduced NextGenerationEU, an 800 billion EUR economic stimulus package funded by external borrowing. The largest part of the package is the Recovery and Resilience Facility (RRF), a funding resource granted to member states, with 312.5 billion EUR in grants and 360 billion EUR in loans (in 2018 prices) (European Commission, 2021b). Green transition is the most important of the RRF's three pillars, and the Facility has emerged as the dominant source of funding for related measures in the EU. In total, the 27 national plans approved under the RRF include 10 billion EUR in hydrogen-related funding (European Commission, 2023d).

Russia's invasion of Ukraine in February 2022 has served as yet another accelerator for Europe's hydrogen ambitions. In May 2022, the European Commission presented its REPowerEU proposal, aimed at fully phasing out Europe's dependence on Russian fossil fuels by 2027. This is to be achieved through a combination of demand reduction, diversification of supplies, and accelerated energy transition, including green hydrogen development. The REPowerEU plan has substantially increased hydrogen targets (albeit non-binding): 10 million tons annually in domestic production and the same amount in imports by 2030. All of a sudden, hydrogen came to be viewed as not only essential for reaching net zero, but as an important element of bolstering Europe's energy security.

With its Hydrogen Strategy and REPowerEU, the EU has articulated an ambitious vision of a future place of hydrogen in its economy. Yet the EU's status as a hydrogen frontrunner is being challenged as global competition accelerates. In August 2022, the United States adopted the landmark Inflation Reduction Act (IRA), complete with an estimated 369 billion USD in financial incentives for energy and climate-related measures. The IRA offers a generous tax credit for clean hydrogen production (up to 3 USD per kg), with the level of remuneration contingent on the amount of lifecycle GHG emissions avoided; a second tax credit targets blue hydrogen production (from natural gas with CCUS). The sheer scale and simplicity of the US scheme have sparked the interest of hydrogen investors worldwide. Faced with a tangible danger of seeing hydrogen investment and technology manufacturing relocate to the US, the EU has swayed towards a more assertive industrial policy.

On 1 February 2023, the Commission unveiled its proposed "Green Deal Industrial Plan for the Net-Zero Age" (GDIP), followed by the legislative proposal on the "Net-Zero Industry Act" (NZIA) in March 2023. With the GDIP, the EU aims to promote investment in strategically important projects, facilitate skill development, speed up permitting for new clean energy production sites, and use trade measures to promote diversified and resilient supply chains (Stolton, 2023). In contrast to previous EU initiatives, NZIA directly focuses on the EU's "net-zero manufacturing capacity" (European Commission, 2023b), setting domestic manufacturing targets for selected "strategic" clean technologies. (Under pressure from the Council and the European Parliament, the Commission's original list of supported technologies has been significantly expanded to include, controversially, nuclear technologies and carbon capture and storage (CCS)). According to the Commission's proposal, at least 40% of clean energy technologies must be produced domestically by 2030, including an electrolyzer capacity to produce 100 GW of green hydrogen. In an effort to motivate the European cleantech industry to keep its production facilities in the EU, NZIA also envisions anti-relocation incentives (to be granted by Member States from their budgets) such as grants, loans or tax breaks or even, in "exceptional cases," subsidies fully matching those offered by third countries. The downside is that the proposed GDIP and NZIA would not be equipped with new money from the EU but would rely mainly on repurposing existing EU-level funds (such as the Recovery and Resilience Facility and the Innovation Fund), as well as making it easier for Member States to grant state aids as described above.

The success of the EU's hydrogen vision to a large extent hinges on the actions of Member States. With limited resources of its own, the EU will not be able to jumpstart a thriving hydrogen economy on its own. On the other hand, excessive reliance on Member States (meaning, in practice, on wealthy member states) carries its own risks, such as entrenching structural inequalities in the EU and undermining competition as a fundamental principle of the European single market. The EU's hydrogen vision will need to be integrated into national strategic planning at the Member State level, with the Commission acting as a coordinator. Indeed, the EU Hydrogen Strategy states that the Commission will "exchange" with Member States on their hydrogen plans using a dedicated informal platform run by DG ENER (European Commission, 2020, p. 9), yet little coordination has taken place so far. Moving forward, however, Member States will be expected to include information on hydrogen-related developments as part of the updates in their National Energy and Climate Plans (NECPs), which would help the Commission to better steer the alignment process.

3 Policies for Domestic Hydrogen Development

The EU has been developing a mix of policy instruments to support a hydrogen economy. It has a strong track record of funding RD&D on hydrogen and has also launched a dedicated industrial alliance to promote hydrogen projects along the entire value chain. While seeking to strengthen carbon pricing as a key market mechanism, the EU is in parallel designing a dedicated regulatory framework for hydrogen and other renewable gases. This includes both supply-side and demand-side measures as well as infrastructure regulations. While there is a consensus in the EU on the importance of the green transition and the need to position Europe better in the global net-zero economy, the three EU institutions involved in the legislative procedure—the European Commission, the European Parliament, and the Council of Ministers—often hold differing views on the design of relevant policies. Some points of disagreement have included the definition of renewable hydrogen, key sources of funding for the EU's green industrial policy (state-level versus EU-level), and rules for a future hydrogen infrastructure.

Aware of the intensifying global cleantech race, the EU and Member States have made billions of EUR available for the transition to net zero and for hydrogen specifically. However, navigating the EU's complex and fragmented funding landscape is a serious challenge for businesses. Permitting and licensing delays are another problematic issue for hydrogen investors, which the European Commission is currently trying to address.

This section provides a detailed review of the EU's key policies targeted at developing a hydrogen economy: RD&D funding measures, the evolving regulatory framework, and various investment support measures and subsidies both at the EU and Member State level. It will also highlight the challenges the EU has faced in each of these areas.

3.1 Research and Innovation

Funding research and innovation has been, for years, the most established and visible way in which the EU has supported hydrogen development. In doing so, the EU has put an emphasis on close cooperation with the private sector. Already in 2002, amidst an earlier wave of interest in hydrogen, the EU established the Fuel Cells and Hydrogen Joint Undertaking (FCH JU), which remained active even when the interest ebbed away elsewhere in the world. In November 2021, the EU's Clean Hydrogen Partnership (also known as the Clean Hydrogen Joint Undertaking) was launched as a flagship European public–private partnership and a successor of the FCH JU. Its members include the European Commission and two partners: the fuel cell and hydrogen industrial grouping called Hydrogen Europe and the research community called Hydrogen Europe Research. The Partnership's goal is to promote R&I in hydrogen technologies, with a special focus on their development, commercialization, and scaling, while applying an integrated European value chain approach. The Clean Hydrogen Partnership (CHP) has been allocated 1 billion EUR in public funding from the EU's Horizon research funding framework for the period from 2021 until 2027; this amount will be supplemented with an additional 1 billion EUR from the private partners. The Partnership puts a strong focus on developing so-called "hydrogen valleys" in the EU, i.e. regional hydrogen ecosystems combining several hydrogen uses and often involving research as well. Hydrogen valleys offer a way to pool hydrogen demand and match it with on-site supply, which is especially important for early projects in the absence of a hydrogen transport infrastructure.

The Commission's REPowerEU package of proposals presented in May 2022 dedicates an additional 200 million EUR in funding to the CHP to double the number of hydrogen valleys in Europe.

3.2 Regulation

Internationally, the EU has emerged as a pioneer in developing a comprehensive regulatory framework for a hydrogen economy, a process that is still ongoing and faces frequent delays due to the presence of conflicting interests. The EU's approach features a combination of stringent sustainability requirements for renewable and other types of low-carbon hydrogen, sectoral quotas, a reform of its emissions trading system, and rules for the future renewable gas infrastructure.

3.2.1 Defining and Certifying Clean Hydrogen

Consistent with its ambitious climate agenda, the EU has chosen early on to prioritize renewable hydrogen both for domestic production and imports. Other types of low-carbon hydrogen (e.g. blue or nuclear-based) can play a role too, provided they can offer a minimum of 70% reduction in GHG emissions compared to unabated fossil-based hydrogen based on steam methane reforming, but they have until now received less political attention compared to green hydrogen.

Having a clear definition under what conditions hydrogen qualifies as renewable is essential for investors as they plan projects and prepare for final investment decisions (Hydrogen Europe and Renewable Hydrogen Coalition, 2022). For a long time, EU institutions and key hydrogen stakeholders wrangled over a compromise on how to design rules and safeguards to ensure that renewable hydrogen is genuinely climate-friendly. Supported by leading climate NGOs, the Commission has consistently called for a stringent set of rules and requirements based on three principles: additionality, temporal correlation and geographic correlation. Additionality requires that renewable hydrogen is produced with new renewable energy capacity, rather than with existing installations, in order to prevent cannibalizing much-needed green electricity from other important decarbonization uses. Temporal correlation refers to matching hydrogen production with the availability of renewable electricity in the grid; in its strictest form it would translate into running electrolyzers only when the wind is blowing or the sun is shining. Finally, geographic correlation requires that the electrolyzer and the renewable energy facility are located close to each other.

The requirements are meant to ensure hydrogen's positive impact on lowering emissions. The downside, however, is that tighter requirements translate into fewer hours electrolyzers can run, lowering their load factor and thereby raising the levelized cost of renewable hydrogen (LCOH). Moreover, proving compliance with the requirements—for example, documenting the provenance of electricity used for hydrogen production on an hourly basis—imposes a significant bureaucratic and economic burden on firms (Parkes, 2022). For this reason, the hydrogen industry and the European Parliament repeatedly voiced concerns—especially after the US adopted the IRA in 2022—that introducing too stringent rules from the outset would overburden the nascent industry and slow down market ramp-up, risking the flight of potential investors. They thus called for laxer rules and/or longer transition periods.

After a constant back-and-forth between the Commission and the Parliament and repeated delays, the compromise was finally achieved in July 2023 with the adoption of two delegated acts[1] amending Articles 27 and 28 of the Renewable Energy Directive (RED II) on so-called RFNBOs. The acronym stands for "renewable liquid and gaseous fuels of non-biological origin" and includes renewable hydrogen and its derivatives such as green ammonia and green methanol. The delegated acts contain a detailed description of conditions when electrolytic hydrogen is recognized as renewable (see Text Box 1), as well as a methodology for calculating related emissions savings. Importantly, these requirements will also apply to hydrogen imported into the EU.

Agreeing on the definition of renewable hydrogen was an important step. However, at the moment the EU still lacks a harmonized certification scheme for green hydrogen which businesses in Europe (and abroad) could use to prove compliance with the requirements set out in the delegated act. Nor are there any certification bodies accredited by the European Commission to provide such services. The hydrogen industry has been calling for such a scheme, as well as for EU guidance for certification and accreditation bodies (Parkes, 2023). Currently, two schemes are applying for accreditation with the Commission, including CertifHy, a voluntary European scheme developed by the Clean Hydrogen Partnership. In the meantime, some Member States like Denmark are moving forward with their own national schemes. Harmonized standards and transparent certification mechanisms are also important for international hydrogen projects that are expected to cater to the European market.

Text Box 1 Defining renewable hydrogen in the EU

Green hydrogen, by definition, is produced with renewable electricity. The EU's Delegated Acts to the Renewable Energy Directive Recast (RED II) that came into force in July 2023 set out conditions when electrolytic hydrogen is recognized as renewable (a "RFNBO"). In some cases, proving this attribute is straightforward. This is, for example, when the electrolyzer is directly connected to a dedicated renewable energy facility. Another option is to source grid electricity in a bidding zone where the share of renewables exceeds 90%. However, very few countries in the European Economic Area can satisfy this condition: this mainly applies to Sweden and Norway. In most cases, electrolyzers will have to be connected to the grid in countries with a large share of

---

[1] A delegated act is a "non-legislative act adopted by the Commission to supplement or amend certain non-essential elements of a legislative act" Delegated acts have the advantage of being easier to adopt compared to a full-scale revision of a legislative act.

non-renewable power generation. In this case, hydrogen producers will have to comply with the requirements of additionality, temporal correlation and geographic correlation to have their hydrogen recognized as a RFNBO.

The additionality requirement, which will enter into force in January 2028, is satisfied if the hydrogen producer concludes a power purchase agreement (PPA) with a new renewable energy facility (defined as not older than 36 months). Early movers that begin producing hydrogen before 2028 have the advantage of only needing to comply with the additionality requirement starting from 2038. There is also a special exception for low-carbon bidding zones (mainly nuclear-dominated) with an emission intensity below 18 gCO_2eq/MJ. In such zones, hydrogen producers will not have to prove additionality but they will still need to sign renewable PPAs and comply with temporal and geographic correlation.

Temporal correlation requires that hydrogen is produced only when there is sufficient renewable electricity available from the installation with which it has signed the PPA. In a hard-won compromise, the delegated act introduces monthly matching until 1 January 2030 and hourly matching afterwards. Finally, geographic correlation is satisfied when the electrolyzer and the renewable energy facility are located in the same electricity bidding zone (in a majority of cases in the EU, bidding zones are defined by national borders), in a neighboring zone with a higher electricity price, or an offshore bidding zone.

3.2.2 Carbon Pricing

Putting a price on CO_2 emissions in a cap-and-trade system is a fundamental element of the EU's approach to promoting decarbonization. Currently, the EU's Emissions Trading System (EU ETS) covers some 40% of the EU's emissions. Revenues from the auctioning of ETS allowances are channeled to the EU Innovation Fund and the Modernization Fund (see sections below), which are important sources of funding for many of the EU's energy transition and hydrogen-related actions. In its present form, however, the EU ETS cannot yet serve as a sufficient incentive for hydrogen investments. Even though the CO_2 price temporarily exceeded the 100 EUR threshold in February 2023, it is still not high enough to encourage capital-intensive investments into clean hydrogen in industries like steel or fertilizers. Secondly, the producers and consumers of unabated hydrogen in the EU (e.g. in refineries or ammonia and methanol production) are currently benefiting from free emission allowances issued to protect them from "carbon leakage", i.e. competition from jurisdictions that do not use carbon pricing. Carbon contracts for difference (CCfDs) planned in the EU and already launched in the Netherlands and Germany, are designed to compensate the difference between the current ETS price and the real cost of abatement, strengthening the incentive to decarbonize.

Moving forward, however, carbon pricing will become more important for hydrogen development in the EU. As part of the wide-reaching reform of the EU ETS agreed in May 2023, free allowances for energy-intensive industry will be gradually phased out, while the total number of allowances in the system overall will be steeply reduced, resulting in higher CO_2 prices. In parallel, in October 2023, the EU launched the first phase of the carbon border adjustment mechanism (CBAM) for a number of energy-intensive imports, including iron and steel, fertilizers, aluminium, and hydrogen. It mandates EU importers to report on embedded emissions in their products and, from 2026, purchase ETS allowances to compensate for these embedded emissions. The ETS will also be expanded to include maritime shipping (a hard-to-abate sector that is expected to use hydrogen-derived clean fuels in the future), and a separate emissions trading system will be set up for buildings, road transport, and fuels, increasing the pressure to decarbonize.

3.2.3   Incentivizing Hydrogen Demand

Stimulating demand for clean hydrogen is one of the most difficult challenges at this early stage of hydrogen development. The EU has been ahead of other global actors in developing several policies aimed at incentivizing demand in industry and transport. A central instrument are sector-specific RFNBO quotas for industry and transport, which are included in the new Renewable Energy Directive (RED III) adopted in October 2023. For industry, the quotas are a 42% share of RFNBOs in total hydrogen use (both as a fuel and feedstock) by 2030 and 60% by 2035, with some caveats.[2] In the transport sector, member states will be obliged to implement a combined quota of 5.5% of biofuels and RFNBOs (of which at least 1% must be RFNBOs) by 2030.

Despite growing evidence that hydrogen mobility is unlikely to compete with battery electric passenger vehicles in the future, the EU is also introducing mandatory deployment targets for hydrogen refueling infrastructure. The Alternative Fuels Infrastructure Regulation (AFIR), which became law in September 2023, requires member states from 2030 onwards to deploy hydrogen refueling infrastructure serving both cars and trucks in all urban nodes and every 200 km along the entire European TEN-T core network. By providing a sufficiently dense refueling network across the EU, AFIR is expected to raise consumers' interest in hydrogen vehicles and thereby in hydrogen-derived fuels.

The EU has also adopted policies that are likely to incentivize demand in the maritime and aviation sectors. The FuelEU Maritime Regulation, which entered into

---

[2] It is possible for member states to discount the contribution of RFNBOs in industry use by 20 per cent under two conditions: (a) the member state has contributed to the EU's overall renewable energy target of 42.5 per cent by 2030; and (b) if the share of hydrogen from fossil fuels consumed in the member state does not exceed 23 per cent in 2030 and 20 per cent in 2035. This caveat has been introduced to give nuclear-based hydrogen a role in industrial decarbonization.

force in October 2023, sets carbon intensity reduction targets for vessels above 5000 tons calling at European ports (which are responsible for 90% of total GHG emissions from the maritime shipping sector) from January 2025. The targets progressively increase from 2% by 2025 to 80% by 2050. In addition, the regulation introduces a 2% renewable fuels usage target by 2034.[3] In turn, the ReFuelEU Aviation regulation aims at decarbonizing aviation through biofuels and synthetic fuels (also known as e-fuels, such as e-kerosene made by combining carbon with green hydrogen). The regulation introduces an obligation for aviation fuel suppliers to ensure that all fuel made available to aircraft operators at EU airports contains a minimum share of sustainable aviation fuels (SAF—currently consisting primarily of biofuels) from 2025 including a minimum share of synthetic fuels. SAF quotas progressively increase until 2050, from 2% by 2025 to 70% by 2050, and synthetic fuel shares in the blend should reach 1.2% in 2030, progressively increasing until reaching 35% in 2050.

3.2.4 Regulating Future Hydrogen Infrastructure

If the EU is to use hydrogen at scale, developing some form of a cross-border hydrogen transport infrastructure will be necessary in order to connect renewables-rich areas, e.g. Spain, with industrial demand centers, such as those in northwest Europe. This will most likely be done as a combination of repurposing some natural gas pipelines, building new dedicated ones for hydrogen and adapting sea ports for handling clean fuels. There is a choice to be made between a leaner, "no-regret" infrastructure accompanied by a strong focus on regional hydrogen valleys that would bundle demand, and a dense—and costlier—network that intersects the entire European continent. The latter vision is supported by European gas TSOs, which are understandably interested in securing a future use for their gas pipeline network assets and have launched a "European Hydrogen Backbone" initiative (currently uniting 33 members, including those in the European neighborhood) mapping hydrogen corridors in Europe up until 2040.

Infrastructure regulation is a highly complex terrain, as the Commission has to navigate the diverging interests of Member States, gas TSOs and users. Important and cost-heavy decisions will have to be taken in the absence of clarity on the expected volumes of future hydrogen supply and demand as well as the possibility that some industries will relocate closer to hydrogen production sites. For businesses, on the other hand, not knowing whether there will be hydrogen infrastructure available in a given region has a significant impact on the project design and final investment decisions. There are difficult political choices to be made, such as on how to deal with the valuation and decommissioning of no longer needed gas pipelines, given the expected fall in gas demand in the EU in net-zero scenarios. Whereas to date electricity and gas regulation has been designed in silos, hydrogen development requires energy system integration, which will pose a number of difficult governance

---

[3] If the Commission reports that in 2031 RFNBO amount to less than 1% in fuel mix.

questions and require skillful EU-level coordination. Also, unlike the EU's ambitious climate target horizons reaching until 2050, infrastructure development planning has shorter timespans: gas network planning is set out in so-called "ten-year network development plans" (TYNDPs) that are developed by gas TSOs. The 2022 edition of the TYNDP included, for the first time ever, plans for hydrogen infrastructure (ENTSOG, 2022).

The Commission's Hydrogen and Decarbonised Gas Package proposal, which was first presented in December 2021 and adopted by the Council in May 2024, is supposed to bring clarity to most of the key questions related to hydrogen network regulation. The package offers incentives for expanding hydrogen infrastructure throughout the EU, such as freeing cross-border hydrogen flows from taxes. It also allows the blending of up to 5% of hydrogen into natural gas networks, which is likely to encourage the integration of hydrogen into the gas infrastructure but has been criticized for promoting an inefficient use of the limited hydrogen that is available. Importantly, the package also proposes opening LNG terminals and gas storage facilities for hydrogen, in the EU's bid to make sure its fossil fuel infrastructure is "hydrogen-ready" (European Commission, 2021a). The policy package envisions the creation of the European Network of National Hydrogen Operators (ENNOH), similar to the existing network of gas transmission system operators (ENTSO-G). ENNOH would be responsible for coordinating hydrogen infrastructure on the basis of Ten-Year Network Development Plans (TYNDPs). In the transitional period until 1 January 2027, as ENNOH is being set up, ENTSO-G will be responsible for such planning.

In essence, the Commission's approach is to regulate the hydrogen infrastructure in the same way as natural gas markets are regulated today. This includes an important set of consumer rights, unbundling of transmission and distribution, open third-party access, integrated network planning, and independent regulatory authorities. Critics point out, however, that this model, developed over the span of decades, may be overly restrictive for the fledgling hydrogen system (Stam et al, 2023). There need to be sufficient incentives for future hydrogen network operators, as well as ample space for learning and experimentation.

3.3 Investment Support and Production Subsidies

In addition to research funding and regulation, the EU has focused on investment support, production subsidies and promoting an industrial alliance for a hydrogen economy. Currently, two key EU-level sources of funding for the green transition are revenues from trade in emissions allowances within the EU ETS, as well as the Recovery and Resilience Facility, the centerpiece of the NextGenerationEU pandemic recovery package. Funds from the RRF are allocated to and administered by Member States. The EU also offers budget guarantees to leverage private capital for investment in hydrogen. Furthermore, the EU determines conditions under which Member States are allowed to subsidize hydrogen projects from their national budgets though so-called "state aids".

With the expected rise in CO_2 prices, revenues from emission allowances are likely to increase in the future. RRF-based funding, however, is scheduled to run out by the end of 2026, with no possibility of new borrowing in the absence of a major crisis. (Some remaining funds from the RRF are likely to be integrated into new support schemes though). There is thus an open question of how to sustain a sufficiently high level of green investment in the EU beyond 2026, with analysts estimating the resulting gap in green investment at 34–35 billion EUR annually until 2030 (Pisani-Ferry et al., 2023).

In gauging hydrogen investment needs, the EU has put an emphasis on close cooperation with the industry. In July 2020, the Commission launched the Clean Hydrogen Alliance, which brings together a wide range of hydrogen stakeholders from around Europe with the aim to "promote investments and stimulate clean hydrogen production and use" (European Commission, n.d.). The Alliance has identified a large pipeline (750+) of viable hydrogen projects in Europe ready for investment. In 2022, it also organized the important Electrolyser Summit, where European manufacturers pledged to increase the combined electrolyzer manufacturing capacity tenfold, to 17.5 GW by 2025, to put the EU on the path to meeting its REPowerEU targets (European Commission, 2022d).

3.3.1 EU-Level Support Schemes

Among existing EU-level instruments, some target hydrogen directly and others can support hydrogen-related initiatives as part of their broader mandate to promote the energy transition. The Innovation Fund is the most important EU-level funding instrument for low-carbon technologies, including hydrogen. It is also one of the world's largest support programs of its kind. The Innovation Fund was launched in 2020 and has been endowed with the revenues from the auctioning of 450 million EU ETS allowances between 2020 and 2030. Its budget thus depends on the future development of the CO_2 price in the EU. Current estimates put it at ca. 40 billion EUR until 2030, a figure that might increase in the future. The Fund focuses on commercializing innovative clean technologies that are past the research stage but are not yet bankable; the EU's contribution is capped at 60% of the CAPEX+OPEX costs. As of May 2024, the Innovation Fund has issued four large-scale calls, each of them explicitly open to hydrogen projects. Supported areas have included renewable hydrogen production, hydrogen solutions for industrial decarbonization, and manufacturing of components for electrolyzers and fuel cells. The Innovation Fund is open to all Member States, as well as Norway and Iceland.

The Innovation Fund is also the main source of funding for EU hydrogen production subsidies. The European Hydrogen Bank, a newly established vehicle with an initial budget of 3 billion EUR, will grant ten-year subsidies to green hydrogen produced in the EU, Norway or Iceland (and in the future, to imported hydrogen as well). The pilot Renewable Hydrogen Auction was held between November 2023 and

February 2024, attracting as many as 132 bids from 17 countries. Seven projects (from Spain, Portugal, Norway and Finland), planning to produce 1.58 tonnes of renewable hydrogen over a period of 10 years, were selected as winners, to be awarded 720 million EUR in total. While the auction offered a fixed-premium subsidy of up to 4.5 EUR/kg, the winning bids requested a much lower level of subsidy, ranging between 0.37 EUR and 0.48 EUR/kg (European Commission, 2024a). The second round of hydrogen auctions is planned for late 2024, with a budget of 1.2 billion EUR (Parkes, 2024). While European hydrogen stakeholders have welcomed the long-awaited launch of the auctions, there are fears that successful bidders may turn out to be those that use lower-cost Chinese electrolyzers. This would result, in essence, in subsidizing the Chinese electrolyzer industry. There is an ongoing debate in the EU on whether to introduce additional ESG requirements for equipment in order to compensate for the unfair advantage of Chinese elelectrolyzer makers, who have lower labor costs and less stringent environmental requirements (Collins, 2023b).

Another important planned subsidy to be funded out of the Innovation Fund are Carbon Contracts for Difference (CCfDs) announced in the REPowerEU package. The goal of CCfDs is to derisk investment in climate-friendly technologies and processes in hard-to-abate sectors such as steel processing, cement production, and the chemical industry, including green ammonia and green methanol production. CCfDs offer-long term contracts to compensate for the difference between the actual CO_2 abatement cost[4] and the reference CO_2 price in the EU ETS that companies would have to pay if they kept using conventional fuels. As such, CCfDs facilitate the switch from conventional to renewable hydrogen in industrial processes and promote the deployment of clean hydrogen and its derivatives to decarbonize industry. While EU-level CCfDs have not been introduced yet, some member states such as the Netherlands and Germany have already launched their own CCfD-type schemes.

Most projects that have received support from the Innovation Fund are imple-mented in the more developed Member States, reflecting their better capacity for innovation. Addressing such structural imbalances in the EU is one of the aims of the Modernization Fund, which was set up in 2020 with the goal of supporting lower-income Member in their transition to climate neutrality. The Fund's beneficiaries include Bulgaria, Croatia, Czechia, Estonia, Greece, Hungary, Latvia, Lithuania, Poland, Portugal, Romania , Slovakia and Slovenia. The Modernization Fund's budget is funded with revenues from the auctioning of 2% of EU ETS allowances between 2021 and 2030, as well as additional allowances transferred by five bene-ficiary states (Croatia, Czechia, Lithuania, Romania and Slovakia). A wide range of hydrogen-related actions are eligible for funding of up to 70% of project costs, including renewable hydrogen production and use, hydrogen mobility and the repur-posing of natural gas infrastructure for low-carbon or renewable hydrogen. The investments out of the Modernization Fund are managed by the beneficiary Member States themselves and are thus subject to state aid clearance by the Commission. Support can be granted in the form of grants, premiums, guarantee instruments,

---

[4] The abatement cost is the actual cost of deploying climate-friendly fuels or technologies and is fixed in the contract as the CO_2 strike price.

loans or capital injections, depending on the decision of the beneficiary Member States.

The Just Transition Fund, in turn, focuses on regions most affected by the transition to climate neutrality, especially those with a strong presence of the coal industry. The Fund has a budget of 17.5 billion EUR for 2021–2027 (in 2018 prices; this includes 10 billion EUR from NextGenerationEU) and offers grants, loans and guarantees to promote economic diversification and retraining. It is open to all member states but its main beneficiaries have been Poland, Germany, Spain, and Eastern European Member States. To access support from the Fund, Member States need to identify regions most in need of support and draw up territorial just transition plans up to 2030. Hydrogen-related projects can be supported by the Fund if they meet other eligibility requirements, such as assisting in the transition of industries or businesses to a greener model.

Yet another support instrument is InvestEU, a strategic blended finance program. It facilitates the EU's transition to climate neutrality by offering repayable loans to riskier and more innovative projects, with the European Investment Bank acting as the main financial partner. The program offers an EU budget guarantee of 26.2 billion EUR with the aim of leveraging 372 billion EUR in public and private investment. Under its sustainable infrastructure window, InvestEU can support a wide range of hydrogen projects including hydrogen production, supply and storage, hydrogen refueling infrastructure, and others. One prominent example is the EU-Catalyst Partnership, which is managed on the EU side by InvestEU and is funded by Horizon Europe and the Innovation Fund. The partnership promises to mobilize up to 820 million EUR for EU-based projects in four areas, including clean hydrogen (European Commission, 2021c).

In the area of infrastructure, a key strategic investment instrument is the Connecting Europe Facility (CEF). Under the TEN-E (Trans-European Networks for Energy) regulation, cross-border energy infrastructure projects in the EU recognized as Projects of Common Interest (PCI) are eligible for EU funding from the Facility (European Commission, 2021d).[5] The revised TEN-E regulation, which entered into force in 2022, also introduces a new category called Projects of Mutual Interest (PMI) to promote infrastructure links with third countries. Supported types of infrastructure include hydrogen, electrolyzers, smart gas and electricity grids, CO_2 networks and energy storage.

The amount of EU funding available for infrastructure projects is relatively limited: the Connecting Europe Facility (CEF), which has an energy pillar for financing PCIs, has a budget of only 5.84 billion EUR for the energy sector for the years 2021–2027. For this reason, obtaining a PCI or PMI status is no guarantee of receiving CEF funding; in fact, most PCIs in the past were implemented without CEF support. However, PCI/PMI projects still benefit from other important

---

[5] To meet this requirement, they need to fulfil the following requirements: have a significant impact on at least two member states and contribute to promoting market and network integration, enhancing security of supply, promoting energy market competition, and/or contribute to the sustainability transition.

advantages, such as political support and faster permitting and environmental assessment procedures. In November 2023, the European Commission published a list of 166 infrastructure projects selected for the PCI/PMI status, including 65 hydrogen and electrolyzer projects (European Commission, 2023g). Some examples include the proposed H2Med hydrogen pipeline connecting Portugal, Spain and France, the AquaDuctus pipeline that aims to bring green hydrogen from North Sea offshore wind farms to German industrial consumers (Buljan, 2023), as well as various electrolyzer facilities, hydrogen storage projects, and cross-border hydrogen valleys in Denmark, France, Germany, the Netherlands, Spain, and other countries (European Commission, 2023f). Following endorsement by the Parliament and the Council, the list became final in April 2024.

On the whole, the amounts of EU funding available are dwarfed by the gargantuan investment needs for a future hydrogen infrastructure. The Commission estimated that achieving the REPowerEU 2030 goals on hydrogen would require investments in the range of 28–38 billion EUR for EU pipelines and 6–11 billion for storage (European Commission, 2022b, p. 7). Further into the future, the projected costs are even higher. In 2022, the European Hydrogen Backbone Initiative estimated the investment needs for a 53,000 km backbone by 2040 at 80 to 143 billion EUR. This would include some 60% repurposed pipelines and 40% new ones (European Hydrogen Backbone, 2022). In a November 2023 report based on updated data, the EHB revised the estimates upwards, citing the rising costs of all components (Martin, 2023b). Given the sheer scale of the investment needed, the EU's task of developing a clear regulatory framework and creating the right incentives becomes all the more crucial.

Finally, the EU has worked on bolstering its industrial policy and promoting cleantech manufacturing. In June 2023, the Commission presented a proposal for the Strategic Technologies for Europe Platform (STEP) to support the uptake and manufacturing of key strategic technologies. STEP focuses on three priority areas: deep and digital technologies, clean technologies (including hydrogen) and biotech. In cleantech, STEP states its aim as supporting the rapid "development and deployment of home-grown clean energy technologies, energy storage system innovations and decarbonisation solutions in the EU" (European Commission, 2023a). Initially, this was imagined as the "European Sovereignty Fund" and was expected to become a major EU-level facility open to all Member States and a key tool of European industrial policy. In the shape proposed in June 2023, however, STEP has not lived up to the expectations. It does not come with significant amounts of new funding but rather aims at reshuffling available instruments to use them more efficiently. In practice, this mostly applies to the remaining funds left from NextGenerationEU. It also calls on Member States to provide a top-up contribution of 10 billion EUR, which would go to key instruments like InvestEU. The hope is that this public financing will be able to attract up to 160 billion EUR in private capital (Bourgery-Gonse, 2023).

3.3.2 Leveraging Member State Support

Due to the limitations of EU-level funding, much of the public financing for hydrogen-related measures in the EU will have to come from Member State budgets. Acknowledging this, the EU has made it easier for Member States to grant national subsidies (known as "state aids") in the area of green transition. Member States have also made active use of funds from the Recovery and Resilience Facility (RRF). In fact, it was in the national Recovery and Resilience Plans developed to access RRF funds that many member states included support for hydrogen development (some of them for the first time ever), with the combined amount reaching 10 billion EUR (European Commission, 2023d).

Important Projects of Common European Interest (IPCEI), introduced in 2014, are a prominent strategic instrument to support R&D and first industrial deployment of strategically important technologies as well as to promote infrastructure links. IPCEI projects are required to have a strong European component and involve at least four Member States. Obtaining the coveted IPCEI status makes projects eligible for state aids, which has the potential to unlock a significant amount of additional private funding. IPCEI undertakings typically bundle tens of individual projects united by a common theme; each project involves participating companies from several member states. As of May 2024, only ten undertakings in total have succeeded in obtaining the IPCEI status, and four of them involve hydrogen. The large-scale hydrogen IPCEI, Hy2Tech (2022; 5.4 billion EUR), Hy2Use (2022; 5.2 billion EUR), Hy2Infra (2024; 6.9 billion EUR), and Hy2Move (2024; 1.4 billion EUR) involve a large number of participating Member States, companies and projects (European Commission, 2024b). Together, they are expected to unlock an additional 24.5 billion EUR in private investment. The four IPCEI support a wide range of projects along the entire hydrogen value chain, including production, transport, storage, infrastructure, and end uses. Yet IPCEI are no silver bullet: applying for an IPCEI status requires a great deal of time and administrative effort, and chances of success are limited. In addition, even when selected, project developers often face long waiting times in accessing the funds to be disbursed by national governments (Martin, 2023a).

Going beyond IPCEI, in reaction to the subsidies announced under the US's Inflation Reduction Act in 2022, the EU has significantly relaxed constraints on granting national subsidies to green transition projects. In March 2023, the European Commission adopted a new Temporary Crisis and Transition Framework (TCTF) to enable Member States to provide support measures in key sectors of the transition to net zero. It introduces the strongest loosening of state aid rules in the EU to date. The new TCTF extends and amends the Temporary Crisis Framework originally adopted in March 2022 (and subsequently prolonged) to help Member States deal with the economic consequences of Russia's war on Ukraine. The TCTF allows state aid for the production and use of both renewable hydrogen and renewable hydrogen-derived fuels. In addition to prolonging support for renewable energy and energy storage and industrial decarbonization (both of which are relevant to renewable hydrogen), the TCTF introduces a new component in the form of support for the manufacturing of a wide range of green technologies, including electrolyzers. TCTF enables Member

States to provide aid in the form of direct grants or, alternatively, tax incentives, loans or guarantees up to a certain percentage of eligible costs. It is likely to become a key instrument guiding "green" state aid in EU countries. The TCTF allows grant support measures until 31 December 2025.

In parallel, the Commission has also revised the General Block Exemption Regulation (GBER). Under GBER, Member States can implement certain aid measures without going through the time-consuming notification and approval procedure. GBER allows Member States to support key sectors for the net-zero transition, including hydrogen, carbon capture and storage, zero-emission vehicles and energy performance of buildings. In addition, it has increased the limits for state aid for undertakings in less-developed regions in the EU.

The loosening of state aid rules has been controversial. A number of Member States—including Central and Eastern European states, but also Belgium, Denmark, Finland, Ireland, the Netherlands, and Sweden—have called on the Commission to avoid a subsidy race and emphasized the importance of preserving a competitive single market (Euractiv, 2023). Indeed, these measures bear the risk of creating a multi-speed transition, where stronger and larger economies use their resources to support their industry and harness the benefits of decarbonization, while poorer states fall behind (Quitzow et al., 2023).

4 External Dimensions of EU Hydrogen Policy

International cooperation is one of the cornerstones of the EU's vision of a future hydrogen economy. The EU and the European Commission (as well as individual Member States) have sought to shape the rules of engagement in the future international hydrogen market and have actively participated in various multilateral governance fora. In the REPowerEU package of proposals, the EU explicitly acknowledged the need for future large-scale imports of clean hydrogen, both for decarbonization and energy security purposes. It also put in place a (non-binding) import target of up to 10 million tons of hydrogen annually by 2030. In this vein, the EU has begun to develop clean hydrogen partnerships with a range of third countries; this includes not only those countries that are well-positioned to emerge as early exporters but also those with a high renewable energy potential and export ambitions but which find themselves at a very early stage of hydrogen development.

4.1 Multilateral Governance

At the global level, the European Union and the European Commission have sought to actively participate in a number of international bodies and fora promoting a hydrogen economy. The EU is a member of the prominent Mission Innovation global initiative, which is a global forum for the governments of leading economies committed to speeding up the energy transition. Mission Innovation puts a strong emphasis on public–private partnerships and has a dedicated work stream on hydrogen ("Clean Hydrogen Mission") where the EU plays a key role as a co-lead (together with Australia, Chile, the UK and the US). The Clean Hydrogen Mission has several key priorities, including R&D promotion, reducing the final costs of clean hydrogen to 2 USD per 1 kg by 2030, and reaching 100 hydrogen valleys globally by 2030. The EU, through its Clean Hydrogen Partnership R&D initiative discussed above (in the section on Research and Innovation), plays a particularly important role in the latter effort. It co-funds the Mission Innovation Hydrogen Valley platform, which showcases flagship hydrogen projects around the world to promote global cooperation (Clean Hydrogen Partnership, n.d.). The EU itself is an undisputed leader in the number of developing hydrogen valleys compared to other global regions.

As a member of the Group of 7 (G7), the EU is now also a party to the Hydrogen Action Pact adopted in May 2022. The Pact has the goal of facilitating joint action in the area of Power-to-X, promoting the use of hydrogen and its derivatives and streamlining the implementation of existing multilateral initiatives. The EU also participates in the newly launched G20's International Hydrogen Economy Initiative. In July 2023, G20 members adopted "G20 High Level Voluntary Principles on Hydrogen", where they pledge to promote collaboration on the global harmonization of standards and certification, technological acceleration, information sharing, mobilization of investment and finance, and hydrogen trade in alignment with WTO rules (G20 Energy Ministers, 2023, pp. 15–16).

Furthermore, the European Commission is one of the 25 contracting parties in the International Energy Agency's influential Hydrogen Technology Collaboration Programme (Hydrogen TCP), which dates back to 1977 and facilitates and coordinates information exchange and innovative R&D activities in the area of hydrogen (IEA, n.d.-b). The Commission is also a member of the International Partnership for Hydrogen and Fuel Cells in the Economy (IPHE), established in 2003, which works on accelerating market deployment of hydrogen and fuel cells as well as policy and regulatory activities. IPHE has two task forces dedicated to hydrogen: the Hydrogen Production Analysis Task Force and the Hydrogen Trade Rules Task Force. In the past years, IPHE has issued several working papers developing a methodology for determining greenhouse gas emissions associated with hydrogen production. This is likely to have an impact on international clean hydrogen standards. The studies were designed and directed by several individuals, including Tudor Constantinescu, Principal Adviser to DG Energy at the European Commission (IPHE, 2021, 2023, p. 11). Finally, the European Commission is a member of the Clean Energy Ministerial Hydrogen Initiative (CEM H2I), a voluntary multi-government initiative launched in 2019 and coordinated by the International Energy Agency. CEM H2I aims at promoting international collaboration "on policies, programs and projects to accelerate the commercial deployment of hydrogen and fuel cell technologies across all sectors of the economy" (IEA, n.d.-a). Some examples of areas of work include the Global Ports Hydrogen Initiative, a global forum to promote clean hydrogen use in industrial port areas.

4.2 Bilateral Hydrogen Cooperation Within EU Climate and Energy Diplomacy

The EU's plans to import hydrogen has led the bloc to assign this clean energy carrier a more prominent place in its international partnerships. The EU's nascent hydrogen diplomacy is informed and shaped by the Union's wider climate and energy diplomacy. It is also influenced by the Union's interest in greater energy security and diversification of suppliers, reflecting rising geopolitical tensions, growing supply chain risks and the accelerating global competition on clean technologies and manufacturing.

As a self-identified global climate leader, the EU has declared climate and energy diplomacy a "core component" of its foreign policy (Council of the EU, 2023a). As noted in the European Council's conclusions in 2021, "the coherent pursuit of external policy goals is crucial for the success of the European Green Deal" (Council of the EU, 2021). The overarching goal of the EU's climate and energy diplomacy is to accelerate the global energy transition and keep the global temperature rise to under 1.5 degrees Celsius by facilitating the uptake of renewable energy and energy efficiency technologies as well as assisting other nations with climate readiness and adaptation and advocating the phaseout of coal (Council of the EU, 2021, 2023a). In line with these efforts, the EU, together with its member states, has emerged as the world's leading provider of public climate finance (Council of the EU, 2023b). In addition, the EU is planning to use Horizon Europe funds to fund research and innovation cooperation in the area of renewable energy and green hydrogen in partnership with the Union of the Mediterranean and the African Union.

In its External Strategy for Engagement with Energy Partners, adopted in May 2022, the European Commission announced its intention "to conclude hydrogen partnerships with reliable partner countries to ensure open and undistorted trade and investment relations for renewable and low carbon fuels" (European Commission, 2022a). Yet for now, hydrogen diplomacy is still a novel field of engagement for the EU, where it lags behind proactive Member States like Germany or the Netherlands, which have signed a slew of hydrogen accords with nations around the globe (see Nunez & Quitzow, 2023; Stam et al., 2023). Still, the EU has been expanding its cooperation on hydrogen with other states. In November 2022, at COP27 the EU signed memorandums of understanding on hydrogen development with the host nation Egypt, as well as with Namibia and Kazakhstan. (With the latter two, parallel accords on cooperation on critical raw materials were signed as well). In February 2023, the EU signed a Memorandum of Understanding with Ukraine on cooperation in the area of renewable gases, including hydrogen and biomethane. Further agreements were signed in 2023, including with Tunisia, Argentina and Uruguay. Notably, all these partner countries find themselves at very early stages of clean hydrogen development and the energy transition more generally, lack developed infrastructural transport links to Europe or are facing serious geopolitical challenges, most notably Ukraine. Given these obstacles, these partnerships are thus unlikely to result in hydrogen exports to Europe in the near future but have longer time horizons.

In parallel, the EU has also undertaken some efforts to develop cooperation with international hydrogen frontrunners. In 2021, the EU and Japan set up "the Green Alliance", aiming at promoting a green economy, including hydrogen development. This was expanded in December 2022, when the two signed a Memorandum of Understanding to "spur innovation and develop an international hydrogen market". However, until now, there have been few activities on that front (European Commission & Government of Japan, 2022). A "Green Partnership" with South Korea followed in May 2023, with hydrogen identified as one of priority areas.

Furthermore, the External EU Energy Strategy mentions developing cooperation on hydrogen and other renewable gases with today's natural gas exporters (European Commission, 2022a). This would allow the EU to build on the existing cooperation and political ties and contribute to the green transition in those fossil fuel-dependent countries. It might also involve the use of (repurposed) fossil fuel infrastructure, such as natural gas pipelines, LNG terminals (Riemer et al., 2022), and storage facilities, thus avoiding stranded assets. However, progress in this area has been limited so far. In May 2022, the European Commission launched a "Strategic Partnership with the Gulf Cooperation Council", which emphasized the region's role as a potential supplier of renewable hydrogen to Europe. Indeed, the Gulf monarchies like Saudi Arabia, Oman and the UAE are increasingly likely to emerge as early hydrogen exporters. Yet in practice, the EU has actually been criticized for "de-prioritising" relations in the area of the energy transition with these countries, failing to engage in serious discussion on hydrogen partnerships with any of them (Bianco, 2023). For now, the GCC countries have mainly focused on developing exports to Asia as well as on efforts to build local green industries, as part of their hydrogen development plans.

Yet even closer to home, hydrogen cooperation with neighbors has uncertain prospects. The UK has an ambitious hydrogen strategy and good gas infrastructure links to Europe, yet it has prioritized the development of a domestic hydrogen market and production facilities, with little attention paid to potential exports to the EU. That said, devolved administrations such as Scotland have shown strong interest in supplying the EU with green hydrogen or derivatives in the future (Weko, 2023). Norway may be a more promising partner, even though its preference for exporting blue hydrogen while using green hydrogen domestically (Skjærseth et al., 2023; see also Kilpeläinen et al., 2023) is somewhat at odds with the EU's emphasis on renewable hydrogen. In April 2023, the EU and Norway launched a "Green Alliance", where hydrogen and CCUS technologies represent one of the areas of cooperation. In

November 2023, the planned RWE-Equinor pipeline aiming to bring Norwegian blue hydrogen (and later possibly green hydrogen) to Germany was granted the status of a Project of Mutual Importance (PMI) by the European Commission. This was the only hydrogen infrastructure project in the list of 10 PMIs selected by the Commission (European Commission, 2023f).

Canada is another potentially significant future hydrogen producer. Soon after Russia's invasion of Ukraine, the European Commission and the Canadian Government set up a dedicated working group on the green transition and liquefied natural gas (LNG) (European Commission & Government of Canada, 2023). The group also aims at "lay[ing] the foundation for the development of reliable hydrogen supply chain between Canada and the EU as well as to develop common approaches to standards and the certification of hydrogen" (European Commission, 2023c). However, Canada currently does not have the infrastructure in place to transport LNG or hydrogen to Europe and thus is unlikely to emerge as an early hydrogen supplier to Europe.

As for the United States, its relationship with the EU in the area of the energy transition has been complex, marked both by rivalry and cooperation. Some points of disagreement included President Trump's 2018 decision to impose import tariffs on EU steel and aluminium (the tariffs were scrapped in 2021), the US's criticism of the Carbon Border Adjustment Mechanism, or the EU's criticism of the Inflation Reduction Act's protectionist elements. Cooperation, by contrast, has tended to center on issues where the US needs the EU as an ally to counterbalance China, or where the EU needs the US to bolster its energy security. Since 2021, the EU and the US have been involved in complicated negotiations aimed at establishing a transatlantic market for green steel and aluminium, with import tariffs linked to CO_2 emissions—a market that is likely to disadvantage Chinese producers. In 2021, they also established the EU-US Trade and Technology Council (TTC), a forum aimed at deepening transatlantic trade and securing critical supply chains, especially for semiconductors. While the TTC does not directly focus on hydrogen, it was within its framework that the US and EU launched the "Transatlantic Initiative on Sustainable Trade" and the "Clean Energy Incentives Dialogue", which can serve as potential platforms to discuss a transatlantic hydrogen market in the future. Finally, in October 2022, the EU and US set up a joint Task Force on the IRA in order to address the EU's concerns about the Act's trade-distorting nature. At the same time, the IRA can also benefit Europe: by providing generous subsidies for hydrogen production, it may transform the US into an attractive supplier of competitively priced clean hydrogen to European offtakers, helping the latter achieve their climate targets (Urbasos, 2023). This would require close cooperation between the EU and US on aligning rules, standards and certification mechanisms. In the end, however, it remains uncertain whether the US will actually export subsidized hydrogen to Europe in the future.

4.3 Cross-Border Hydrogen Trade

While the EU has set a very ambitious hydrogen import target for 2030, a European framework for trade with third countries is still in the making. There is a benefit to having a coordinated, European-level approach: as Europe's experience during the gas crisis of 2022 painfully demonstrated, the scramble to individually secure energy supplies by Member States led to ballooning prices. In this spirit, the EU is planning to include hydrogen imports in the EU Energy Platform, initiated in April 2022 for joint energy purchases (European Commission, 2022c). The platform is intended to promote security of supply, pool demand and coordinate infrastructure use, with the first-ever EU tender for natural gas successfully taking place in May 2023. However, the timeline for including hydrogen in this mechanism is still unclear.

Another proposed measure is Europeanizing and expanding Germany's H2 Global mechanism, a scheme for subsidizing hydrogen imports from third countries. H2Global features double-sided auctions—for producers of hydrogen and its derivatives on the one end and for hydrogen offtakers on the other. The price difference between the two is compensated by the specially created intermediary company, Hint.co. H2Global offers ten-year guaranteed offtake contracts, helping to derisk investments in hydrogen production and supply to the European market. H2Global's first tender, for green ammonia, took place in November 2022. In June 2023, the EU Commissioner for Energy Kadri Simson and Germany's H2Global foundation announced plans to link the European Hydrogen Bank and H2Global to organize auctions open to funding from all Member States for hydrogen imports from third countries. The details of the scheme are currently being determined. While H2 Global's procedures have been criticized for being excessively bureaucratic, it will play an important role in testing and streamlining processes that may be later introduced in the EU.

4.4 Investing in the Hydrogen Economy in Third Countries

As a major provider of climate and development finance, the EU engages in a series of bilateral and multilateral initiatives in support of investments in the decarbonization of the energy sector. The EU's Neighbourhood, Development and International Cooperation Instrument—Global Europe (NDICI-GE) framework launched in 2021 is the main instrument for financing the EU's external actions in third countries and includes grants, guarantees and blended finance (Rizzi & Varvelli, 2023). A prominent part of this instrument is the European Fund for Sustainable Development Plus (EFSD+), which may also finance hydrogen-related projects.

The EU's Global Gateway, a major connectivity initiative launched in December 2021 as a response to China's Belt and Road Initiative (Lau & Moens, 2022), is expected to play a stronger role in EU support for hydrogen development abroad as well. Climate and Energy are one of the five priority areas of the initiative, which has been described as a "geopolitical instrument for long-term, sustainable infrastructure investments" (German Federal Foreign Office, 2023). It is supposed to offer partner countries—especially in the Global South—long-term sustainable partnerships that go beyond resource extraction or aid. Aiming to mobilize 300 billion EUR in private and public investment between 2021 and 2027, the Global Gateway adopts a Team Europe approach, pooling the resources of the EU, the Member States, and key European and national financial institutions such as the EIB, the EBRD or KfW (European Commission, 2021e). To what extent the goal of leveraging large-scale funds will be successful remains uncertain, however. Past efforts have shown that the amounts of private sector finance that EU facilities have leveraged in partner countries is extremely modest (Prontera & Quitzow, 2023).

Global Gateway has recently added a focus on investing in hydrogen economy in partner countries in various regions, although many of the announced projects are still at a very early stage. It is also uncertain how exactly these projects will be financed, although green bonds to be issued by the European Investment Bank and the European Central Bank are reportedly under consideration (Collins, 2023c). In Africa, flagship projects include a hydrogen power plant in Morocco, as well as green hydrogen production in Namibia based on the construction of an 85 MW solar plant equipped with electrolyzers (Team Europe, 2023a). According to a joint declaration signed with the Namibian government, the European Investment Bank (EIB) will potentially provide a loan of up to €500 m to finance renewable hydrogen and renewable energy investments (Green Hydrogen Organization, 2023, p. 9). In Kenya, the EU is planning to invest 3.4 billion EUR in climate and nature, with 12 million EUR in grants committed to leverage investment into green hydrogen. Another partner is Mauritania: in October 2023, the EU launched a Team Europe initiative in the country, with the goal of supporting its ambition to become a green hydrogen hub. The initiative is expected to fund capacity building and green energy infrastructure, as well as promote job creation and help put in place an appropriate "legal and fiscal framework" (Shumkov, 2023). In particular, the EU Technical Assistance Facility is assisting the country with developing a legal framework for hydrogen called the "hydrogen code" (Weko et al., 2023). In Latin America, the EU is planning to support the production of green hydrogen in Chile, Uruguay, Costa Rica, Argentina and Paraguay (Team Europe, 2023b). As part of these plans, in June 2023, the EU launched two cooperation initiatives on hydrogen with Chile. The first is the Team Europe Renewable Hydrogen Funding Platform for Chile, where the EIB and the German development bank KfW pooled 200 million EUR to support Chile's renewable hydrogen industry. The second is the Team Europe technical assistance program that will strengthen the conditions for the promotion of the renewable and sustainable hydrogen economy in the country. Finally, at the European Hydrogen Week in November 2023, the European Commission announced plans to make 2 billion EUR available under the Global Gateway initiative, with a share of this money going to support the development of the hydrogen value chain in Brazil. This would involve the construction of a 10 GW green hydrogen and ammonia production facility in the Brazilian state of Piaui, to be shipped to industrial offtakers in southeastern Europe (European Commission, 2023e).

India, with its much-touted hydrogen potential, is expected to become another important partner. In February 2023, the European Investment Bank officially joined the India Hydrogen Alliance, promising to support large-scale hydrogen hubs and other green hydrogen industrial projects throughout the country with up to 1 billion EUR in indicative investment and a dedicated credit facility (EIB, 2023).

Finally, the EU has also joined forces with other major donors in engaging with a number of fossil fuel-dependent emerging economies on hydrogen and the energy transition as part of so-called Just Energy Transition Partnerships (JETPs). JETPs are a financing mechanism offering grants and loans to promote green transformation in partner economies, especially those that are highly dependent on coal or other fossil fuels. The most prominent example is the Just Energy Transition Partnership (JETP) with South Africa. It represents an 8.5 billion USD agreement between the South African government and France, Germany, the US, UK, and the EU to promote the decarbonization of the South African energy sector. In line with the JETP priority areas, the South African government has published a JETP Investment Plan, which includes an investment plan for green hydrogen infrastructure and development projects. The EU is also a partner donor in several further JETPs, with Indonesia, Vietnam, and Senegal; some of these (e.g. the agreement with Vietnam) mention green hydrogen development among other measures (European Commission, 2022e).

4.5   International Repercussions of EU Hydrogen Policies

In addition to the EU's targeted international engagement on hydrogen, its policies and regulations in themselves play an important role in shaping hydrogen developments globally. Sustainability requirements for hydrogen are a prominent example. While in the EU, the prolonged debate preceding the adoption of the definition of renewable hydrogen has been viewed as problematic, the very same concerns about ensuring hydrogen's positive impact on climate are now fueling the debate in the United States. Leading US climate and energy think tanks like Energy Innovation Policy & Technology, the Princeton and MIT research centers (Bergman, 2023), as well as climate NGOs and some business groups, all call for the implementation of strict sustainability requirements. They insist on the same principles that the EU has implemented: additionality, time matching (temporal correlation in EU parlance), and deliverability (analogous to geographic correlation in the EU) (Esposito et al., 2023). In one form or another, the three principles are likely to become part of the US guidelines to be issued by the Treasury as well. In fact, the G20 Voluntary Principles on Hydrogen adopted in July 2023, which mention the harmonization of hydrogen standards, suggest that the US might be moving towards embracing EU rules (Collins, 2023a). Emerging economies like Brazil are also strongly drawing on the EU's sustainability requirements as they design their own regulatory framework for hydrogen.

In addition, given that the EU is expected to become a major import market for clean hydrogen, complying with its rules and regulations will be essential for any potential exporters. In the delegated acts on RFNBOs, the EU has made it clear that the strict definition of renewable hydrogen will apply to imported hydrogen as well. In fact, given that many potential hydrogen exporters have highly fossil-reliant grids (e.g. the Gulf States, Morocco or Australia), compliance with the additionality requirement will be particularly important in these contexts to ensure a positive climate impact (Bellona, 2023). Another oft-overlooked requirement is that grid-connected renewable energy facilities supplying an electrolyzer may not have received any other type of production subsidy (King & Spalding, 2023). Thus there are some concerns that the burden of complying with EU requirements may lead potential exporters to opt for other, less demanding, destinations, such as Asia.

Finally, the Carbon Border Adjustment Mechanism (CBAM), in itself the first instrument of its kind globally, has a clear external dimension and will impact international trade flows. The CBAM entered its transitional phase in October 2023. This means that European importers of hydrogen are now obliged to report embedded emissions in their products. In the next stage, from 2026 onward, they will have to pay an EU ETS price for emissions in their imported products (including hydrogen and several industrial goods using hydrogen). This is expected to incentivize non-EU producers to reduce the carbon footprint of their goods to remain competitive.

5   Conclusion

The European Union has invested a great deal of effort in promoting an ambitious vision of a European hydrogen economy. Major shocks like the Covid-19 pandemic and Russia's invasion of Ukraine, but also the US's adoption of the Inflation Reduction Act in 2022, have all served as strong accelerators for the EU's hydrogen plans. The EU is developing a pioneering regulatory framework for hydrogen, complete with stringent sustainability standards, production subsidies and demand quotas for industry and transport. The impacts of the EU's energy transition and hydrogen policies reverberate far beyond its borders. At the same time, intra-European divisions on issues like the definition of renewable hydrogen, sources of funding for European industrial policy or the future shape of the hydrogen infrastructure have repeatedly delayed the legislative process, affecting investor confidence. There are also fears that with the introduction of hydrogen auctions, European production subsidies may end up supporting Chinese electrolyzer manufacturers. In addition, there is a large looming gap in green transition investment once the RRF funding runs out in 2026. As EU-level funding is limited, the financial backing for the emerging hydrogen economy hinges to a significant degree on Member States' readiness to support it. The intensifying global cleantech race has led the EU to significantly relax its restrictions on Member State-level subsidies in the area of green transition. At the EU level, however, the bloc has not been able to agree on an ambitious new industrial policy instrument. The hydrogen funding landscape in Europe is fragmented and difficult to navigate, and accessing the funds comes with a protracted bureaucratic process. It remains to be seen whether the EU will be able to compete with other global players like the US, which offer sizeable and easily accessible support schemes.

Internationally, the European Commission and the EU are actively involved in the emerging multilateral hydrogen governance. Key focal points have included the harmonization and mutual recognition of hydrogen standards, as well as facilitating the creation of hydrogen valleys worldwide. When it comes to hydrogen diplomacy, however, the EU as a bloc lags behind its proactive Member States like Germany or the Netherlands. The EU's stated ambition is to go beyond securing hydrogen supplies by promoting the "green transformation" of the partner countries' economies, capacity building and local value addition. Focusing on co-benefits and supporting local socioeconomic development is important for attracting partners, especially in the Global South. However, to move from declarations to reality, the EU will need to prove its ability to mobilize large volumes of public and private financing. These efforts are still in their very early phase, and much more will have to be done. The EU will also need to decide on how to best address the IRA-related developments in the US in order to nurture a mutually beneficial energy transition and hydrogen partnership.

Finally, the EU's ambition to import 10 million tons of hydrogen by 2030 has been criticized for being unrealistic and misguiding expectations in partner countries. For large-scale imports to materialize, the EU would need to engage much more proactively with prospective exporter nations, including both emerging frontrunners, like the Gulf states, and future exporters in the Global South. While the latter may have excellent renewable energy endowments, they are faced with much higher costs of capital compared to Europe, the difficulties of mobilizing investment into hydrogen exports in the absence of an international hydrogen market, and a skills gap. Infrastructure for large-scale hydrogen imports to Europe is missing, too. Moving forward, potential exporters will also need to comply with the EU's stringent hydrogen sustainability requirements, which will require close cooperation with the EU on developing and harmonizing or accrediting hydrogen certification schemes. In addition, a number of high-potential hydrogen hopefuls are increasingly interested in downstream value creation, using green hydrogen or derivatives to manufacture industrial goods domestically. This potential reconfiguration of green industrial value chains is likely to have implications for the EU's hydrogen import plans and, more generally, for its green industrial strategy on the whole.

Acknowledgements Research for this chapter was financially supported by the German Federal Foreign Office within the framework of the project "Geopolitics of the Energy Transformation—Implications of an International Hydrogen Economy" (GET Hydrogen), funding reference number AA4521G125.

References

Bergman, A. (2023, August 2). 45V Hydrogen tax credit in the inflation reduction act: Comparing hourly and annual matching. *Resources for the Future*. https://www.rff.org/publications/issue-briefs/45v-hydrogen-tax-credit-in-the-inflation-reduction-act-comparing-hourly-and-annual-matching/

Bellona Europe. (2023). *Risks and challenges of importing hydrogen to Europe*. https://bellona.org/news/energy-systems/2023-09-risks-and-challenges-of-importing-hydrogen-to-europe

Bianco, C. (2023, June 16). Renewable relations: A strategic approach to European energy cooperation with the Gulf states. *European Council on Foreign Relations*. https://ecfr.eu/publication/renewable-relations-a-strategic-approach-to-european-energy-cooperation-with-the-gulf-states/

Bourgery-Gonse, T. (2023, June 21). Commission 'annihilated symbolic value' of EU Sovereignty Fund, leading MEP says. *Euractiv*. https://www.euractiv.com/section/economy-jobs/news/commission-annihilated-symbolic-value-of-eu-sovereignty-fund-leading-mep-says/

Bradford, A. (2019). *The Brussels effect: How the European union rules the world*. Oxford University Press.

Buljan, A. (2023, January 24). Partners apply for 'Project of Common Interest' status for aqua ductus green hydrogen pipeline. *OffshoreWIND.Biz*. https://www.offshorewind.biz/2023/01/24/partners-apply-for-project-of-common-interest-status-for-aquaductus-green-hydrogen-pipeline/

Clean Hydrogen Partnership. (n.d.). *Mission innovation hydrogen valley platform*. Retrieved 23 October, 2023, from https://h2v.eu/

Collins, L. (2023a, September 11). G20 leaders' declaration on hydrogen suggests US will adopt similar rules to EU on green H2 production. *Hydrogen Insight*. https://www.hydrogeninsight.com/policy/g20-leaders-declaration-on-hydrogen-suggests-us-will-adopt-similar-rules-to-eu-on-green-h2-production/2-1-1516033

Collins, L. (2023b, November 24). EU bosses want to stop hydrogen subsidies going to Chinese electrolyser makers, but Germany objects. *Hydrogen Insight*. https://www.hydrogeninsight.com/electrolysers/exclusive-eu-bosses-want-to-stop-hydrogen-subsidies-going-to-chinese-electrolyser-makers-but-germany-objects/2-1-1559837

Collins, L. (2023c, November 28). European Commission is considering Contracts for Difference for green hydrogen offtakers. *Hydrogen Insight*. https://www.hydrogeninsight.com/policy/exclusive-european-commission-is-considering-contracts-for-difference-for-green-hydrogen-offtakers/2-1-1561729

Council of the EU. (2021, January 25). *Council adopts conclusions on climate and energy diplomacy*. https://www.consilium.europa.eu/en/press/press-releases/2021/01/25/council-adopts-conclusions-on-climate-and-energy-diplomacy/

Council of the EU. (2023a). *Council conclusions on Climate and Energy Diplomacy*. https://www.consilium.europa.eu/media/62942/st07248-en23.pdf

Council of the EU. (2023b, October 2). *Financing the climate transition*. https://www.consilium.europa.eu/en/policies/climate-finance/

European Commission. (n.d.). *European clean hydrogen alliance. European commission*. Retrieved 23 October 2023, from https://single-market-economy.ec.europa.eu/industry/strategy/industrial-alliances/european-clean-hydrogen-alliance_en

European Commission. (2020). *A hydrogen strategy for a climate-neutral Europe* (COM/2020/301 final). https://eur-lex.europa.eu/legal-content/EN/TXT/?uri=CELEX:52020DC0301

European Commission. (2021a). *Hydrogen and decarbonised gas market package*. https://energy.ec.europa.eu/topics/markets-and-consumers/market-legislation/hydrogen-and-decarbonised-gas-market-package_en

European Commission. (2021b, June 16). *NextGenerationEU: Questions and answers on the recovery and resilience facility*. European Commission: Press Corner. https://ec.europa.eu/commission/presscorner/detail/en/qanda_21_3014

European Commission. (2021c, November 2). *Q&A: EU-Catalyst Partnership*. European Commission: Press Corner. https://ec.europa.eu/commission/presscorner/detail/en/qanda_21_5647

European Commission. (2021d, November 19). *Q&A on the fifth list of energy Projects of Common Interest (PCIs)*. European Commission: Press Corner. https://ec.europa.eu/commission/pressc orner/detail/en/qanda_21_6093

European Commission. (2021e, December 1). *Global Gateway: Up to €300 billion for the European Union's strategy to boost sustainable links around the world*. European Commission: Press Corner. https://ec.europa.eu/commission/presscorner/detail/en/ip_21_6433

European Commission. (2022a). *EU external energy engagement in a changing world* (JOIN/2022/23 final). https://eur-lex.europa.eu/legal-content/EN/TXT/?uri=JOIN%3A2022%3A23% 3AFIN&qid=1653033264976

European Commission. (2022b). *REPowerEU Plan* (COM/2022/230 final). https://eur-lex.europa. eu/legal-content/EN/TXT/?uri=COM%3A2022%3A230%3AFIN&qid=1653033742483

European Commission. (2022c). *EU energy platform*. https://energy.ec.europa.eu/topics/energy-sec urity/eu-energy-platform_en

European Commission. (2022d, May 5). *Hydrogen: Commission supports industry commitment to boost by tenfold electrolyser manufacturing capacities in the EU*. European Commission: Press Corner. https://ec.europa.eu/commission/presscorner/detail/en/ip_22_2829

European Commission. (2022e, December 14). *Political Declaration on establishing the JETP with Viet Nam*. European Commission: Press Corner. https://ec.europa.eu/commission/presscorner/ detail/en/statement_22_7724

European Commission. (2023a). *Strategic Technologies for Europe Platform: Target investment areas*. https://commission.europa.eu/strategy-and-policy/eu-budget/strategic-technologies-eur ope-platform/target-investment-areas_en

European Commission. (2023b, February 1). *Q&A: Green deal industrial plan for the net-zero age*. European Commission: Press Corner. https://ec.europa.eu/commission/presscorner/detail/ en/qanda_23_511

European Commission. (2023c, March 7). *Joint press release on Canada-EU relations*. European Commission: Press Corner. https://ec.europa.eu/commission/presscorner/detail/en/ip_23_1486

European Commission. (2023d, May 11). *Speech of Timmermans at the World Hydrogen Summit 2023*. European Commission: Press Corner. https://ec.europa.eu/commission/presscorner/det ail/en/speech_23_2704

European Commission. (2023e, November 20). *Keynote speech by President von der Leyen at the European Hydrogen Week 2023, via video message*. https://ec.europa.eu/commission/pressc orner/detail/en/speech_23_5907

European Commission. (2023f, November 28). *Annex to the Commission Delegated Regulation (EU) …/… amending Regulation (EU) No 2022/869 of the European Parliament and of the Council as regards the Union list of projects of common interest and projects of mutual interest*. https://energy.ec.europa.eu/system/files/2023-11/Annex%20PCI%20PMI%20list.pdf

European Commission. (2023g, November 28). *Commission proposes 166 cross-border energy pro-jects for EU support to help deliver the European Green Deal*. https://ec.europa.eu/commis sion/presscorner/detail/en/ip_23_6047

European Commission. (2024a, April 30). *European Hydrogen Bank auction provides €720 million for renewable hydrogen production in Europe*. https://ec.europa.eu/commission/presscorner/det ail/en/ip_24_2333

European Commission. (2024b). *Approved integrated Important Projects of Common European Interest (IPCEI)*. https://competition-policy.ec.europa.eu/stateaid/ipcei/approved-ipceis_en

European Commission & Government of Japan. (2022). *EU-Japan Memorandum of Cooperation on Hydrogen*. https://energy.ec.europa.eu/publications/eu-japan-memorandum-cooper ation-hydrogen_en

EIB. (2023, February 8). *India: EIB backs green hydrogen deployment and joins India Hydrogen Alliance.* https://www.eib.org/en/press/all/2023-045-eib-backs-green-hydrogen-dep loyment-in-india-and-joins-india-hydrogen-alliance

ENTSOG. (2022). *The hydrogen and natural gas ten-year network development plan (TYNDP).* European Network of Transmission System Operators for Gas (ENTSOG).

Esposito, D., Gimon, E., & O'Boyle, M. (2023). Smart design of 45V hydrogen production tax credit will reduce emissions and grow the industry. *Energy Innovation.* https://energyinnovation.org/wp-content/uploads/2023/04/Smart-Design-Of-45V-Hyd rogen-Production-Tax-Credit-Will-Reduce-Emissions-And-Grow-The-Industry.pdf

Euractiv. (2023, February 15). *Eleven EU countries urge 'great caution' in loosening state aid rules.* https://www.euractiv.com/section/economy-jobs/news/eleven-eu-countries-urge-great-caution-in-loosening-state-aid-rules/

European Hydrogen Backbone. (2022, April 5). *European Hydrogen Backbone grows to meet RE-PowerEU's 2030 hydrogen targets.* https://ehb.eu/newsitems#european-hydrogen-backbone-grows-to-meet-repowereu-s-2030-hydrogen-targets

European Commission & Government of Canada. (2023, March 8). *Joint Cooperation Committee Report on the State of the EU-Canada Relationship (2020–2022).* Government of Canada: International. https://www.international.gc.ca/world-monde/international_relations-relations_intern ationales/can-eu_agreement-accord_can-ue-2022.aspx?lang=eng

European Parliament & Council of the EU. (2022). *Trans-European Networks for Energy (TEN-E) Regulation* (PE/2/2022/REV/1). http://data.europa.eu/eli/reg/2022/869/oj/eng

G20 Energy Ministers. (2023). *G20 energy transitions ministers' meeting outcome document and chair's summary.* https://www.g20.org/content/dam/gtwenty/gtwenty_new/document/G20_ Energy_Transitions_Ministers%E2%80%99_Meeting_Outcome_Document_and_Chair% E2%80%99s_Summary.pdf

German Federal Foreign Office. (2023, July 5). *EU Global Gateway: Global partnerships for democratic and sustainable standards.* https://www.auswaertiges-amt.de/en/aussenpolitik/eur ope/eu-global-gateway--globale-partnerschaften-fuer-demokratische-und-nachhaltige-standa rds/2607028

Green Hydrogen Organization. (2023). *Getting the right blend: Innovative development finance for the large-scale renewable and green hydrogen economy.* https://gh2.org/sites/default/files/ 2023-04/GH2_Getting%20the%20Right%20Blend_2023_digital.pdf

Hancock, A. (2023, October 25). Campaigners warn EU over funds for hydrogen infrastructure. *Financial Times.* https://www.ft.com/content/d78b5048-008b-4364-8b61-a6d7d505571e

Hydrogen Council & McKinsey. (2023). Hydrogen Insights 2023: The state of the global hydrogen economy, with a deep dive into renewable hydrogen cost evolution. https://hydrogencouncil. com/wp-content/uploads/2023/12/Hydrogen-Insights-Dec-2023-Update.pdf

Hydrogen Europe & Renewable Hydrogen Coalition. (2022, October 6). *Industry Letter on RFNBO Delegated Act.* https://renewableh2.eu/wp-content/uploads/2022.10.6_Industry-letter-on-RFNBO-DA.pdf

IEA. (n.d.-a). *CEM hydrogen initiative.* International energy agency. Retrieved October 23, 2023, from https://www.iea.org/programmes/cem-hydrogen-initiative

IEA. (n.d.-b). *Hydrogen TCP—research and innovation in hydrogen technology by IEA.* International Energy Agency. Retrieved 23 October, 2023, from https://www.ieahydrogen.org/

IPHE. (2021). *Methodology for Determining the Greenhouse Gas Emissions Associated with the Production of Hydrogen* (Version 1; A Working Paper by the IPHE Hydrogen Production Analysis Task Force). International Partnership for Hydrogen and Fuel Cells in the Economy. https:// www.iphe.net/_files/ugd/45185a_ef588ba32fc54e0eb57b0b7444cfa5f9.pdf

IPHE. (2023). *Methodology for determining the greenhouse gas emissions associated with the production of hydrogen* (Version 3; A Working Paper by the IPHE Hydrogen Production Analysis Task Force). International Partnership for Hydrogen and Fuel Cells in the Economy.

Kilpeläinen, S., Quitzow, R., & Tsoumpa, M. (2023). *Hydrogen in the Nordics: Drivers of European cooperation?* Friedrich-Ebert-Stiftung. https://library.fes.de/pdf-files/a-p-b/20629.pdf

Lau, S., & Moens, B. (2022, December 20). EU to launch Global Gateway projects, challenging China's Belt and Road. *POLITICO.* https://www.politico.eu/article/global-gateway-european-union-launch-china-belt-and-road/

Martin, P. (2023a, July 7). European green hydrogen projects are being delayed due to torturously slow EU subsidy processes, say developers. *Hydrogen Insight.* https://www.hydrogeninsight.com/policy/european-green-hydrogen-projects-are-being-delayed-due-to-torturously-slow-eu-subsidy-processes-say-developers/2-1-1483308

Martin, P. (2023b, November 24). Europe's 'hydrogen backbone' of cross-border pipelines will cost billions more euros than initial estimates. *Hydrogen Insight.* https://www.hydrogeninsight.com/production/europes-hydrogen-backbone-of-cross-border-pipelines-will-cost-billions-more-euros-than-initial-estimates/2-1-1560429

King & Spalding. (2023, September 5). EU regulatory challenges complicating the development of international green hydrogen projects. *King & Spalding: News & Insights.* https://www.kslaw.com/news-and-insights/eu-regulatory-challenges-complicating-the-development-of-international-green-hydrogen-projects

Nunez, A., & Quitzow, R. (2023). *Germany's Hydrogen Strategy: Securing Industrial Leadership in a Carbon-Neutral Economy.* RIFS Discussion Paper. https://publications.rifs-potsdam.de/rest/items/item_6002778_4/component/file_6002779/content

Parkes, R. (2022, May 24). Proposed stringent EU rules on green hydrogen 'would put the brakes on development'. *Recharge.* https://www.rechargenews.com/energy-transition/proposed-stringent-eu-rules-on-green-hydrogen-would-put-the-brakes-on-development-/2-1-1223746

Parkes, R. (2023, July 14). Give us the means of proving that our green hydrogen is truly renewable, H_2 industry urges EU. *Hydrogen Insight.* https://www.hydrogeninsight.com/policy/give-us-the-means-of-proving-that-our-green-hydrogen-is-truly-renewable-h2-industry-urges-eu/2-1-1486766

Parkes, R. (2024, June 12). Second round of European Hydrogen Bank subsidy auction allocated €1.2bn—€1bn less than expected. *Hydrogen Insight.* https://www.hydrogeninsight.com/policy/second-round-of-european-hydrogen-bank-subsidy-auction-allocated-1-2bn-1bn-less-than-expected/2-1-1659860

Pisani-Ferry, J., Tagliapietra, S., & Zachmann, G. (2023). *A new governance framework to safeguard the European Green Deal.* Bruegel Policy Brief.

Prontera, A., & Quitzow, R. (2023). Catalytic Power Europe: Blended finance in European external action. *JCMS: Journal of Common Market Studies, 61*(4), 988–1006. https://doi.org/10.1111/jcms.13442

Quitzow, R., Triki, A., Wachsmuth, J., Fragoso Garcia, J., Kramer, N., Lux, B., & Nunez, A. (2023): Mobilizing Europe's full hydrogen potential: Entry-points for action by the EU and its member states. *HYPAT Discussion Paper No 5/2023.* https://publica-rest.fraunhofer.de/server/api/core/bitstreams/6eb07f2f-f7c9-42dc-8b8c-41f09b415eab/content

Rayner, T., Szulecki, K., Jordan, A. J., & Oberthür, S. (2023). The global importance of EU climate policy: An introduction. In *Handbook on European union climate change policy and politics* (pp. 1–21). Edward Elgar Publishing. https://www.elgaronline.com/edcollchap-oa/book/9781789906981/book-part-9781789906981-11.xml

Riemer, M., Schreiner, F., & Wachsmuth, J. (2022). Conversion of LNG terminals for liquid hydrogen or Ammonia: Analysis of technical feasibility und economic considerations. *Fraunhofer Institute for Systems and Innovation Research ISI*

Rizzi, A., & Varvelli, A. (2023). Opening the global gateway: Why the EU should invest more in the southern neighbourhood. *ECFR Policy Brief*. https://ecfr.eu/wp-content/uploads/2023/03/Opening-the-Global-Gateway-Why-the-EU-should-invest-more-in-the-southern-neighbourhood.pdf

Shumkov, I. (2023, October 25). EU launches new initiative to develop green hydrogen in Mauritania. *Renewables Now*. https://renewablesnow.com/news/eu-launches-new-initiative-to-develop-green-hydrogen-in-mauritania-837706/

Skjærseth, J. B., Eikeland, P. O., Inderberg, T. H. J., & Larsen, M. L. (2023). Norway's internal and external hydrogen strategy. *RIFS Discussion Paper*. https://www.rifs-potsdam.de/en/output/publications/2023/norways-internal-and-external-hydrogen-strategy

Stam, R., van der Linde, C., & Stapersma, P. (2023). The netherlands as a future hydrogen hub for Northwest Europe: Analysing domestic developments and international engagement. *RIFS Discussion Paper*. https://doi.org/10.48481/rifs.2023.013

Stolton, S. (2023, January 17). Von der Leyen announces Net-Zero Industry Act to compete with US subsidy spree. *POLITICO*. https://www.politico.eu/article/von-der-leyen-announces-net-zero-industry-act-to-compete-with-us-subsidy-spree/

Team Europe. (2023a). *EU-Africa flagship projects for 2023 (Global Gateway Projects)*. https://international-partnerships.ec.europa.eu/system/files/2023-09/EU-Africa-flagship-projects-sep2023.pdf

Team Europe. (2023b). *EU-Latin America and the Caribbean flagship projects for 2023 (Global Gateway Projects)*. https://international-partnerships.ec.europa.eu/system/files/2023-05/EU-LAC-flagship-projects-for-2023-v05.pdf

Urbasos, I. (2023, October 17). *EU hydrogen diplomacy 2.0: aligning climate ambition and energy security*. https://www.realinstitutoelcano.org/en/analyses/eu-hydrogen-diplomacy-2-0-aligning-climate-ambition-and-energy-security/

Vogler, J. (2023). Global dimensions of EU climate, energy and transport policies. In *Handbook on European union climate change policy and politics* (pp. 144–157). Edward Elgar Publishing. https://www.elgaronline.com/edcollchap-oa/book/9781789906981/book-part-9781789906981-22.xml

Weko, S., Farrand, A., Fakoussa, D., & Quitzow, R. (2023). The Politics of Green Hydrogen Cooperation: Emerging Dynamics in Morocco, Algeria and Mauritania. *RIFS Study*. https://publications.rifs-potsdam.de/rest/items/item_6003183_2/component/file_6003196/content

Weko, S. (2023). The UK hydrogen strategy: Falling behind in the green race? *RIFS discussion paper*. https://www.rifs-potsdam.de/en/output/publications/2023/uk-hydrogen-strategy-falling-behind-green-race

Yana Zabanova is a research associate at the Research Institute for Sustainability—Helmholtz Centre Potsdam (RIFS Potsdam). Her interests include EU hydrogen policy and the geopolitics and geoeconomics of the energy transition and hydrogen more broadly. She is also a PhD candidate at the University of Groningen, where she focuses on renewable energy and hydrogen development in Eurasian hydrocarbon-rich countries from a comparative perspective.

Germany's Hydrogen Strategy: Securing Industrial Leadership in a Carbon–Neutral Economy

Almudena Nunez and Rainer Quitzow

Abstract This chapter provides a review of Germany's import-oriented hydrogen strategy. It places the German policy approach in the context of its broader energy transition strategy, aimed not only at a transition of Germany's energy and industrial system to carbon neutrality by 2045 but also at the promotion of the German *Energiewende* approach abroad. The chapter begins by providing a short review of the German *Energiewende* policy legacy, relating it to it emerging hydrogen policy. On this basis, it provides a comprehensive review of Germany's national hydrogen strategy with a particular focus on its outward-oriented elements. The chapter closes with discussion of key strengths and weaknesses and the broader geopolitical and geoeconomic implications of the strategy.

1 Introduction

In June 2020, Germany joined a group of hydrogen frontrunners, including Japan, France, South Korea, Australia, Netherlands and Norway, to become the seventh country to adopt an official government strategy for the promotion of hydrogen. Along with Korea and Japan, it stands out from this group, due to the important role that the strategy assigns to the development of hydrogen imports (Albrecht et al., 2020). Since launching the strategy, the German government has put in place not only a large portfolio of measures targeting its domestic hydrogen sector, but also for the development of international hydrogen partnerships and supply chains.

This chapter provides a review of Germany's ambitious import-oriented hydrogen strategy. It places the German policy approach in the context of its broader Energiewende (energy transition) strategy, aimed not only at a transition of Germany's energy and industrial system to carbon neutrality by 2045 but also at

A. Nunez · R. Quitzow (✉)
Research Institute for Sustainability (RIFS), Helmholtz Centre Potsdam, Berliner Str. 130, 14467 Potsdam, Germany
e-mail: rainer.quitzow@rifs-potsdam.de

R. Quitzow
Technische Universität Berlin, Straße des 17. Juni 135, 10623 Berlin, Germany

© Helmholtz-Zentrum Potsdam, Deutsches GeoForschungsZentrum GFZ 2024
R. Quitzow and Y. Zabanova (eds.), *The Geopolitics of Hydrogen*, Studies in Energy, Resource and Environmental Economics, https://doi.org/10.1007/978-3-031-59515-8_3

the promotion of the German Energiewende approach abroad. The chapter begins by providing a short review of the German Energiewende policy legacy, relating it to it emerging hydrogen policy. On this basis, it provides a comprehensive review of Germany's national hydrogen strategy with a particular focus on its outward-oriented elements. It discusses Germany's external hydrogen policy along the following five dimensions: political dialogue and diplomacy (both bilateral and multilateral); interventions aimed at building international supply chains; cooperation in research and innovation; capacity building and skill development; and activities aimed at addressing questions of sustainability. The chapter closes with a brief discussion of the geopolitical and geoeconomic implications of Germany's hydrogen strategy.

2   Energiewende as Industrial Modernization Strategy: The Emerging Role of Hydrogen

Accounting for 27% of the EU's industrial production in 2021 (Eurostat, 2021), Germany retains a large and robust industrial sector. Led by its automotive sector, this includes Germany's famous Mittelstand as well as a range of industries supplying basic materials. The former consists of highly specialized, small and medium-sized enterprises (SMEs), many of them engaged in the manufacturing of machinery and production equipment. The latter comprises, among other things, the production of chemicals and related derivatives as well as mineral- and metal-based materials and products. Together these basic materials account for over 20% of German industrial output (Destatis, 2020).

Germany's strong industrial performance is matched by a pioneering climate and energy policy, known as the German *Energiewende*. Aimed at promoting decarbonization while retaining German industrial and technology leadership, this policy approach was initiated by the so-called Red-Green coalition government, consisting of the Social Democrats and the Green Party in the period from 1998 to 2005. At the time, this policy was driven by the Greens' historic mission to phase-out nuclear energy, the struggle that originally gave rise to the party in the late 1970s (Quitzow et al., 2016). To realize this goal while fighting climate change, the government engaged in a market and industry support program for wind and solar energy. Underpinned by the concept of ecological modernization (Mol & Jänicke, 2009), this policy vision built on the idea that Germany could translate environmental policy success into an export-oriented strategy for innovation and industrial development (Quitzow & Thielges, 2020; Quitzow et al., 2016).

Since this initial period, this strategy has developed into an increasingly comprehensive approach for the transformation of the energy and, more recently, the industrial sector. Based on the continued deployment of renewable energy, the *Energiewende* is now the backbone of a strategy aimed at carbon neutrality by 2045. Its current interim climate targets, adopted in the Climate Change Act passed in 2021, aim at a reduction of greenhouse gases by 65% by 2030 and 88% by 2040 (compared to the 1990 levels) (BMF, 2021). These are underpinned by legally binding targets for the deployment of renewables, enshrined in the Renewable Energy Act. In its most recent revision, following Russia's invasion of Ukraine, the law stipulates that 80% of Germany's electricity demand should come from renewable sources by 2030 (BMWK, 2022a). To achieve this goal, Germany will have to more than double the generation of renewable electricity from 240 TWh annually in 2021 to approximately 600 TWh.

The elimination of GHG emissions from German industrial production will require the use of significant volumes of hydrogen, most prominently in the steel and chemical sectors (IEA, 2019). Together these sectors account for close to 50% of industrial greenhouse gas (GHG) emissions or over 10% of Germany's total emissions. In this vein, Germany has launched so-called Carbon Contracts for Difference (CCfDs) ("*Klimaschutzverträge*") as a tool for assisting the transition of the industrial sector. The instrument is designed to provide funding to the industry to enable it to switch to carbon neutral production processes targeting emission-intensive industrial sectors, including steel, cement, and chemicals. Energy-intensive industries are compensated for the additional costs associated with building and operating with more climate-friendly alternatives under these contracts. By doing so, they buffer 15 years of additional CAPEX and OPEX expenditures and protect companies from price fluctuations. Climate-friendly industries can develop expertise, infrastructure, and markets by building the first plants using CCfD funding. Preliminary procedures for CCfDs began in the summer of 2023, and bidding procedures (submission of application documents and competitive bids) began in March 2024 (BMWK, 2023b).

Safeguarding German industrial production in these sectors while meeting its increasingly ambitious climate targets represents a major motivation for Germany's hydrogen strategy. In addition to stakeholders in these hard-to-abate sectors, the strategy is supported by an emerging electrolyzer industry as well as firms engaged in the operation of its natural gas grid. The latter stand to gain from the development of hydrogen-related transport and storage infrastructure, while maintaining few direct stakes in the extraction of natural gas. Parts of the German automotive industry are also keen on the development of fuel cell vehicles as well as hydrogen-based synthetic fuels. The former offer the promise of larger shares of domestic value creation and employment than battery-electric vehicles, while the latter represent a potential lifeline for high-end vehicles based on the traditional internal combustion engine.

This constellation of economic interests combined with Germany's well-developed *Energiewende* policy approach have provided fertile ground for an ambitious hydrogen strategy to support industrial decarbonization along with leadership in hydrogen-related technologies and components. The original strategy (Federal Government of Germany, 2020) proposed 38 measures in support of the following strategic objectives:

- Fostering the competitiveness of hydrogen produced from renewable energy and establish it as a viable option for replacing fossil-based energy carriers and industrial feedstocks

- Developing Germany's domestic hydrogen market, including related transport, distribution and quality assurance infrastructure, that is able to absorb imports
- Strengthening German industry and securing global market opportunities for German firms in the sector
- Fostering science and skilled labor
- Establishing international markets and cooperation

The updated hydrogen strategy included a slightly revised framework aimed at the following four fields of action (BMWK, 2023a):

- "Ensuring availability of sufficient hydrogen" including domestic and imported hydrogen and hydrogen derivatives production
- "Developing an efficient hydrogen infrastructure" encompassing both national and European hydrogen infrastructure and the necessary infrastructure for imports
- "Implementing hydrogen applications" with a particular focus on the decarbonization of industry
- "Creating effective framework conditions" by enhancing research, innovation, and training of professional personnel, as well as establishing operational procedures, standards, and certifications

The German government foresees two main phases for the development of the hydrogen economy: the first aims at establishing the foundation and ramp-up of a functioning market, and the second one, expected to start in 2024, is focused on the consolidation of the domestic market and the increased usage of European and international hydrogen resources for the German economy. While the original strategy in 2020 proclaimed an exclusive focus on promoting green hydrogen, the latest update (summer 2023) provides scope for indirect support for other types of low-carbon hydrogen. It provides the option of supporting it via its demand-side support instruments during the ramp-up phase of the market to ensure access to sufficient quantities of hydrogen. This includes fossil-based hydrogen production coupled with carbon capture and storage (blue hydrogen) in certain applications, taking into account ambitious GHG reduction targets (BMWK, 2023a). Nonetheless, the focus remains strongly on the promotion of green hydrogen. Similarly, the strategy clearly prioritizes so-called hard-to-abate sectors but does leave open the possibility of supporting other "niche applications", for instance in the transport sectors. In sum, Germany has articulated a focus on the use of green hydrogen for reducing GHG emissions in hard-to-electrify sectors, while leaving space for accommodating pragmatic arrangements with international partners as well as domestic industry groups.

3 Germany's Outward-Oriented Hydrogen Strategy

According to its 2023 strategy, the government expects that Germany will require between 95 to 130 TWh of hydrogen for the uses targeted by the strategy by 2030 (BMWK, 2023a). To cover part of this demand, the government aims at establishing up to 10 GW of domestic production capacity, which corresponds to 28 TWh of renewable hydrogen production and 40 TWh of renewable energy generation. It is estimated that 50–70% (45–90 TWh) of the 95–130 TWh demand forecast for 2030 will be met by imports from abroad, in the form of hydrogen and hydrogen derivatives (BMWK, 2023a). This means that Germany aims to import large volumes of hydrogen by 2030 and foresees international cooperation as a key enabler for this. Of the 9 billion euros originally assigned to the implementation of the strategy, 2 billion euros were reserved for international partnerships (BMWK, 2020b). In its strategy update, the government has announced that it plans to develop a dedicated import strategy, providing further weight to the issue moving forward.

The outward orientation has been supported by the coalition government, composed of Social Democrats, Greens and the Liberal Party, which took office in December 2021. The coalition decided not only to transfer the domestic climate portfolio from the Ministry for the Environment (Bundesministerium für Umwelt, Naturschutz, nukleare Sicherheit und Verbraucherschutz, BMUV) to the expanded Ministry for Economic Affairs and Climate Protection (Bundesministerium für Wirtschaft und Klimaschutz, BMWK). It also created a consolidated portfolio for climate foreign policy in the Federal Foreign Office (Auswärtiges Amt, AA) and has developed an official climate foreign policy strategy. The agenda has been backed up by an expanded staff for international climate and energy policy at both BMWK and AA and a newly appointed Special Representative for International Climate Policy within AA. The various federal ministries have a large portfolio of activities around the globe achieving a presence in 50 non-EU countries (see Fig. 1 for a map indicating the presence of the five ministries in partner countries). Increasingly, the German government is also seeking to overcome its tradition of strict ministerial boundaries and is attempting to coordinate its international climate and energy policy more actively across key ministries. Labelled the „Team Germany " approach (Federal Foreign Office, 2022), BMWK, AA, BMUV and the Ministry for Economic Cooperation and Development (Bundesministerium für wirtschaftliche Zusammenarbeit und Entwicklung, BMZ) have established an interministerial coordination mechanism for this purpose. Similarly, the hydrogen strategy is coordinated at the level of State Secretaries (vice-ministers), including all ministries.

3.1 From Energy Partnerships to Hydrogen Partnerships

At the political level, Germany's external hydrogen policy can build on a well-developed set of bilateral partnerships focused on energy-related matters, developed

Fig. 1 Engagement of German federal ministries in partner countries outside Europe. *Source* Author's own, based on official government sources (Status: April 2024)

and expanded to more than twenty partner countries over the course of the past fifteen years. They include major industrialized and emerging economies as well as a number of countries in the Middle East and North Africa. Known as Energy Partnerships and Dialogues, depending on their degree of formalization, these partnerships are in most cases led by BMWK and provide an institutionalized forum for dialogue and exchange at both the political and the expert-level (Quitzow & Thielges, 2020; Quitzow et al., 2019). Generally organized along thematic working groups, these partnerships provide the German government with an entry point for cooperation with a series of potential hydrogen exporters. In that vein, a number of existing partnerships have added hydrogen-related questions as an important pillar of their thematic portfolio, including countries like Australia, the United Arab Emirates or Chile. In addition, new partnerships have been formed with a particular—though not exclusive—focus on hydrogen cooperation. Countries in this group include Canada, Saudi Arabia and Namibia (see annex for a full list). The government has placed particular attention on the latter, appointing former State Secretary Rainer Baake as Special Commissioner of the BMWK for German-Namibian Climate and Energy Cooperation. Notably, despite Germany's strong stance in favor of green hydrogen, these diplomatic efforts have also included support for the import of blue hydrogen and related derivatives in cooperation with partners from the United Arab Emirates and Norway (BMWK, 2022c, 2022d).

These activities, led mainly by BMWK, have been supplemented by the H2 Diplo initiative, sponsored by AA. With offices in Angola, Colombia, Kazakhstan, Kenya, Nigeria, Saudi-Arabia and Ukraine, it focuses primarily on dialogue with fossil fuel exporting countries, aiming to utilize hydrogen and related industrial opportunities as an entry-point for discussing their transition to a carbon–neutral future. This is part of a larger initiative aimed at enhancing awareness and understanding of geopolitical dimensions of the energy transition. To this end, AA has initiated two flagship reports under the umbrella of the International Renewable Energy Agency (IRENA), first on the geopolitics of the energy transformation with a focus on renewables (IRENA, 2019) and second on hydrogen (IRENA, 2022b).

3.2 Multilateral Engagement

Germany is also engaging in political dialogue on hydrogen-related issues in the EU and in a number of multilateral fora. At the EU-level, Germany's most prominent engagements, launched during its Presidency of the European Council, have been an initiative to support so-called Important Projects of Common European Interests (IPCEI) on Hydrogen Technologies and Systems and an Agenda Process Green Hydrogen for the development of a Strategic Research and Innovation Agenda (SRIA). The former was officially kicked-off with the signing of a Manifesto for the development of a European "Hydrogen Technologies and Systems" value chain in December 2020 and has yielded 62 projects in Germany with funding commitments in excess of 8 billion euros from federal and state-level budgets (BMWK, 2021). The SRIA, published in March 2022 (Expert groups of the agenda process, 2022), sets out a research agenda to be supported in joint activities by EU member state governments.

In addition, to these EU-level processes, the German government has also engaged in targeted dialogue with a number of its European neighbors with coastal areas in the North Sea. In the "Esjberg Declaration on the North Sea as a Green Power Plant of Europe" of May 2022, the leaders of Belgium, Denmark, Germany and the Netherlands have committed to develop at least 65 GW of off-shore wind and 20 GW of green hydrogen production in the North Sea by 2030. Similarly, Germany has supported the call for accelerated hydrogen development made in a joint declaration by the Pentalateral Energy Forum, which encompasses the Benelux countries, France, Austria, Switzerland and Germany.

Furthermore, Germany launched a number of hydrogen infrastructure development projects in 2023 to leverage the import potential of its European neighborhood. Firstly, it has adopted a law for the development and financing of a core hydrogen network (*Wasserstoff-Kernnetz*), which aims to begin transporting hydrogen by 2025 and will connect ports, industry, power plants, and storage facilities by 9,700 km of pipelines. The network envisions links to a future European hydrogen grid (Collins, 2023). During the winter of 2023 Germany and Italy have agreed to expand their energy cooperation to strengthen supply security and transition to carbon neutrality, focusing on the development of gas and hydrogen pipelines across the Alps. With the signing of the German-Italian Action Plan, both governments support the diversification of energy supplies and enhancement of energy connectivity in Europe. Plans include the construction of new gas and hydrogen pipelines between Germany and Italy, especially along the "South Central Corridor" connecting the two countries to North Africa (Wettengel, 2023).

Beyond Europe, the German government has promoted the market ramp-up for green hydrogen via both the UN and the G7. During its G7 presidency, it supported the launch of the G7 Hydrogen Action Pact and sponsored an IRENA report to define possible areas of cooperation in that context (IRENA, 2022a). It is also a founding member of the Green Hydrogen Catalogue, an initiative launched within the UN High-level Dialogue on Energy. The initiative provides a forum for publishing hydrogen-related commitments, thereby aiming to motivate countries and other stakeholders to contribute to the green hydrogen market. The German government is also engaged in working groups of the International Partnership on the Hydrogen Economy (IPHE) on Regulations, Codes, Standards, & Safety (RCSS), Trade Rules, and Production Analysis. The most tangible result, coming from the Production Analysis Task Force, has been the development of a methodology for determining the greenhouse gas emissions associated with the production of hydrogen (IPHE, 2022). Here, however, German participation was limited to an observer role, reflecting Germany's focus on green hydrogen so far.

3.3 Promoting International Hydrogen Supply Chains

The political efforts outlined above are underpinned by a number of programs aimed at supporting the development of European and international hydrogen supply chains. In terms of total funding volume, the 62 projects (as part of IPCEI) (BMWK, 2021) supported by the German government represent the largest funding effort in this regard. They include nineteen projects in support of a total of 2 GW of electrolysis capacity in Europe, fifteen projects in support of storage and transport infrastructure, sixteen projects for green hydrogen use in industry (most prominently in steel but also chemical production and refineries) and twelve projects for hydrogen use in the mobility sector. All these projects are located in Germany, albeit with participation of companies from other member states, as stipulated by the EU rules governing IPCEIs.

Outside of the EU, the BMWK is promoting not only the development of production capacities for green hydrogen in potential export countries but also the supply chains needed to deliver them to the German market. For this purpose, the BMWK has supported the creation of the H2 Global Foundation for the implementation of an auction mechanism to support investments in both the supply of hydrogen (and its derivatives) and their use. The foundation operates an auction mechanism, in which the so-called Hydrogen Intermediary Network Company (HINT.CO) enters into long-term (10-year) contracts for the purchase of hydrogen or hydrogen derivatives and short-term (2-year) sales contracts with corresponding users. The supply contracts are based on a price for its delivery to a port in the Netherlands, Belgium or Germany, thus stimulating private investments along the entire supply chain. The H2 Global mechanism represents a system of Contracts for Difference (CfD's) where the difference between the prices offered by both the producers and the users are compensated by HINT.CO with funding from the BMWK. The first round of auctions focuses on the procurement of hydrogen-related products from partner countries outside the EU and the European Free Trade Association (EFTA) (Bollerhey et al., 2022). For this, H2 Global counts on 900 million euros of public financing to be expanded by an additional 3,5 billion (BMWK, 2022b). The first auction for green ammonia was launched in November 2022, and two others for green methanol and sustainable aviation fuel (SAF) followed in 2023 (HINTCO, 2023). Moreover, Germany and the EU have agreed that other EU governments who are interested in implementing international hydrogen tenders can contribute to H2Global as part of the "Team Europe" approach of the European Hydrogen Bank's international pillar (BMWK, 2023c). Further, using public funds worth €600 million, the German and Dutch governments intend to hold a joint auction for renewable hydrogen imports through H2 Global's platform (Martin, 2023).

In addition to H2 Global, the German government operates a number of additional initiatives and funding schemes to support hydrogen-related investments outside the EU. The BMZ has sponsored the Business Alliance Energy (Unternehmensallianz Energie) to engage German firms in a dialogue on how to best support investments in the Power-to-X value chain in developing and emerging economies. It is also building on long-standing energy cooperation with Morocco and Tunisia to launch so-called Hydrogen Alliances for development of the sector. Moreover, the funding program "Internationale Wasserstoffprojekte" (International Hydrogen Projects) has provided grant-based funding of up to 15 million euros per project delivered cooperatively by the BMBF and the BMWK (BMWK & BMBF, 2021) for hydrogen projects outside the EU along the entire supply chain, i.e. production, storage, transport and use. Jointly administered by BMWK and BMBF, it has not only provided funding for the projects themselves but also for related research activities (see below for more details on research and innovation-related activities). The BMWK's International Hydrogen Ramp Up program (H2Uppp) has supported SMEs from Germany and other EU Member States with the preparation of hydrogen projects in developing and emerging countries. In coordination with the German Energy Solutions Initiative, H2Uppp has provided technical advice for the identification, preparation and implementation of hydrogen projects. Finally, BMWK has supported a number of larger scale efforts. For instance, BMWK is supporting Siemens Energy's Haru Oni project in Chile for the integrated production of hydrogen-based synthetic fuels with 8.23 million euros (BMWK, 2020a). It has also committed US$968 million in the form of grants and concessional lending for energy-related investments within the multi-donor Just Energy Transition Partnership with South Africa, which will partially go to green hydrogen projects (International Partners Group, 2022).

3.4   International Cooperation in Hydrogen-Related Research

Complementing its efforts to stimulate the ramp-up of an international hydrogen market and supply chain, the German government is promoting international cooperation to promote research and innovation. This includes support for technology development as well as analysis to support broader knowledge development related to a future hydrogen economy. The BMBF is the central ministry in this field, with additional activities in applied research and development sponsored by the BMWK. The BMBF has placed particular attention on research collaborations with other European frontrunners as well as Australia. Cooperation with Australia has been the most prominent, receiving support via the high-profile projects HySupply which counts on 1.7 million euros of funding provided by the BMBF (BMBF, 2021a) and HyGate with 50 million euros of funding (BMBF, 2021a, 2021b, 2021c), two collaborative research efforts related to the development of a hydrogen supply chain between the two countries. As mentioned above, the BMBF has also sponsored an agenda process for defining a joint research program for the European Research Area. In addition to this, the ministry has supported a joint call for research proposals with the Netherlands as well as a multi-country call for proposals with a number of EU countries and Canada under the EUREKA umbrella, a public network for research collaboration. The BMBF has also launched hydrogen-related research cooperation initiatives with Ukraine and Central Asian countries, Canada, New Zealand, South Korea, Japan, and Namibia. The latter is part of a larger scheme to forge a hydrogen partnership with the country, led by the BMBF. In this vein, the two countries signed a Memorandum of Understanding to form a partnership for research and development on green hydrogen in 2021. In addition, under the framework of the International Future Labs, BMBF has funded scholars from around the world to conduct research visits in Germany on hydrogen-related topics, and it has created the Redefine H2 Economy project to bring international researchers to Germany to conduct research on high-temperature electrolyzers, gasification methods as well as the synthesis of basic chemicals and energy sources, providing 5 million euros of funding for the latter project (Technical University of Munich, 2021).

Complementing these BMBF-led activities, the BMWK has sponsored a number of large-scale R&D projects with private sector participation. These include include thyssenkrupp's project "Element One" in Saudi Arabia for the development of electrolyzer technologies supported by BMWK with 1.5 million euros of funding (BMWK, 2020a, 2020b, 2020c), and a joint project for the development of transport solutions based on liquid organic hydrogen carriers (LOHC), involving among others Uniper, the Emirati oil company ADNOC (Abu Dhabi National Oil Company) and Japan's largest power generation company, Jera.

3.5 Capacity Building and Skill Development

In addition to engagement on cutting-edge research and innovation, the German government has also supported capacity and skill development in developing and emerging countries. This includes activities sponsored by BMZ within its broad portfolio of energy-related cooperation activities as well as activities led by the BMBF and the BMWK. The latter is the sponsor of the International PtX Hub, a knowledge and training center on Power-to-X technologies and hydrogen. It has developed training modules on green hydrogen basics, production processes, logistics, economic aspects, regulation, and sustainability. It has country-based activities to support knowledge and skill development in countries across Asia, Africa, the Middle East and South America. The BMBF is also supporting capacity and skill development with a focus on African countries. Among the most relevant initiatives are a Master's Graduate Program in West Africa on green hydrogen technologies, based on collaboration between research institutions in Germany and the West African Science Service Centre on Climate Change and Adapted Land Use (WASCAL) graduate schools, providing funding of 8 million euros (Jühlich Aachen Research Alliance, 2021). It has also funded a research project for the assessment of hydrogen production potentials in Africa, starting with West Africa. Additionally, the partnership activities in Namibia, which builds on 40 million euros of funding from BMBF (BMBF, 2021a, 2021b, 2021c), include training and skill development. Finally, BMZ is funding capacity building within its Hydrogen Alliances with Morocco and Tunisia as well as in Brazil and South Africa.

3.6 Promoting a Sustainable Hydrogen Economy

The German government is a very vocal supporter of green hydrogen as its preferred solution for building a climate-friendly hydrogen economy. As a result, its activities to ensure the sustainability of a future hydrogen economy are focused on the sustainability of green hydrogen rather than alternative forms of low-carbon hydrogen production, i.e. hydrogen from natural gas in combination with CCS technologies or "blue hydrogen". Questions of sustainability have been one of the main pillars of work at the PtX Hub. It has developed a conceptual framework for PtX sustainability, which distinguishes four sustainability dimensions: Environmental, Economic, Governance, and Social. It has incorporated the related issues in its training programs and has promoted them as the possible basis for sustainability standardization and certification. Moreover, the first auction launched under the H2 Global scheme included not only requirements to ensure that the electricity utilized for the production of hydrogen is from renewable sources but also additional sustainability requirements pertaining to the use of environmental management systems, water and land use, biodiversity, labor standards as well as local skill development.

Beyond these focused engagements, the German government has not engaged in any major effort to promote the development of internationally recognized sustainability standards or certification schemes. Rather, it has been revealed that the BMWK intervened at the EU-level to request a more gradual phase-in of so-called additionality rules (i.e. rules proposed by the European Commission to ensure that the renewable energy used for the production of green hydrogen would be additional to existing renewable energy capacities in the power mix) (Kurmayer, 2022). Rather than the proposed 100% by 2027, the German government requested a much lower share of 20, 25 and 30% by 2026, 2028 and 2030, respectively. Moreover, as mentioned above, the German government has not actively engaged in the definition of standards for the production of blue hydrogen, despite the fact that it has already concluded agreements for the import of blue hydrogen from UAE, Canada (Quitzow et al., 2023) and Norway.

4 Conclusion

Germany's external hydrogen strategy represents a comprehensive effort to promote an international hydrogen economy, focused on securing German technology leadership as well as imports to meet future hydrogen demand. It is seeking to achieve these goals by supporting cooperation with international hydrogen technology leaders, including Australia, Canada and a number of European countries, as well as potential exporters, most notably Namibia, UAE, Chile and Morocco. In doing so, Germany has assumed a very prominent international leadership role on green hydrogen. Its H2 Global initiative in particular represents an innovative and highly visible effort in this regard. Similarly, its partnership with Namibia has attracted significant international attention. Within EU policy-making, Germany has taken a more passive role in comparison, focusing efforts primarily on smaller groupings of like-minded countries, such as the set of North Sea countries, rather than advancing more ambitious schemes to promote hydrogen production across the EU as a whole. German leadership and visibility at the international level paired with a more cautious approach at the EU-level is reminiscent of past efforts around renewable energy. Although German renewable energy policy was highly successful at inspiring other countries to follow its policy model, its European policy engagement was mainly focused on defending its domestic support schemes under pressure from the Commission (Solorio et al., 2014). Nevertheless, Europe's hydrogen strategy is well-aligned with Germany's import-oriented ambitions, though the EU still lacks a strong framework for implementation. As a result, initial hydrogen production capacities in Europe are likely to be clustered in those Member States with the needed financial capacity to support investments, rather than financially weaker Member States with relatively abundant renewable energy resources. This means that potential imports from other EU Member States may remain underexploited, relative to their potential.

Another pronounced characteristic of Germany's outward-oriented hydrogen strategy is its strong focus on promoting the production of hydrogen (and derivatives like ammonia and synthetic fuels) in partner countries. The strategy does not explicitly aim to build broader industrial development partnerships based on value-added products, such as hydrogen-based direct-reduced iron as an input for steel making. While it is in the interest of Germany's incumbent industries as well as the corresponding industrial work force to retain value creation in Germany, there may also a be a case for exploring broader green industrial partnerships, especially with countries in the European Neighborhood. Such partnerships could provide the basis for the development of climate-friendly industrial value chains centered on the EU market. The strong focus on hydrogen imports may distract German policy makers and businesses from building broader industrial alliances. These are not only important for securing German and European industrial leadership and influence in a future carbon neutral economy. They may also be the key to sustaining the interest among potential hydrogen exporters, including those in the EU. It would also provide opportunities for a more active promotion of German and European standards in these emerging green industrial value chains, an area that has not figured prominently in the strategy to date.

Another challenge for Germany is its ambiguous stance on the role of blue hydrogen. Despite its vocal support for green hydrogen, it has moved towards increased support for blue hydrogen, including agreements in support of blue hydrogen imports from Norway and the UAE. In light of this, it is essential that the German government develops a more nuanced position on blue hydrogen and supports the build-up of related analytical capacities. This is needed to engage with partners and ensure that investments in blue hydrogen are aligned with German and international climate targets. A more open yet nuanced and principled approach to blue hydrogen would also provide the basis for engaging more actively with the US and other EU partners, offering opportunities to influence global developments.

Lastly, a decision made by the country's constitutional court in November 2023 regarding the remaining budget to combat the Covid-19 pandemic resulted in a €60 billion cut to the Climate and Transformation Fund, which was intended to finance investments in the field of climate and energy. With a lower budget available in 2024, investments in the hydrogen sector and industrial decarbonization may suffer, though specific impacts remain uncertain at the time of writing (Packroff, 2023).

Acknowledgements Research for this chapter was financially supported by the German Federal Foreign Office within the framework of the project "Geopolitics of the Energy Transformation— Implications of an International Hydrogen Economy" (GET Hydrogen), funding reference number AA4521G125.

Annex

Germany's official Energy Partnerships and Dialogues with a focus or component on hydrogen (Status: April 2024).

Country	Year	Type of partnership and role of hydrogen cooperation
Algeria	2015	Energy partnership, with hydrogen-related activities and dialogue
Angola	2011	Energy partnership led by federal foreign office with an additional hydrogen diplomacy office
Australia	2017	Energy partnership with an additional "Hydrogen Accord" signed in 2021 and German-Australian hydrogen innovation and technology incubator
Brazil	2008/ 2017*	Energy partnership, with hydrogen-related activities and dialogue
Canada	2021	Energy partnership including memorandum of understanding to create a hydrogen alliance
Chile	2019	Energy partnership, with hydrogen-related activities and dialogue, and support from the government's PtX Hub
China	2006	Energy partnership with hydrogen-related activities and dialogue
Colombia	2023	Memorandum of understanding signed for Energy and Climate partnership and additional support from the Government's PtX Hub
India	2006	Energy partnership with a planned task force on hydrogen and support from the government's PtX Hub
Japan	2019	Energy partnership with a working group on "clean hydrogen"
Jordan	2016/ 2019*	Energy partnership with hydrogen-related activities and dialogue, and additional support from the Government's PtX Hub
Kazakhstan	2022	Energy dialogue, including hydrogen-related activities and dialogue, and a hydrogen diplomacy office sponsored by the federal foreign office
Korea	2019	Energy partnership, including hydrogen-related activities and dialogue within the working group on "new green technologies"
Morocco	2012	Energy partnership, including an additional memorandum of understanding for the development of an "Alliance for the Development of the Power-to-X Sector" and additional support from the government's PtX Hub
Mexico	2019	Energy Partnership, including hydrogen-related activities and dialogue at the sub-national level
Namibia	Planned	Memorandum of understanding for the creation of a hydrogen partnership to be led by BMBF and additional support from the government's PtX-Hub
Nigeria	2008	Energy partnership led by federal foreign office with an additional hydrogen diplomacy office
Norway	2023	Climate, renewable energy and green industry partnership including activities on hydrogen

(continued)

(continued)

Country	Year	Type of partnership and role of hydrogen cooperation
Oman	2017	Energy dialogue with hydrogen-related activities and dialogue
Qatar	2022	Energy dialogue with hydrogen-related activities and dialogue
Saudi-Arabia	2019	Energy partnership with an additional memorandum of understanding on hydrogen cooperation and a hydrogen diplomacy office sponsored by the federal foreign office
South Africa	2013	Energy partnership including a focus on green hydrogen and additional support from the government's PtX-Hub and financial support committed in the just energy transition partnership investment plan
Tunisia	2021	Energy partnership, including hydrogen-related activities and dialogue
Turkey	2012/ 2018*	Energy partnership with a focus on hydrogen in the working group on "sector coupling"
Ukraine	2020	Energy partnership with a working group on hydrogen and a hydrogen diplomacy office sponsored by the federal foreign office
United Arab Emirates	2017/ 2022*	Energy and climate partnership with hydrogen-related activities and dialogue
United Kingdom	2023	Energy and climate partnership including activities on hydrogen
United States of America	2019/ 2021*	Climate and energy partnership including activities and dialogue on hydrogen
Vietnam	Planned	Planned energy dialogue with hydrogen-related activities and dialogue supported by the government's PtX-Hub

Sources Information based on https://www.bmwk.de/Navigation/DE/Wasserstoff/Internationale-Wasserstoffzusammenarbeit/internationale-wasserstoffzusammenarbeit.htmlBMWK (2022) Fortschrittsbericht zur Umsetzung der Nationalen Wasserstoffstrategie; Quitzow and Thielges (2020, p. 11) and internal government documents.
*If more than one year is listed, the energy cooperation has experienced a relaunch or upgrading to a more formalized or more comprehensive partnership.

References

Albrecht, D. U., Bünger, D. U., Michalski, D. J., Raksha, T., Wurster, R., & Zerhusen, J. (2020). *International hydrogen strategies*. World Energy Council.

BMWK & BMBF. (2021). Bekanntmachung der Förderrichtlinie für internationale Wasserstoffprojekte im Rahmen der Nationalen Wasserstoffstrategie und des Konjunkturprogramms: Corona-Folgen bekämpfen, Wohlstand sichern, Zukunftsfähigkeit stärken. https://www.bmwk.de/Redaktion/DE/Downloads/F/20210410-pm-sachstand-foerderrichtlinie.pdf?__blob=publicationFile&v=10

BMBF. (2021). 2021 Federal government report on energy research. https://www.bmwk.de/Redaktion/EN/Publikationen/Energie/federal-government-report-on-energy-research-2021.pdf?__blob=publicationFile&v=5

BMBF. (2021). Karliczek: Germany and Namibia form partnership for green hydrogen. https://www.bmbf.de/bmbf/shareddocs/pressemitteilungen/de/2021/08/172_namibia_eng.pdf?__blob=publicationFile&v=1

BMBF. (2021). Karliczek: Mit HyGATE eine deutsch-australische Lieferkette für Grünen Wasserstoff aufbauen. https://www.bmbf.de/bmbf/shareddocs/pressemitteilungen/de/karliczek-mit-hygate-eine-deut-r-gruenen-wasserstoff-aufbauen.html

BMF. (2021). Immediate climate action programme for 2022. https://www.bundesfinanzministerium.de/Content/EN/Standardartikel/Topics/Priority-Issues/Climate-Action/immediate-climate-action-programme-for-2022.html

BMWK. (2020a). 'Haru Oni' PtX project Minister Altmaier hands over first approval notice for inter-national green hydrogen project. https://www.bmwk.de/Redaktion/EN/Pressemitteilungen/2020/12/20201202-haru-oni-ptx-project-minister-altmaier-hands-over-first-approval-notice-for-international-green-hydrogen-project.html (Accessed 17 March 2023)

BMWK. (2020b). The National Hydrogen Strategy. https://www.bmwk.de/Redaktion/EN/Publikationen/Energie/the-national-hydrogen-strategy.pdf?__blob=publicationFile&v=6

BMWK. (2020). PtX-Projekt „Element One": Altmaier übergibt Förderbescheid für internationales Projekt für grünen Wasserstoff in Saudi-Arabien. https://www.bmwk.de/Redaktion/DE/Pressemitteilungen/2020/12/20201216-altmaier-uebergibt-foerderbescheid-fuer-internationales-projekt-fuer-gruenen-wasserstoff.html

BMWK. (2021). "Wir wollen bei Wasserstofftechnologien Nummer 1 in der Welt werden": BMWi und BMVI bringen 62 Wasserstoff-Großprojekte auf den Weg. https://www.bmwk.de/Redaktion/DE/Pressemitteilungen/2021/05/20210528-bmwi-und-bmvi-bringen-wasserstoff-grossprojekte-auf-den-weg.html

BMWK. (2022a). Federal minister Robert Habeck says Easter package is accelerator for renewable energy as the Federal Cabinet adopts key amendment to accelerate the expansion of renewables. https://www.bmwk.de/Redaktion/EN/Pressemitteilungen/2022/04/20220406-federal-minister-robert-habeck-says-easter-package-is-accelerator-for-renewable-energy.html (Accessed 13 March 2023).

BMWK. (2022b). Federal ministry for economic affairs and climate action launches first auction procedure for H2Global – €900 million for the purchase of green hydrogen derivatives. https://www.bmwk.de/Redaktion/EN/Pressemitteilungen/2022/12/20221208-federal-ministry-for-economic-affairs-and-climate-action-launches-first-auction-procedure-for-h2global.html (Accessed 13 March 2023).

BMWK. (2022a). Federal Minister Robert Habeck: "Expand Cooperation on Hydrogen with United Arab Emirates", https://www.bmwk.de/Redaktion/EN/Pressemitteilungen/2022/03/20220321-federal-minister-robert-habeck-expand-cooperation-on-hydrogen-with-united-arab-emir-ates.html&sa=D&source=docs&ust=1663696730911095&usg=AOvVaw1G0gS2BrNRfGA5K1cYkPAy (Accessed 13 March 2023).

BMWK. (2022d). Joint statement Germany—Norway. https://www.bmwk.de/Redaktion/DE/Downloads/J-L/20220316-joint-statement-norway.html (Accessed 13 March 2023).

BMWK. (2023a). National Hydrogen strategy update. https://www.bmwk.de/Redaktion/EN/Publikationen/Energie/national-hydrogen-strategy-update.pdf?__blob=publicationFile&v=2

BMWK. (2023b). Klimaschutzverträge. https://www.bmwk.de/Redaktion/DE/Downloads/V/vorstellung-klimaschutzvertrage.pdf?__blob=publicationFile&v=4

BMWK. (2023c). Wichtige Etappe für globalen Markthochlauf für grünen Wasserstoff: Bundesregierung und EU -Kommission machen H2Global zum europäischen Wasserstoff-Projekt. https://www.bmwk.de/Redaktion/DE/Pressemitteilungen/2023/06/20230601-bundesregierung-und-eu-kommission-machen-h2global-zum-europaeischen-wasserstoff-projekt.html

Bollerhey, T., Exenberger, M., Geyer, F. & Westphal, K. (2022). H2 Global-Idea, Instrument and Intentions. http://files.h2-global.de/H2Global-Stiftung-Policy-Brief-01_2022-EN.pdf

Collins, L. (2023). German hydrogen pipeline network will begin transporting H2 in 2025, with 9700km in place by 2032, says government. https://www.hydrogeninsight.com/policy/german-hydrogen-pipeline-network-will-begin-transporting-h2-in-2025-with-9-700km-in-place-by-2032-says-government/2-1-1554455

Destatis. (2020). Key data of enterprises in manufacturing. Destatis—German Federal Statistical Office.

Eurostat. (2021). Industrial Production Statistics. https://ec.europa.eu/eurostat/statistics-explained/index.php?title=Industrial_production_statistics#Industrial_production_by_country

Expert groups of the agenda process. (2022). Strategic Research and Innovation Agenda. Key findings and conclusions of the agenda process for the European research and innovation initiative on green hydrogen. Final version. Secretariat of the agenda process. https://www.bmbf.de/SharedDocs/Downloads/files/SRIA_green_hydrogen.pdf?__blob=publicationFile&v=4

Federal Foreign Office. (2022). Team Germany on the conclusion of COP27. Press Release, Federal Foreign Office, 20.11.2022.

Hydrogen Action Pact. International Renewable Energy Agency (IRENA). IRENA. (2022b). Geopolitics of the Energy Transformation: The Hydrogen Factor. International Renewable Energy Agency (IRENA).

IEA. (2019). *The future of hydrogen.* International Energy Agency (IEA).

International Partners Group (IPG). (2022). Joint statement: South Africa just energy transition investment plan. *European Commission, 7*(11), 2022.

IPHE. (2022). Methodology for determining the greenhouse gas emissions associated with the production of Hydrogen, Version 2. Methodology for determining the greenhouse gas emissions associated with the production of hydrogen.

IRENA. (2019). A new world: The geopolitics of the energy transformation. International Renewable Energy Agency (IRENA).

IRENA. (2022a). Accelerating hydrogen deployment in the G7: Recommendations for the HINTCO. 2023. About HINTCO. https://www.hintco.eu/

Jülich Aachen Research Alliance. (2021). Shaping the energy supply of the future. https://www.jara.org/en/research/energy/news/detail/Master-program-imp-egh-green-hydrogen

Kurmayer, N. (2022). Revealed: How Germany stepped in to delay EU's 'green' hydrogen rules. *Euractiv, 31*(10), 2022.

Martin, P. (2023). Germany and Netherlands plan €600m joint auction for green hydrogen early next year. https://www.hydrogeninsight.com/policy/germany-and-netherlands-plan-600m-joint-auction-for-green-hydrogen-early-next-year/2-1-1554704

Mol, A., & Jänicke, M. (2009). The origins and theoretical foundations of ecological modernisation theory. In A. P. J. Mol, D. A. Sonnenfeld, & G. Spaargaren (Eds.), *The Ecological Mod-ernisation Reader: Environmental Reform in Theory and Practice* (pp. 7–27). Routledge.

Packroff, J. (2023). Germany solves budget spat by cutting climate fund and increasing energy taxes. https://www.euractiv.com/section/economy-jobs/news/germany-solves-budget-spat-by-cutting-climate-fund-and-increasing-energy-taxes/

Quitzow, R., & Thielges, S. (2020). The German energy transition as soft power. *Review of International Political Economy, 29*(2), 598–623. https://doi.org/10.1080/09692290.2020.1813190

Quitzow, R., Röhrkasten, S., & Jänicke, M. (2016). The German energy transition in international perspective. In IASS Study. Institute for Advanced Sustainability Studies.

Quitzow, R., Thielges, S., & Helgenberger, S. (2019). Deutschlands Energiepartnerschaften in der internationalen Energiewendepolitik (IASS Diskussionspapier, März 2019). Institute for Advanced Sustainability Studies.

Quitzow, R., Mewes, C., Thielges, S., Tsoumpa, M., & Zabanova, Y. (2023). Building Partner-ships for an International Hydrogen Economy: Entry-Points for European Policy Action. Fried-rich-Ebert-Stiftung.

Solorio, I., Öller, E., & Jörgens, H. (2014). Solorio, Israel. In A. Brunnengräber & M. R. Di Nucci (Eds.), *Im Hürdenlauf zur Energiewende* (pp. 189–200). Springer Fachmedien.

Technical University of Munich. (2021). TUM to coordinate Future Lab for Green Hydrogen. https://www.tum.de/en/news-and-events/all-news/press-releases/details/tum-koordi niert-zukunftslabor-fuer-gruenen-wasserstoff

Wettengel, J. (2023, November 23). Germany, Italy push for new pipeline across Alps. *Clean Energy Wire*. https://www.cleanenergywire.org/news/germany-italy-push-new-pipeline-across-alps

Almudena Nunez is Energy & Climate Policy Advisor International Relations at VNG AG in Leipzig, Germany. Prior to joining VNG in January 2024, she was a Research Associate at the Research Institute for Sustainability (RIFS Potsdam), Helmholtz Centre Potsdam in the research group "Geopolitics of Energy and Industrial Transformation" where she focused on the policy and regulatory aspects of green hydrogen as part of the Global Hydrogen Potential Atlas project (HyPat). She holds a bachelor's degree in International Relations with a specialization in Business and International Finances from Anahuac University in Mexico, her home country, and a Master's degree in Public Policy with a specialization in International Political Economy and European Public Policy from the Willy Brandt School of Public Policy at the University of Erfurt in Germany.

Rainer Quitzow leads the research group "Geopolitics of Energy and Industrial Transformation" at the Research Institute for Sustainability (RIFS Potsdam), Helmholtz Centre Potsdam. His research focuses on climate and energy policy with a particular focus on questions of industrial policy and their geoeconomic implications within emerging green sectors. Rainer Quitzow is also Professor of Innovation and Sustainability at the Technische Universität Berlin. Before his career as a researcher, he worked at the World Bank in Washington, DC. He holds a PhD in Political Science from the Freie Universität Berlin.

France's Hydrogen Strategy: Focusing on Domestic Hydrogen Production to Decarbonise Industry and Mobility

Ines Bouacida ⓘ

Abstract France was one of the European frontrunners in formulating policies to develop hydrogen for decarbonisation, releasing its first hydrogen plan in 2018. This was followed by a larger, €9-billion strategy in 2020 (to be updated in 2024), hot on the heels of plans released by the European Commission and Germany. The French strategy for hydrogen deployment focuses in particular on applications where hydrogen is key for deep decarbonisation, including refineries and the chemical industry as well as steel production, and the mobility sector. The country aims to have a head start on European and world competitors thanks to its large electricity resources from the existing nuclear fleet and by building new nuclear capacity. Additionally, it relies on several existing innovation hubs specialising in hydrogen, as well as on the support of many local governments involved in hydrogen development and a relatively structured hydrogen industry. The French strategy for hydrogen includes few ambitions at the international level beyond scientific and technological cooperation within the European Union. The political priority is to develop a domestic industry sized to meet national demand, which is seen as a more secure sourcing strategy than relying on imports. This comes in contrast with the positions of France's neighbours, notably Spain, Portugal and Germany, which are pushing to enable cross-border trade of hydrogen as early as possible. This situation has generated political tensions within the European Union and in particular in the Franco-German relationship.

1 Introduction

France has been one of the leading European Member States in hydrogen development, publishing its first hydrogen strategy in 2018 in the midst of policy discussions on the future of the power sector. This initial strategy formed the foundation of a

I. Bouacida (✉)
Institute for Sustainable Development and International Relations (IDDRI), Sciences Po, Paris, France
e-mail: ines.bouacida1@sciencespo.fr

© Helmholtz-Zentrum Potsdam, Deutsches GeoForschungsZentrum GFZ 2024
R. Quitzow and Y. Zabanova (eds.), *The Geopolitics of Hydrogen*, Studies in Energy, Resource and Environmental Economics, https://doi.org/10.1007/978-3-031-59515-8_4

more ambitious, €9-billion hydrogen strategy adopted in 2020. Hydrogen technologies are considered a key driver for the decarbonisation of the industry and transport sectors as well as a building block to (re)build France's industrial competitiveness.

This case study explores key characteristics of French national hydrogen policy, the factors influencing French engagement in international hydrogen trade, and the external dimensions of French hydrogen development. While France has ambitious plans for the development of a national hydrogen industry and can count on the support of regional governments, its lack of political enthusiasm for hydrogen imports sets it at odds with many of its European partners. Despite that, France is involved in hydrogen diplomacy both at the European and global levels and the 2023 proposition for the update of the hydrogen strategy suggests a possible strengthening of French diplomacy.

2024 should bring updates to the policy framework for hydrogen in France, with a final, more specific version of the 2023 revised hydrogen strategy and of the 2028 and 2050 energy and climate planning reference documents (Multi-Annual Energy Planning and Low-Carbon National Strategy). These processes could clarify further the role of hydrogen in the French energy transition and sectoral implementation targets.

2 The French Hydrogen Economy Today

In 2019, France produced around 800 kt of hydrogen (approx. 26 TWh), including about half as a side product from other industrial processes, and half as dedicated production, which makes it the sixth largest producer of hydrogen in Europe (it neither imports nor exports hydrogen). Hydrogen is mainly used for refineries (almost half) and for ammonia production (one third) (see Fig. 1). Dedicated hydrogen production only relies on steam methane reforming (SMR) using natural gas (AFHYPAC et al., 2020).

Additionally, the impulse to develop "decarbonised" hydrogen in 2020 built on a previous plan from 2018 (MTES, 2018). Although the spending did not exceed €100 million (Ministère de l'Économie, 2020), it enabled the development of 11 mobility projects and 5 industrial projects using locally-produced hydrogen (ADEME, 2019, 2020).

The 2018 strategy had two aims: first, to decarbonise existing uses of hydrogen in the industry, starting with those close to commercial viability (refineries, ammonia); then to develop hydrogen in mobility as a "complement" to battery electric mobility (heavy road transport especially at the local level and shipping and aviation). It also mentioned eventually using hydrogen as system storage to replace natural gas and to store variable renewable electricity, subject to the findings of further research. The objective was to reach 10% of "decarbonised" hydrogen in the industry by 2023 and 20–40% by 2028. Modest targets for the transport sector were also formulated: 5000 light-duty vehicles by 2023; although at the end of 2022, only around 500 such vehicles are in service (France Hydrogène, 2022).

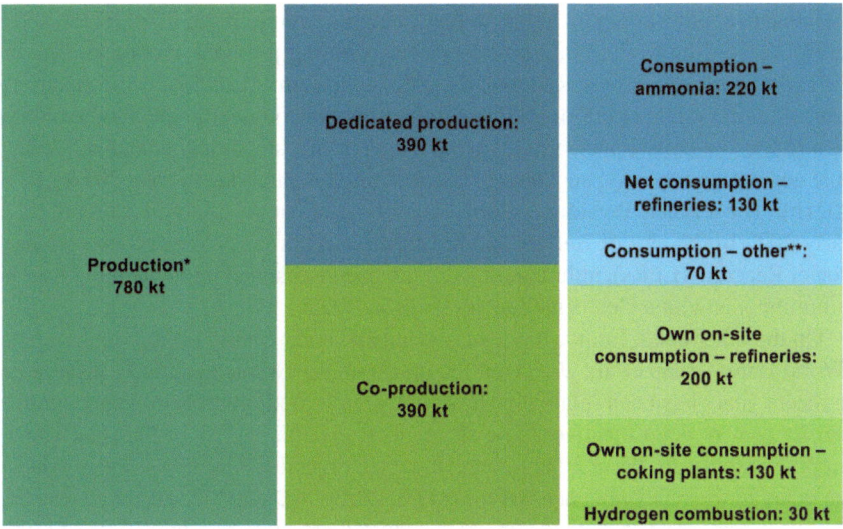

* production: hydrogen from processes generating pure H2 or mixed with other gases
** of which hexamethylenediamine: 40kt; surface treatment of metals: 10 kt; hydrogen peroxide: 7 kt.

Fig. 1 Hydrogen production and consumption in France in 2019. *Source* AFHYPAC et al. (2020) drawing on Hinicio & EY (2020). Reproduced with permission (slightly modified)

France's commitment to funding research and innovation projects around hydrogen over the past decades has made the Paris region one of the 10 leading global hydrogen innovation clusters according to an IEA ranking (IEA & European Patent Office, 2023). It is the top European cluster on the list, ranking highest after several Japanese and US zones and before German ones. France is also one of the world leaders in hydrogen patenting, with 6% of all hydrogen-related international patent families originating in the country (IEA & European Patent Office, 2023). Several French private actors are world leaders, including Air Liquide (the world's second-largest producer of hydrogen), EDF (a world leader in electricity technologies), Alstom (company with the first commercial hydrogen fuel cell-powered passenger train in operation) and McPhy (a leading producer of alkaline electrolysers).

3 The Role of Hydrogen in French Climate and Energy Policy

Today, the French power system has surplus,[1] low-emission power generation capacity, partly thanks to its large nuclear fleet (about 60 GW), despite reduced availability in winters 2021–2022, 2022–2023 and 2023–2024 due to regular planned

---

[1] In other words, generation capacity structurally exceeds national consumption.

maintenance, unforeseen technical defects, and the ageing of the fleet (RTE, 2021a, 2023). According to the 2022 power-system planning political announcements by the French President of the Republic and 2021 long-term scenarios by the electricity transmission system operator RTE, the power system could still have large volumes of surplus, low-emission electricity in the longer term, using both nuclear and renewable generation capacity, depending on actual implementation (Elysée, 2022c; RTE, 2021b). The French President of the Republic Emmanuel Macron announced in early 2022 in the city of Belfort that the state would order 6–14 new Evolutionary Power Reactors (EPRs), although that has not been translated into concrete financial commitments with a clear timeline yet (Elysée, 2022c).

On the other hand, France has a large wind and solar energy production potential (World Bank Group et al., 2023) but the uptake of renewable electricity installation in recent years has been relatively slow compared to both other European countries and its own policy objectives. The share of renewable energy in the final energy supply was only 20.7% in 2022—while the 2020 objective was 23%—mostly due to the delay in the renewable heat sector (MTE, 2023a; SDES, 2022). France installed 2 GW of wind power capacity in 2022, while Germany installed 7 GW. Further renewable buildout faces barriers: slow. Administrative procedures, strong local opposition to onshore wind and solar panels, and insufficient cooperation between national and regional level (IEA, 2022). Given that several studies have shown that the cheapest hydrogen is produced from renewables, the slowness of renewable energy deployment might hinder hydrogen development in France (Bouacida & Berghmans, 2022).

Hydrogen features in existing energy and climate policy plans, but because these plans were formulated before the 2020 hydrogen strategy, its role is smaller than what the recent hydrogen strategy and political announcements suggest. French climate policy is laid out in the 2020 National Low-Carbon Strategy (in French, SNBC), which defines a pathway to climate neutrality by 2050, and the 2020 Multi-Annual Energy Programming Law (in French, PPE), which represents the legal basis for the implementation of energy transition targets until 2028. The Ministry for the Energy Transition was meant by law to provide an Energy and Climate Planning Law for the 2030 and 2035 horizon by July 2023 (Code de l'énergie Article L100-1A, 2023), but political discussions were delayed and will continue in 2024, although the detailed timeline and legislative tools (law, decree, ordinance) are unknown. The Ministry published in November 2023 some non-binding objectives for 2030 for consultation. Delays are attributed by commentators to a lower political profile of climate policy as well as political disagreements relating to the use of biomass in the 2030–2025 time horizon, and to political tensions in the French Parliament where there is no absolute majority (Contexte, 2023; Goar & Mouterde, 2023).

The 2024 revisions of French climate policy should include recent agreements to increase ambition at the European level. In particular, the 2022 revision of the Renewable Energy Directive sets targets for the incorporation of renewable hydrogen and derived fuels (so-called Renewable Fuels of Non-Biological Origin, RFNBOs) in the industry (42% by 2030, 60% by 2035) and in the transport sector (5.5% of RFNBOs and advanced biofuels by 2030).

Lastly, achieving energy independence has long been a prominent feature in energy policy and still is a core objective of long-term climate and energy policy and political discourse (Andriosopoulos & Silvestre, 2017; Elysée, 2022c; MTES, 2019). In practice, French energy import dependency is relatively high at 45 percent and only slightly below the EU average (55 percent). Additionally, political discourse by the current President of the Republic has suggested that France should aspire to *technology leadership* in green technologies in order to achieve "industrial" and "ecological sovereignty", meaning to maintain industrial jobs and gain economic competitiveness (Elysée, 2021, 2022b, 2022c, 2023a). This dual narrative has played in favour of the development of hydrogen technologies as well as nuclear power which would provide the needed "low-carbon" electricity for hydrogen production. In practice, the government has launched France 2030, a €54 bn, 5-year investment plan aiming at developing key technologies for competitiveness and decarbonisation, including for hydrogen (Ministère de l'Economie, 2022), and adopted in October 2023 a Green Industry Law aiming to make France a technological leader, as a response to the American IRA, although its true efficacy in fostering industrial activities is put to doubt and its true impact on the hydrogen economy is unclear (Ministère de l'Economie, 2022, 2023; Roux-Goeken, 2023).

4   Hydrogen Strategies in France: Main Objectives and Implementation

4.1   Three Government Priorities for 2030 and Several French Companies' Strategies

The French hydrogen strategy was published in September 2020 and features three priorities: decarbonise the industry, develop hydrogen mobility, support research, innovation and capacity building. It initially earmarked €7.2 bn until 2030, including €2 bn from the recovery plan launched in 2020.

One year later, the hydrogen plan received an additional €1.9 bn from the France 2030 plan, including €1.7 bn dedicated to financing Important Projects of Common European Interest (IPCEI). IPCEIs are large transnational innovation and infrastructure projects that, upon approval by the European Commission, become eligible for state aid from Member States. The breakdown of the initial €7.2 bn for hydrogen is shown in Fig. 2; the details of France 2030's budget is unknown. The 2023 revision of the hydrogen plan did not earmark any additional funding. It is open for consultation between December 2023 and January 2024 and its final version should be released in early 2024.

Although the 2020 French hydrogen strategy mobilises far more financial resources than its 2018 counterpart (€9 bn v. €100 million), the political priorities are similar. The strategy takes a cautious approach with respect to the end uses of hydrogen and focuses on decarbonising existing uses of fossil hydrogen in industry

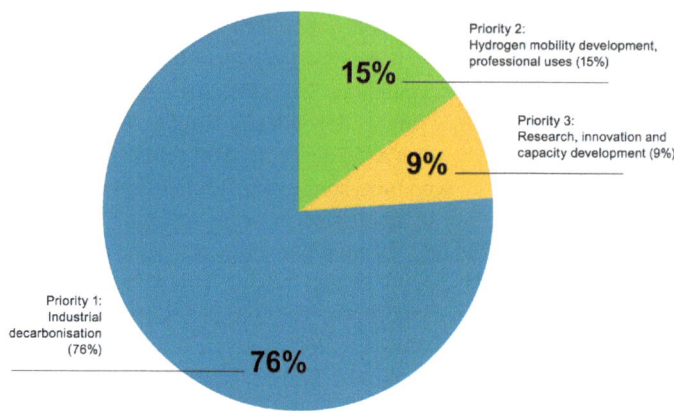

Fig. 2 Breakdown of hydrogen strategy budget 2020–2030. *Source* Ministère de la Transition Ecologique (2021). Reproduced with permission (slightly modified)

(refineries and ammonia), with mobility applications (mainly heavy- and light-duty vehicles) coming second. Part of the funding is directed to research and innovation into potential other hydrogen uses. Some end-uses frequently featured in other European hydrogen strategies, such as heat in buildings and passenger cars, are not mentioned. Aviation and maritime transport are mentioned within the innovation strategy for hydrogen.

The December 2023 proposed revision gives a stricter hierarchy among uses in the transport sector, only using hydrogen in the segments where there are no alternatives, especially where direct electrification or biofuels are not sufficient. It clearly states that in road transport, hydrogen should only be developed in "intensive" use cases and is not relevant for most segments, including passenger cars.

In the existing SNBC, hydrogen from electrolysis is developed up to 40 TWh by 2050, which is slightly more than current hydrogen consumption in France. It is only produced domestically, and it is used mostly in industry (20 TWh by 2050), followed by the power sector (15 TWh). Little indication is available regarding hydrogen uses by sub-sector. The consumption of electrolytic hydrogen happens for the largest part after 2030 (MTES, 2019). The PPE's main objectives for hydrogen are that 20–40% of hydrogen used in industry is renewable by 2028 and the deployment of hydrogen light- and heavy-duty road vehicles, namely up to 50,000 light-duty vehicles and up to 2000 heavy-duty vehicles by 2028, although targets for the latter are quite modest compared to the size of the whole fleet (MTES, 2020).

French hydrogen policy only mentions electrolysis to produce hydrogen, so far excluding fossil-based hydrogen production combined with carbon capture and storage. Government policies since 2020 focus on "decarbonised" hydrogen, a term that refers to hydrogen produced using electrolysis, either from nuclear or renewable electricity. The 2020 strategy sets a 6.5 GW electrolyser objective for 2030, which has not been updated since. So far, the French electrolyser project pipeline is quite advanced, with over thirty projects at the final investment decision (FID) stage or beyond, with most projects exceeding the 1 MW threshold (IEA, 2023).

France also aims to control other parts of the hydrogen supply chain: in 2022, the French Prime Minister announced co-financing for ten "gigafactories" for the production of electrolysers by 2030, of which one was already approved by the European Commission, as well as factories for fuel cells and other hydrogen equipment.

Implementation of these strategic priorities has been relatively slow, however. So far, only a small part of the budget has been disbursed and much of this spending has been directed at the mobility sector (in particular urban vehicles) and renewable hydrogen production (ADEME, 2019; Ministère de la Transition énergétique, 2023). A €4 bn contracts-for-difference scheme for electrolytic hydrogen production was announced in September 2023; first auctions should take place in 2024, although the details of the mechanism are not known yet (France Hydrogène, 2023a; MTE, 2023b).

Following the publication of the 2020 French government strategy, Engie, a partly state-owned gas utility, and EDF, the predominantly state-owned electricity utility, also formulated global hydrogen strategies. Though their objectives are relatively vague, they seem to align with government objectives, although they formulate targets for the world level. For example, Engie aims to develop 4 GW of renewable electrolysis, 700 km of dedicated hydrogen pipelines, and 1 TWh of storage worldwide by 2030 (Engie Hydrogène, 2020). EDF, which owns the entire French nuclear fleet and became entirely state-owned in 2023, aims to promote "low-carbon" hydrogen, including both nuclear- and renewables-based hydrogen production. It aims to install 3 GW of electrolysis capacity by 2030 for uses matching those specified in the government strategy of 2020 (light-duty vehicles, existing industrial uses, synthetic fuels, and, in the longer term, power production) (EDF, 2022).

4.2   Strategies Formulated by French Regions

In addition to the national strategy, many regional-level strategies have been formulated, with most regions developing a roadmap or strategy in the course of 2020. French regions do not have competence regarding energy planning, and the funding they can dedicate to hydrogen development is limited. However, their portfolio of competences includes regional transport, regional economic development, and spatial planning, all of which could play a role in hydrogen development.

The priorities set by regions are quite consistent with national goals, with a stronger emphasis on the transport sector. In some cases, however, they also seek to encourage non-priority uses not mentioned in the national strategy, such as in buildings (Région Pays de la Loire, 2020). Hydrogen is seen as a tool for job creation in local industrial ecosystems and for implementing climate action. In Brittany, which is heavily dependent on energy imports from other French regions, hydrogen has been identified as a means to enhance energy autonomy (Région Bretagne, 2020).

Other regions are keen to position themselves on the international hydrogen market, such as the Grand Est Region, which shares borders with Belgium, Luxembourg, Germany and Switzerland, and the PACA Region, which has strong commercial links with the Mediterranean region (Région Bretagne, 2020; Région PACA, 2020). The Grand Est region was a key stakeholder in the mosaHYc project, a cross-border hydrogen infrastructure project in the Saar-Lorraine-Luxembourg region (GRTgaz & CREOS, 2020).

Regions were also essential in starting the first hydrogen projects, including the Île-de-France region (surrounding Paris), the Rhône Valley, the southwest area (including Toulouse and Bordeaux, on the border with Spain), the Dunkerque area (close to Belgium), the Rhine area (on the border with Germany) (IEA & European Patent Office, 2023). Regional stakeholders, including regional governments and small and medium-sized enterprises will remain key in the implementation of national hydrogen goals.

4.3   Elements on Infrastructure and Cross-border Trade

The French vision for pure hydrogen infrastructure differs slightly from that of many other EU Member States, due to different national features.

The French 2020 strategy is cautious when it comes to the topic of transport infrastructure buildout, only highlighting the need to support research and innovation to prepare for the "hydrogen infrastructure of tomorrow" (MTE, 2020). The 2023 proposed update fleshes out a strategy for the next decade, with a strong focus on the main industrial centres (so-called "hubs"), which should all have access to hydrogen by 2030 thanks to 500 km of hydrogen pipelines connecting consumption to production and storage facilities. A precise layout of the first hydrogen routes should be defined by 2026 (MTE, 2023b).

The French national strategy envisions limited cross-border trade of hydrogen or derivatives. The 2020 version does not mention international trade at all. While the 2023 proposal assumes a global market, that only develops at a later stage (no earlier than 2040) and the wording on France's position in the global market remains vague. In view of the spatial concentration of industrial uses, large-scale pipeline infrastructure is not be needed to decarbonise the industry sector (Agora Energiewende & AFRY Management Consulting, 2021; Bouacida, Wachsmuth, et al., 2022a, 2022b).

However, the French position on infrastructure and cross-border hydrogen trade could shift towards a stronger support for connections between industrial clusters, but only at a later stage (MTE, 2023b). French electricity and gas network operators as well as the hydrogen industry have explored the technical and industrial possibilities around importing hydrogen products in several studies (Amber Grid et al., 2022; RTE, 2022). France has also agreed to "H2Med", a pipeline connecting Barcelona to Marseille and continuing to Germany (Elysée, 2022a).

5   External Dimensions of Hydrogen Development in France

5.1   Favourable Technical and Political Conditions for France to Become a Hydrogen Exporter and a European Hub

France possesses a number of advantages that would enable it to become an exporter of hydrogen to its European neighbours. In particular, the availability of low-emissions electricity (existing and planned) and its location in Europe would allow France to transition from its role as a European transit hub for natural gas (since the decrease of natural gas supply from Russia) to a hydrogen hub (GRTgaz & Teréga, 2022).

Metropolitan France has natural gas interconnections with most its neighbours (Belgium, Germany, Switzerland, Spain), which could in theory be retrofitted to accommodate for pure hydrogen (GRTGaz, 2019). It also possesses three large LNG terminals, which could be converted to process ammonia or liquid hydrogen, although the latter entails both technical and economic uncertainties (Riemer et al., 2022). Additionally, its location between potential hydrogen exporters (Iberian Peninsula, North Africa) and importers (Belgium, Germany, the Netherlands) make France a potential node for European hydrogen transport. Some of France's neighbours, most prominently Spain and Germany, have expressed an interest in implementing such a strategy.

These advantages have been noted in several technical studies. A World Energy Council (2021) study found that France could export up to 0.7 Mt of hydrogen and import up to 0.5 Mt of hydrogen by 2050, depending on actual nuclear and renewable electricity capacity installation (World Energy Council, 2021). Similarly, a study by natural gas network operators found that the transport costs of imported hydrogen would be offset by the opportunity to access cheaper hydrogen resources (European Hydrogen Backbone, 2021). Such technical analyses should be complemented by research on the industrial and environmental conditions for the development of international value chains.

5.2   International Perspectives in French Policy Plans

Despite some favourable technical and political circumstances, French national policy does not identify cross-border hydrogen transport as a priority, and the levels of demand envisioned in existing scenarios could likely be covered through domestic production.

Accordingly, the 2020 strategy does not mention cross-border trade at all. However, the 2023 update announces a diplomatic strategy to be adopted in the first half of 2024. It still prioritises developing a national hydrogen economy over international supply chains, but it contemplates possible trade after 2040. It also announces government support to French companies so that they can export their technologies without however mentioning specific mechanisms.

International competition in the production of hydrogen could pose a threat to French plans to become a leader in industrial hydrogen technologies. Cross-border hydrogen infrastructure is a pre-requisite for large-scale hydrogen imports; slowing down the buildout of cross-border infrastructure could therefore limit the import of cheap hydrogen and derivatives.

The official French position on imports also echoes doubts expressed in several independent analyses regarding the feasibility and desirability (in relation to technical and environmental concerns) of hydrogen imports, especially in the short to medium term (Bouacida, 2022; SCI4climate.NRW, 2021; Wietschel et al., 2020).

A number of private stakeholders in France have expressed more favourable views on imports, including the hydrogen industry association France Hydrogène (Gouty, 2022) and companies directly related to the natural gas sector. France Hydrogène is the national hydrogen industry federation and has offices at the regional level. It includes the main hydrogen private players at the French level, of which multinational companies such as Air Liquide or Engie, as well as smaller actors such as McPhy and Ad Venta. Its membership includes numerous national-level public stakeholders, including the Caisse des Dépôts (French public finance institution) and local governments like Alsace Collectivité européenne and research institutions (CNRS, CEA). This allows France Hydrogène to present a unified political message to influence policy discussions and promote the interests of the hydrogen industry.

Engie—a founding member of the Hydrogen Council and part of France Hydrogène—has announced that three quarters of its targeted electrolyser capacity of 4 GW by 2030 will be produced outside France (Engie Hydrogène, 2020; Le Figaro avec AFP, 2022). GRTgaz and Teréga, the French gas TSOs, are involved in the gas TSO consortium "European Hydrogen Backbone", which is investigating technical options for the development of interconnections for hydrogen imports; the two TSOs consider hydrogen imports to Europe "likely" (GRTgaz & Teréga, 2021).

Several French corporate players are involved in projects abroad that do not envisage exports to France: Engie has partnered with Anglo-American in South Africa, EDF is involved in a renewable hydrogen megaproject in Egypt, Total Eren is developing a large-scale green hydrogen project in Chile and Mauritania, and Lhyfe in Finland. These investments are still in the early stages of development.

The 2022 gas supply crisis has not fundamentally altered the position of French players on international dimensions of hydrogen policy. However, hydrogen policy is evolving to accommodate recent industrial and political developments, especially in the context of updates to energy and climate planning.

Developing electrolytic hydrogen is not a significant lever to decrease European dependence on Russian natural gas imports in the short to medium term because the industry is nascent and because hydrogen could not and should not be developed in most of today's natural gas uses (Bouacida, Rüdinger, et al., 2022). Additionally, France was not strongly dependent on Russian natural gas before the crisis (only 17 percent of its supply came from Russia), meaning that France was less affected than many other European Member States (Ministère de la Transition Energétique, 2021).

5.3   French Hydrogen Diplomacy at the EU Level

French diplomacy on hydrogen at the European Commission and in the Council of the EU has been strong but mostly focused on the criteria for "sustainable" hydrogen in Green Deal texts. As has occurred in contexts such as the EU taxonomy for sustainable activities and the net-zero industry act (Messad, 2023a; Simon & Taylor, 2022), France has pushed for nuclear-based electricity and hydrogen to be included into sustainable criteria, and has treated this as a red line in several cases (Simon et al., 2023). The underlying goal of these diplomatic efforts is to keep the door open for nuclear power capacity to be eligible for public funding and able to contribute to renewable or low-emission energy goals. Challenges around the deployment of renewable electricity in France have added impetus to these efforts (see Sect. 4).

France has quite successfully applied this strategy in several European policy processes, notably the Hydrogen and Decarbonised Gas Package and the Renewable Energy Directive and its two delegated acts. According to the adopted revision of the Renewable Energy Directive (RED), countries whose fossil share in hydrogen consumption is less than 23% in 2030, 20% in 2035 can discount the industry target for renewable hydrogen consumption by 20%., A RED delegated act exempts electrolysis projects connected to power grids with a 90% renewables share or an emission factor under 65 gCO_2/kWh (today, only France and Sweden qualify) to prove they contributed to building additional renewable power capacity (European Commission, 2023; European Union, 2023). These provisions were considered a diplomatic win for France (France Hydrogène, 2023b; Messad, 2023c) as they ease the targets for renewable hydrogen if non-fossil hydrogen is also produced—in this case, hydrogen from nuclear electricity.

To defend nuclear, France cites article 194 of the Treaty on the Functioning of the European Union which protects Member States' *"right to decide the conditions for exploiting its own energy resources, choose between different energy sources and decide the general structure of its energy supply"*. In February 2023, it built an "alliance" aiming to defend nuclear power in climate legislation with other Member States. This included Central and Eastern Europe countries (except Austria and the Baltic countries), Finland, the Netherlands, Sweden, Belgium and Italy with an "observer status". Although this alliance has been relatively loose so far, it constitutes a blocking minority in Council negotiations, and France has had several successes on hydrogen since it was formed. It also sparked the creation of a rival alliance led by Austria in defence of renewables (Messad, 2023b).

Regarding the promotion of hydrogen imports, France has been very cautious in European discussions so far. First, ongoing policy discussions in the EU favour the construction of cross-border hydrogen interconnections as the basis of a trans-European hydrogen network to enable pipeline imports from outside Europe, as articulated in the positions of EU institutions on the current version of the hydrogen and gas markets decarbonisation package. Here, France appears to stand alone with its focus on building local hydrogen ecosystems around industrial hubs over diffuse hydrogen projects. On the other hand, it has not expressed strong opposition to cross-border hydrogen interconnections so far.

Second, in order to gain allies in its support for nuclear energy, France has agreed to back other Member States in their pursuit of fossil-based hydrogen and natural gas. This was the case in taxonomy negotiations in 2021, when France allied with pro-natural gas states to categorise both nuclear and natural gas as sustainable (Hubert, 2022). Similarly, France agreed to end a 20-year-long discussion on the development of a gas inter-connection, initially referred to as the Midcat pipeline, then H2Med, between France and Spain through the Pyrenees. Based on its hydrogen policy, France had little need for a cross-border pipeline, which would enable cheaper Spanish hydrogen to compete with French hydrogen. However, it gave way to pressure from Spain and Germany in exchange for their support for nuclear-based hydrogen in negotiations around the renewable energy directive, while also negotiating an alternative route for the pipeline through the Mediterranean (Euractiv with Reuters, 2023). Similar agreements could emerge in future negotiations to regulate hydrogen imports to the EU, for example.

Supporting European technological and industrial cooperation on hydrogen has been a core part of the French strategy from 2020 (MTE, 2020). Such cooperation is essential to achieve the technological leadership on hydrogen that the EU and France are aiming for.

In particular, France is involved in the European Clean Hydrogen Alliance, which was founded by the European Commission to facilitate partnerships between industry players and Member States, and which aims to set up a pipeline of hydrogen projects covering the whole value chain across Europe.

Additionally, France has participated in Important Projects of Common European Interest (IPCEIs) that allow Member States to jointly propose and finance hydrogen projects subject to the approval of the European Commission. For example, the infrastructure project Masshylia led by Total and Engie near Marseille to incorporate renewable hydrogen to a biofuel refinery was approved as an IPCEI in July 2022.

In particular, France and Germany have built ties for hydrogen development, despite political disputes on the definition of "sustainable" hydrogen. A French-German joint working group for hydrogen was expected to deliver its conclusions by April 2023[2] and both countries have pledged to develop a joint roadmap, although it was never mentioned again (Elysée, 2023b).

5.4   French Diplomacy on the International Stage

France is involved in numerous international initiatives looking to coordinate efforts to foster hydrogen development, such as the International Partnership for Hydrogen and Fuels Cells in the Economy (IPHE) (led by the Frenchman Laurent Antoni),

---

[2] To the author's knowledge, no conclusions of the working group have been published at the time of writing.

IRENA and the IEA. The French 2023 strategy reasserts the importance of international arenas, especially for the definition of standards, so that they do not exclude French hydrogen technologies.

Additionally, major French corporate actors have engaged in global partnerships aiming to coordinate companies at the global level to increase private participation in the hydrogen sector. These include Engie, Total and Air Liquide, who were founding members of the Hydrogen Council.

This contrasts with French foreign policy on hydrogen, which has focused on promoting industrial partnerships for French companies but has not involved in production projects dedicated to exports towards Europe or France. For example, the Indo-French roadmap for developing green hydrogen, signed in late 2022, focuses on fostering information exchange e.g., on regulatory development and scientific research, and on helping industry players form partnerships (Ambassade de France en Inde, 2022). Although France also discussed hydrogen cooperation with Norwegian players, the resulting partnership focused rather on carbon capture and storage (CCS) (Business Norway, 2022; Olje- og energidepartementet, 2022). This approach was reinforced in the 2023 proposed update of the French hydrogen strategy.

France also contributes to energy policy processes in developing countries through Agence Française pour le Développement (AFD) and various development assistance programmes. Climate and environment are a key feature of AFD's objectives and the agency has dedicated significant resources (roughly one fifth of its spendings in 2018) to energy transition objectives, with a particular focus on energy access, energy supply decarbonisation and energy efficiency (AFD, 2019). However, the agency is not currently involved in hydrogen projects on a significant scale. This could change after 2024 as the proposed revised strategy of December 2023 mentions AFD loans as a tool to promote French industrial technologies abroad.

6 Conclusion

In terms of political ambition to develop hydrogen technologies, France is among the frontrunners in the European Union. Building on its first national hydrogen strategy of 2018, France adopted ambitious additional targets in 2020, at a time when many other European Member States and the European Commission were still developing hydrogen plans or implementing post-pandemic recovery programmes. Early 2024 should see a final version of the updated national strategy which was proposed in December 2023.

France has several assets in the race to technological leadership: a surplus of nuclear and renewable electricity (which it plans to maintain), world-ranking innovation clusters, strong political support at the national and regional level and several global-scale private players. This partly explains why France has not engaged significantly in building import routes for hydrogen: the political priority is to develop existing assets and lead technological innovation and deployment. This agenda is evident in various projects that have been funded so far in France, with a visible focus on mobility projects and hydrogen valleys, although the implementation of hydrogen projects in industry has been slower. Some French corporations however do project on engaging in imports of hydrogen products and in production projects abroad.

The discrepancy between French hydrogen priorities and those of its European partners raises political issues. France's unwillingness to cooperate on the construction of cross-border hydrogen connections has compromised Spanish and German ambitions to build a pipeline route to distribute hydrogen from renewables-rich Spain to the German steel and chemical industry via metropolitan France. This generated diplomatic tensions at the European level between France and Germany and Spain, which were only partly mitigated by France conceding to the development of a pure hydrogen pipeline connecting Barcelona and Marseille, the so-called "BarMar", and subsequently to a longer "H2Med" pipeline between Spain and Germany. Although this pipeline could enable large-scale trade of hydrogen between the Iberian Peninsula and Germany, many obstacles still stand in the way of its realisation, including technical and implementation challenges. This puts further strain on the French-German relationship on energy issues, adding to tensions relating to the question of whether new nuclear projects should be incentivised in European energy and climate legislation to achieve the energy transition. France and Germany still managed to reach a deal which they both qualified as "a win" on the electricity market design reform in October 2023, despite tensions through the negotiations (Simon & Kurmayer, 2023).

In a scenario where few or no hydrogen interconnectors would be built between France and Spain, Germany would need to achieve its import goals by alternative and possibly more expensive routes, or parts of its industry could have to relocate to regions with a more affordable energy supply. Spain and Portugal would need to adapt their hydrogen policies towards alternative export routes or a larger domestic market.

Acknowledgements Research for this chapter was financially supported by the German Federal Foreign Office within the framework of the project "Geopolitics of the Energy Transformation— Implications of an International Hydrogen Economy" (GET Hydrogen), funding reference number AA4521G125.

References

ADEME. (2019, May 3). *#Hydrogène : #AàP « Ecosystèmes de #mobilité hydrogène », très forte mobilisation des acteurs industriels et des territoires, 11 projets sélectionnés.* Communiqué de Presse. https://presse.ademe.fr/2019/05/hydrogene-aap-ecosystemes-de-mobilite-hydrogene-tres-forte-mobilisation-des-acteurs-industriels-et-des-territoires-11-projets-selectionnes.html

ADEME. (2020). *52 histoires de transition écologique. Mobilité et industrie misent sur l'hydrogène vert.* 52 Histoires Pour s'inspirer. https://52histoires2020.ademe.fr/histoire/mobilite-et-indust rie-misent-sur-l-hydrogene-vert

AFD. (2019). Stratégie 2019–2022 Transition énergétique. *Les Transitions Énergétiques.* https://doi.org/10.2307/j.ctvt1shs7.12

AFHYPAC, Ernst & Young, & Hinicio. (2020). *Etude de la demande potentielle d'hydrogène renouvelable et/ou bas carbone en France à 2030.*

Agora Energiewende, & AFRY Management Consulting. (2021). *No-regret hydrogen. Charting early steps for H2 infrastructure in Europe.* https://www.agora-energiewende.de/en/publicati ons/no-regret-hydrogen/

Ambassade de France en Inde. (2022, October 18). *France and India adopt Joint Roadmap on Green Hydrogen.* https://in.ambafrance.org/France-and-India-adopt-Joint-Roadmap-on-Green-Hydrogen

Amber Grid, Bulgartransgaz, Conexus, CREOS, DESFA, Elering, Enagás, Energinet, Eustream, FGSZ, FluxSwiss, Fluxys Belgium, GasConnect Austria, Gasgrid Finland, Gassco, Gasunie, Gas Networks Ireland, GAZ-SYSTEM, GRTgaz, … Transgaz. (2022). *European Hydrogen Backbone. A European hydrogen infrastructure vision covering 28 countries.*

Andriosopoulos, K., & Silvestre, S. (2017). French energy policy: A gradual transition. *Energy Policy, 106,* 376–381. https://doi.org/10.1016/J.ENPOL.2017.04.015

Bouacida, I. (2022). *Hydrogen imports in Europe: A lever for climate cooperation?* IDDRI Blog. https://www.iddri.org/en/publications-and-events/blog-post/hydrogen-imports-europe-lever-climate-cooperation

Bouacida, I., & Berghmans, N. (2022). Hydrogen for climate neutrality: Conditions for deployment in France and Europe. *IDDRI Study, 2.* https://www.iddri.org/en/publications-and-events/study/hydrogen-climate-neutrality-conditions-deployment-france-and-europe

Bouacida, I., Rüdinger, A., & Berghmans, N. (2022). Sortir de la dépendance au gaz naturel russe : Quelles stratégies pour l'UE et la France ? *Document de Propositions Iddri, 3.* https://www.iddri.org/fr/publications-et-evenements/propositions/sortir-de-la-depend ance-au-gaz-naturel-russe-quelles

Bouacida, I., Wachsmuth, J., & Eichhammer, W. (2022). Impacts of greenhouse gas neutrality strategies on gas infrastructure and costs: Implications from case studies based on French and German GHG-neutral scenarios. *Energy Strategy Reviews, 44*(June 2021), 100908. https://doi.org/10.1016/j.esr.2022.100908

Business Norway. (2022, September 12). *Programme: French-Norwegian Decarbonization forum.* https://norwayevent.com/programme-hydrogen-and-ccs-forum/

Contexte. (2023). Stratégie française énergie-climat : « Notre travail de prospective ne boucle pas complètement aujourd'hui », prévient Sophie Mourlon. *Contexte Energie.* https://www.con texte.com/actualite/energie/strategie-francaise-energie-climat-notre-travail-de-prospective-ne-boucle-pas-completement-aujourdhui-previent-sophie-mourlon-2_177782.html

EDF. (2022). *Le Plan Hydrogène du groupe EDF.* https://www.edf.fr/groupe-edf/espaces-dedies/journalistes/tous-les-communiques-de-presse/le-groupe-edf-lance-un-nouveau-plan-industriel-dedie-a-lhydrogene-100-bas-carbone

Elysée. (2021, November 16). *Devenir le leader de l'hydrogène vert, voilà notre objectif avec France 2030 !* https://www.elysee.fr/emmanuel-macron/2021/11/16/deplacement-beziers-gen via-france-2030

Elysée. (2022a). *Déclaration conjointe à l'occasion du lancement du projet H2Med.* https://www. elysee.fr/emmanuel-macron/2022/12/09/declaration-conjointe-a-loccasion-du-lancement-du-projet-h2med

Elysée. (2022b). *Devenir le leader des industries vertes !* https://www.elysee.fr/emmanuel-macron/2022/11/08/devenir-le-leader-des-energies-vertes

Elysée. (2022c, February 10). *Reprendre en main notre destin énergétique !* https://www.elysee.fr/emmanuel-macron/2022/02/10/reprendre-en-main-notre-destin-energetique

Elysée. (2023a). *Déclaration de M. Emmanuel Macron, président de la République, sur la planifi-cation écologique, à Paris le 25 septembre 2023.* https://www.vie-publique.fr/discours/291196-emmanuel-macron-25092023-planification-ecologique

Elysée. (2023b). *French-German declaration of 22 January 2023.* https://www.elysee.fr/en/emm anuel-macron/2023/01/22/french-german-declaration

Engie Hydrogène. (2020). *L'accélérateur de décarbonation.*

Euractiv with Reuters. (2023). France in new row with Germany and Spain over nuclear-derived hydrogen. *Euractiv.* https://www.euractiv.com/section/energy-environment/news/fra nce-in-new-row-with-germany-and-spain-over-nuclear-derived-hydrogen/

European Commission. (2023). *Commission delegated act to the Renewable Energy Directive establishing a Union methodology setting out detailed rules for the production of renewable liquid and gaseous transport fuels of non-biological origin.* https://eur-lex.europa.eu/legal-con tent/EN/TXT/?uri=uriserv%3AOJ.L_.2023.157.01.0011.01.ENG&toc=OJ%3AL%3A2023% 3A157%3ATOC

European Hydrogen Backbone. (2021). *Analysing future demand, supply, and transport of hydrogen.* https://gasforclimate2050.eu/?smd_process_download=1&download_id=718

European Union. (2023). *Renewable Energy Directive 2023.*

France Hydrogène. (2022). *Chiffres clés - Vig'Hy.* Vig'Hy. https://vighy.france-hydrogene.org/chi ffres-cles/

France Hydrogène. (2023a). *4 milliards pour la production d'hydrogène décarboné.* https://www. france-hydrogene.org/magazine/4-milliards-pour-la-production-dhydrogene-decarbone/?cn-reloaded=1

France Hydrogène. (2023b). Réaction de France Hydrogène à l'adoption des deux actes délégués définissant les carburants renouvelables d'origine non-biologique (RFNBOs). *Communiqué de Presse.* https://www.france-hydrogene.org/reaction-de-france-hydrogene-a-ladoption-des-deux-actes-delegues-definissant-les-carburants-renouvelables-dorigine-non-biologique-rfn bos/?cn-reloaded=1

Goar, M., & Mouterde, P. (2023, September 13). « Il n'y a plus de bande passante » : l'avenir incertain de la loi de programmation sur l'énergie et le climat. *Le Monde.* https://www.lemonde.fr/planete/article/2023/09/13/l-avenir-incertain-de-la-loi-de-pro grammation-sur-l-energie-et-le-climat_6189111_3244.html

Gouty, F. (2022, June 8). Importation d'hydrogène ? « Cela n'a de sens seulement si l'énergie consommée est renouvelable ». *Actu Environnement.* https://www.actu-environnement.com/ ae/news/importation-hydrogene-interview-philippe-boucly-france-hydrogene-energie-renouv eable-39774.php4

Code de l'énergie Article L100–1A. (2023). https://www.legifrance.gouv.fr/codes/section_lc/LEG ITEXT000023983208/LEGISCTA000023985174/#LEGISCTA000023985174

GRTGaz. (2019). *Conditions techniques et économiques d'injection d'hydrogène dans les réseaux de gaz naturel.* https://www.grtgaz.com/sites/default/files/2020-12/Conditions-techniques-eco nomiques-injection-hydrogene-reseaux-gaz-rapport-2019.pdf

GRTgaz, & CREOS. (2020). Hydrogène : Lancement de mosaHYc, un projet de conversion de réseau de gaz transfrontalier au 100% hydrogène. *Communiqué de Presse.* https://www.grtgaz. com/medias/communiques-de-presse/hydrogene-lancement-mosahyc

GRTgaz, & Teréga. (2021). Confirmation d'une demande européenne d'hydrogène de 2300 TWh en 2050, facilitée par la construction de la Dorsale européenne. *Communiqué de Presse.* https://www.grtgaz.com/medias/communiques-de-presse/european-hydrogen-bac kbone-rapport-juin2021

GRTgaz, & Teréga. (2022). Hiver 2022–2023 : Le système gaz français devrait faire face à la demande en s'appuyant sur la gestion prudente des stocks et la sobriété de tous les consomma-teurs. *Communiqué de Presse.* https://www.grtgaz.com/medias/communiques-de-presse/perspe ctives-systeme-gazier-hiver-2022

Hubert, A. (2022). Le nucléaire et le gaz remportent la bataille de la taxonomie verte. *Briefing Énergie.* https://www.contexte.com/article/energie/le-nucleaire-et-le-gaz-remportent-la-bataille-de-la-taxonomie-verte_143809.html

IEA. (2022). *France 2021 Energy Policy Review.* https://www.iea.org/reports/france-2021/execut ive-summary

IEA. (2023). *IEA Hydrogen Production Projects Database.* https://www.iea.org/data-and-statis tics/data-product/hydrogen-production-and-infrastructure-projects-database#hydrogen-produc tion-projects

IEA, & European Patent Office. (2023). *Hydrogen patents for a clean energy future.* https://iea. blob.core.windows.net/assets/1b7ab289-ecbc-4ec2-a238-f7d4f022d60f/Hydrogenpatentsfor acleanenergyfuture.pdf

Le Figaro avec AFP. (2022, November 3). Hydrogène vert : Engie produira principalement à l'étranger pour tenir ses objectifs. *Le Figaro.* https://www.lefigaro.fr/flash-eco/hydrogene-vert-engie-produira-principalement-a-l-etranger-pour-tenir-ses-objectifs-20221103

Messad, P. (2023a). Paris plots response to von der Leyen's 'unfortunate' comments on nuclear. *Euractiv.* https://www.euractiv.com/section/energy-environment/news/paris-plots-response-to-von-der-leyens-unfortunate-comments-on-nuclear/

Messad, P. (2023b, March 29). Nuclear vs renewables: Two camps clash in Brussels. *Euractiv.* https://www.euractiv.com/section/energy-environment/news/nuclear-vs-renewables-two-camps-clash-in-brussels/

Messad, P. (2023c, June 19). *France finally satisfied with EU deal on renewables directive.* Euractiv. https://www.euractiv.com/section/energy-environment/news/france-finally-satisfied-with-eu-deal-on-renewables-directive/

Ministère de l'Economie. (2022). *France 2030 : un plan d'investissement pour la France.* https://www.economie.gouv.fr/france-2030

Ministère de l'Economie. (2023). *Projet de loi industrie verte : découvrir les 15 mesures.* https://www.economie.gouv.fr/industrie-verte-presentation-projet-loi

Ministère de l'Économie. (2020, September 9). *Présentation de la stratégie nationale pour le développement de l'hydrogène décarboné en France.* https://www.economie.gouv.fr/presentat ion-strategie-nationale-developpement-hydrogene-decarbone-france

Ministère de la Transition Energétique. (2021). *Gaz naturel | Chiffres clés de l'énergie - Édition 2021.* https://www.statistiques.developpement-durable.gouv.fr/edition-numerique/chi ffres-cles-energie-2021/14-gaz-naturel

Ministère de la Transition énergétique. (2023, February 1). *"Ecosystèmes territoriaux hydrogène" : Agnès Pannier-Runacher annonce 14 nouveaux lauréats de l'appel à projets.* https://www.ecologie.gouv.fr/ecosystemes-territoriaux-hydrogene-agnes-pannier-run acher-annonce-14-nouveaux-laureats-lappel

MTE. (2020). *Stratégie nationale pour le développement de l'hydrogène décarboné en France. Dossier de presse.* https://www.gouvernement.fr/dossier-de-presse-strategie-nationale-pour-le-developpement-de-l-hydrogene-decarbone-en-france

MTE. (2023a). *Chiffres clés de l'énergie - Édition 2023.* https://www.statistiques.developpement-durable.gouv.fr/edition-numerique/chiffres-cles-energie-2023/

MTE. (2023b). *Stratégie nationale pour le développement de l'hydrogène décarboné en France.* https://www.ecologie.gouv.fr/consultation-sur-nouvelle-strategie-francaise-deploi ement-lhydrogene-decarbone

MTES. (2018). *Plan de déploiement de l'hydrogène pour la transition énergétique.*

MTES. (2019). *Synthèse du scénario de référence de la stratégie française pour l'énergie et le climat.*

MTES. (2020). *Programmation Pluriannuelle de l'Energie 2019–2023, 2024–2028.*

Olje-og energidepartementet. (2022). *Norway and France will strengthen cooperation on CCS.* https://www.regjeringen.no/en/aktuelt/norway-and-france-will-strengthen-cooperation-on-ccs/id2952199/

Région Bretagne. (2020). *Déploiement de l'hydrogène renouvelable : feuille de route bretonne 2030. 20_DCEEB_0.*

Région PACA. (2020). *Plan Régional Hydrogène.* 1–59.

Région Pays de la Loire. (2020). *Feuille de route hydrogène pour les Pays de la Loire.* https://www.paysdelaloire.fr/sites/default/files/2020-09/annexe-4-feuille-de-route-h2_vf.pdf

Riemer, M., Schreiner, F., & Wachsmuth, J. (2022). Conversion of LNG Terminals for Liquid Hydrogen or Ammonia. *Fraunhofer ISI Study.*

Roux-Goeken, V. (2023). Industrie: Des mesures vertes et des pas mûres votées au Parlement | Contexte. *Contexte Energie.* https://www.contexte.com/article/environnement/projet-loi-indust rie-verte-des-mesures-vertes-et-des-pas-mures-votees-au-parlement_176333.html

RTE. (2021a). *Bilan prévisionnel de l'équilibre offre-demande d'électricité en France - principaux enseignements.*

RTE. (2021b). *Futurs énergétiques 2050. La production d'électricité (chapitre 4).* https://bilan-electrique-2020.rte-france.com/

RTE. (2022). *Futurs énergétiques 2050. Le rôle de l'hydrogène et des couplages (chapitre 9) [février 2022].* https://www.rte-france.com/analyses-tendances-et-prospectives/bilan-previsionnel-2050-futurs-energetiques

RTE. (2023). *Bilan prévisionnel. Edition 2023 [principaux résultats].* https://www.rte-france.com/analyses-tendances-et-prospectives/les-bilans-previsionnels

SCI4climate.NRW. (2021). *Wasserstoffimporte. Bewertung der Realisierbarkeit von Wasserstoffimporten gemäß den Zielvorgaben der Nationalen Wasserstoffstrategie bis zum Jahr 2030.*

SDES. (2022). *Chiffres clés des énergies renouvelables édition 2022.*

Simon, F., & Kurmayer, N. J. (2023, October 19). *Deal on EU electricity market reform: What did Paris and Berlin obtain?* Euractiv. https://www.euractiv.com/section/electricity/news/deal-on-eu-electricity-market-reform-what-did-paris-and-berlin-obtain/

Simon, F., & Taylor, K. (2022). LEAK: EU drafts plan to label gas and nuclear investments as green. *Euractiv.* https://www.euractiv.com/section/energy-environment/news/leak-eu-drafts-plan-to-label-gas-and-nuclear-investments-as-green/

Simon, F., Taylor, K., Kurmayer, N. J., Messad, P., & Romano, V. (2023, February 15). The Green Brief: France's pro-nuclear crusade has only just begun. *Euractiv.*

Wietschel, M., Bekk, A., Breitschopf, B., Boie, I., Edler, J., Eichhammer, W., Klobasa, M., Marscheider-Weidemann, F., Plötz, P., Sensfuß, F., Thorpe, D., & Walz, R. (2020). Opportunities and challenges when importing green hydrogen and synthesis products. *Fraunhofer ISI Policy Brief, 3.* https://www.isi.fraunhofer.de/en/presse/2020/presseinfo-26-policy-brief-wasserstoff.html

World Bank Group, ESMAP, VORTEX, & DTU. (2023). Global Wind Atlas - France Mean Wind Speed at 100m. *Global Wind Atlas.* https://globalwindatlas.info/fr/area/France?download=print

World Energy Council. (2021). Decarbonised hydrogen imports into the European Union: challenges and opportunities. *La Revue de l'énergie, Hors-Série.*

Ines Bouacida is a Research Fellow in Climate and Energy at the Institute for Sustainable Development and International Relations (IDDRI) in Paris. She works on the energy transition in France and Europe, in particular on the role of gas in decarbonisation, on the future of gas infrastructure and the development of hydrogen. She holds a Master's degree in Sustainable Development from Utrecht University (the Netherlands) with a major in Energy and Materials.

International Dimension of the Polish Hydrogen Strategy. Conditions and Potential for Future Development

Michał Smoleń and Wojciech Żelisko

Abstract Poland is the third-largest producer of hydrogen in the EU, with around 1 million tonnes generated every year. This grey hydrogen is made almost exclusively from steam methane reforming, based on fossil gas as a feedstock, and utilised primarily by the chemical and petrochemical sectors for ammonia production and for various processes in refineries, respectively. In 2021, the Polish Hydrogen Strategy (PHS) was published as the first official government strategy for low-carbon hydrogen economy development. The document presents an optimistic and ambitious approach, with a focus on domestic production and use of hydrogen in multiple sectors. However, the cost-effective generation of low-carbon hydrogen in Poland can face significant challenges, such as the relatively low availability of clean electricity, reliance on natural gas imports and limited experience with carbon capture technologies. The aspects regarding the emerging global hydrogen market are largely omitted in the PHS. This chapter analyses the possible background of this fact and the factors influencing the future Polish position in this market. Even though Poland is not likely to be a frontrunner in these developments, it is believed to become a pragmatic participant. In fact, several Polish companies, including those state-owned, have already launched some cooperative initiatives at the European level.

1 Introduction

The future of Polish hydrogen has recently sparked significant public interest. In late 2021 the government approved the Polish Hydrogen Strategy until 2030 with an outlook until 2040, the first national document of this kind (Polish Ministry of Climate & Environment, 2021a). Eleven hydrogen valleys are being established to

M. Smoleń (✉) · W. Żelisko
Instrat Foundation, Hoża 51, 00-681, Warsaw, Poland
e-mail: michal.smolen@instrat.pl

W. Żelisko
e-mail: wojciech.zelisko@instrat.pl

© Helmholtz-Zentrum Potsdam, Deutsches GeoForschungsZentrum GFZ 2024
R. Quitzow and Y. Zabanova (eds.), *The Geopolitics of Hydrogen*, Studies in Energy, Resource and Environmental Economics, https://doi.org/10.1007/978-3-031-59515-8_5

support hydrogen cooperation between regional and local governments, large state-owned enterprises, academia and private businesses (Industrial Development Agency JSC, 2023). All Polish coal regions currently undergoing a transition process have mentioned hydrogen in their respective Territorial Just Transition Plans. Hydrogen solutions are being pursued by stakeholders from sectors as diverse as oil and gas, mining, shipping or automotive, as a possible driver for further development in the disruptive era of decarbonisation. Domestic subsidies and industry marketing have already encouraged local administrations to invest in hydrogen buses (Municipal Office in Konin, 2022) to reduce GHG emissions and air pollution from public transport.

Nevertheless, hydrogen is not yet a mainstream topic in Poland, although the introduction of the Polish Hydrogen Strategy prompted some discussion among a broad range of stakeholders. There are several diverse visions for the future national hydrogen economy regarding optimal applications and production modes (zero-carbon versus low-carbon), the shape of the market (centralised versus decentralised) and how best to implement necessary measures (top-down versus bottom-up). For example, some sceptical voices are emerging, raising questions about the environmental sustainability and economic viability of the promoted hydrogen uses or the available supply of low-carbon hydrogen in the coming decades. What is more, the alignment between local visions and concrete regulations being set at the EU level does not appear clear.

The international hydrogen market is seen by expert bodies (IRENA, 2022) and some national hydrogen strategies (German Federal Ministry for Economic Affairs & Energy, 2023) as an important source of low-carbon hydrogen that is needed to decarbonise advanced industrial countries. This international dimension of the hydrogen economy is notably less prominent in Polish discussions. The Polish Hydrogen Strategy is, similarly to analogous documents prepared in other countries in 2020 and 2021, visibly influenced by the economic consequences of the COVID-19 pandemic. Public investment in hydrogen technologies was seen as a way to mobilise domestic industry and the energy sector and to avoid a lasting economic slowdown. While energy security and independence have become key talking points in Polish energy policy discussions after the 2022 Russian invasion of Ukraine, they were initially dominated by the most immediate questions regarding the fossil fuel supply for the 2022/2023 winter and appropriate interventions to shield households and the economy. While there were some early signs of broader reorientation in energy policies such as the March 2022 announcement of assumptions for the energy policy update (Polish Ministry of Climate & Environment, 2022a), the actual strategic and legislative changes were, as of June 2024, carried out only partially, with little focus on the hydrogen economy. There is a possibility for proper reorientation after the October 2023 parliamentary elections gave a majority to liberal committees that stated support for more ambitious energy transition goals.

Against this backdrop, this chapter aims to analyse the external dimensions of the Polish Hydrogen Strategy. Firstly, we provide an overview of the Polish hydrogen economy and identify key factors affecting its future potential. Subsequently, we review the Polish Hydrogen Strategy and its in-depth analytical annex (Kupecki et al., 2021). On this basis, the third part focuses on the role of international engagement—or a relative lack thereof—within the strategy and its possible explanations. Finally, we place the issue in the broader context of Polish international energy policy, including both preliminary assumptions for the Energy Policy of Poland until 2040 update and ongoing policy decisions (both domestic and international), arriving at conclusions for European stakeholders.

2 Polish Hydrogen Economy—State of Play, Potential and Challenges

Poland is in a peculiar starting position in the global push towards a new hydrogen economy (see Fig. 1). It is the third-largest producer of hydrogen in the EU, with around 1 million tonnes[1] produced every year (Polish Ministry of Climate & Environment, 2021a, p. 7). This hydrogen is so-called grey hydrogen, generated almost exclusively from steam methane reforming and utilised by the chemical, petrochemical, steel and food sectors. Existing Polish hydrogen production facilities are for the most part a by-product of relatively developed medium-tech industries, in which hydrogen is predominantly made on-site at large industrial plants. Internal trade is limited in scale and hydrogen imports and exports are moderate, but their net balance is practically zero.[2]

Proponents of the Polish hydrogen economy, including industry stakeholders and policymakers, list several domestic advantages (Kupecki et al., 2021, pp. 455–456), such as the scale of the pre-existing hydrogen economy, steady demand from numerous industries, or developed automotive and rolling stock industries, which can participate in the hydrogen value chains. Moreover, salt caverns[3] could be used for large-scale hydrogen storage and facilitate the creation of hydrogen hubs, and offshore wind power—with a potential that has been assessed at up to 33 GW by the wind industry association (Polish Wind Energy Association, 2022)—for renewable

---

[1] The hydrogen strategy document mentions the value of 1.3 million tonnes per year, yet wrongly, since this higher level applies to the potential of hydrogen production (including, for example, hydrogen, which could theoretically be separated from coke-oven gas), rather than the production capacity of 1.1 million tonnes and the actual production of nearly 0.8 million tonnes (European Hydrogen Observatory, 2023a). The last value applies to the generation of pure hydrogen—it rises to circa 1 million tonnes when hydrogen by-product is included.

[2] Poland engages in a moderate-scale hydrogen trade with other EU members. In 2022, Poland imported almost 0.2 million tonnes of hydrogen mainly from Czechia, the Netherlands and Slovakia and exported virtually all 0.26 million tonnes to Czechia, Slovakia and Hungary (European Hydrogen Observatory, 2023b).

[3] For example, PGNiG, a subsidiary of Orlen, a state-controlled Polish oil and gas conglomerate, presently utilises such caverns for fossil gas storage, which in the more distant future could potentially be repurposed for hydrogen. In general, the total potential of hydrogen storage in Polish onshore salt caverns is estimated at 10 PWh (roughly 300 300 MtH2), the second biggest in the EU (Institute of Power Engineering et al., 2023a, p. 53).

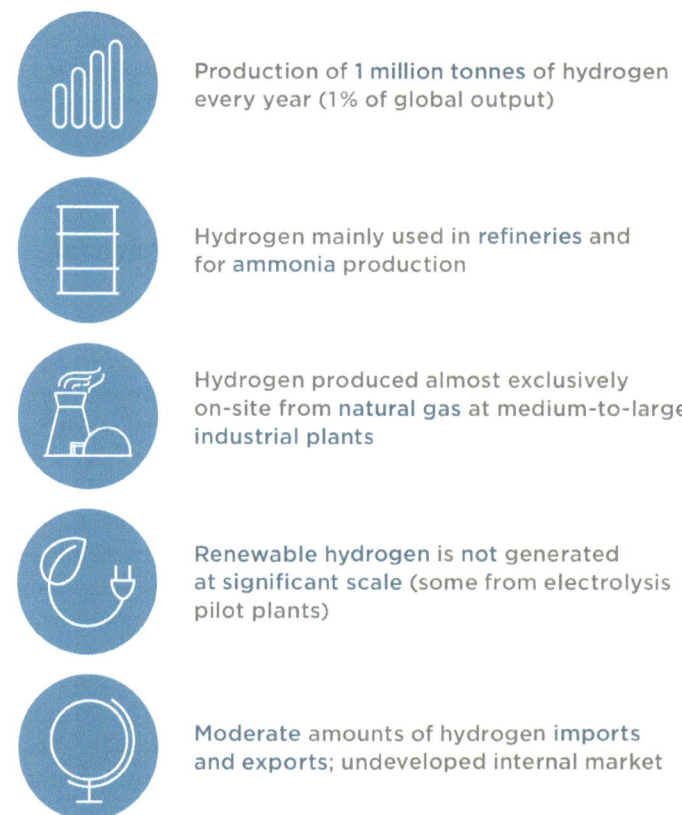

Fig. 1 Polish hydrogen economy at a glance. *Source* Authors' own elaboration based on the data from the Polish Hydrogen Strategy, the European Hydrogen Observatory, the Institute of Power Engineering, and expert knowledge

hydrogen production. Hydrogen itself is touted as a valuable tool for the restructuring of the Polish economy, especially coal-dependent regions.

Nevertheless, various barriers also exist, which can be divided into two loosely defined categories –energy-related and more general barriers. With regard to energy-related issues, Poland lags behind most EU countries in the deployment of renewable energy—the current energy strategy envisions only 32% of renewable energy in electricity production by 2030 (Polish Ministry of Climate & Environment, 2021b, p. 10), though in 2024 that figure may be updated in the new National Energy and

Climate Plan. Onshore wind has suffered[4] after regulations introduced in 2016 made it impossible to initiate new projects (or repower old installations) on 99.7% of the Polish land area (Czyżak et al., 2021), whereas offshore wind development was assessed to be behind schedule in 2022 (Polish Supreme Audit Office, 2022), yet recently next steps have been achieved. Solar power has experienced significant growth since 2019 (Instrat, 2023), but it is mostly in the form of small-scale rooftop PV with relatively low capacity factors due to climate conditions. Renewable electricity will be crucially needed to replace the ageing fleet of coal-fired power plants.[5] Hydrogen produced from biomass, biofuels or municipal waste, promoted by oil and gas incumbents to local administrations as fuel for hydrogen-powered buses, has limited scale-up potential, especially as the same resources will constitute attractive fuels for district heating. Moreover, Poland is unlikely to become a competitive producer of so-called blue hydrogen, i.e. hydrogen generated from natural gas with additional carbon capture and storage (CCS) to reduce CO2 emissions, because it is a gas importer (Statistics Poland, 2023, p. 19) and has very few concrete achievements in the field of CCS technologies (Global CCS Institute, 2023).[6] Ultimately, Polish nuclear power will not be available until the mid-2030s at the earliest.

There are other types of challenges to overcome as well. For example, in comparison with the EU average, Poland's R&D and innovation performance is relatively weak overall (European Commission, 2023a)—and hydrogen-related R&D in the country has been limited (Maj & Szpor, 2020). Furthermore, Poland does not currently produce electrolysers or mine minerals used in their manufacturing, although apparently work focused on creating such electrolyser manufacturing facility is underway. Another obstacle in the first stage of hydrogen economy development is a spatial dispersion of industrial facilities—key industrial centres such as Silesia or central Poland are located far from the Baltic Sea (with its offshore wind

---

[4] These regulations were amended in March 2023. The biggest uncertainty had been their final shape—the initial proposal to modify the minimum required distance between the turbine and its surroundings (mainly houses and environmental protection areas) from the so-called "10H" rule (i.e. ten times the height of the turbine to the tip of its blade; this in practice usually corresponds to ca. 2 km) to 500 m had lost ground in favour of a proposal of 700 m. This will limit the area available for onshore wind turbines by half, and heavily impact the portfolio of projects already prepared for the 500-m limit, thus delaying the significant addition of new capacities perhaps even beyond 2030 (Kopeć, 2023). There is a possibility, though, that the new government will change the regulations to the initially proposed 500-m limit.

[5] Although the deployment of renewables in the Polish power system has been relatively dynamic so far, more and more of them face rejections of connection to the electricity grid. For instance, in 2022, over 51 GW of new capacity (Energy Regulatory Office, 2023, p. 101)—an overwhelming majority of which were renewables—came across this problem, which can be addressed by modernising the distribution grid in the country. Its underinvestment may well negatively affect the possibility of deploying the desired electrolyser capacity in the future and accomplishing the goals of the Polish Hydrogen Strategy.

[6] This may change, however, since Poland has had talks with Norway over the possibility of realising joint initiatives in the field of CCUS technologies (Polish Ministry of Climate and Environment, 2022f) and has amended the Act of Geological and Mining Law, giving a green light to the development of domestic carbon capture projects (Polish Ministry of Climate and Environment, 2023c).

power potential and possible hydrogen import terminals). Finally, some actors also consider the length of administrative procedures and the lack of necessary regulations as well as an adequate financial support system as considerable obstacles.

Despite these constraints, the hydrogen economy has the potential to become an essential part of Poland's decarbonisation pathway—especially in uses with no viable clean alternatives, such as in the steel and chemical industries. However, the abovementioned limitations affect the competitiveness of Polish hydrogen production against, on the one hand, other clean energy solutions, and on the other, that of international hydrogen players. In case of not achieving such satisfactory competitiveness, a great economic risk could come with the relocation of some energy-intensive industries to other parts of the world, more abundant with green energy carriers like renewable hydrogen.

The discussed barriers have not prevented Polish entities, especially state-owned companies, from adopting quite an ambitious approach, in the hope of leading the national hydrogen revolution. Orlen, a state-controlled Polish oil and gas conglomerate, published in 2022 its hydrogen strategy (which acknowledges the goals of the Polish Hydrogen Strategy), stating the objective of 540 MW of low- and zero-carbon hydrogen capacity[7] (both domestic and international in Czechia and Slovakia) for 2030 (Orlen, 2022). Orlen's strategy focuses on different aspects, such as decarbonising its assets with blue hydrogen, supplying hydrogen fuel to the transport sector or utilising it to generate renewable electricity and heat, in line with the national strategy. The company envisages the development of hydrogen production facilities (and hubs) and 100 hydrogen fuel stations across Poland, Czechia, and Slovakia, and explicitly mentions participation "as a supplier and customer" in the European Hydrogen Backbone. Regarding other hydrogen endeavours, Orlen has signed agreements with PESA, the biggest Polish producer of rolling stock, and Alstom, a global leader in the transportation sector, for the development of hydrogen rail and with several local administrations for the implementation of hydrogen public transport.[8] In addition to this, the conglomerate has created a hydrogen academy,[9] a hydrogen hub in Trzebinia (Poland) and an R&D centre in Płock (Poland) (Orlen, 2023). Orlen's recent merger with the Polish state-controlled oil and gas enterprises, Lotos and PGNiG, is likely to further add to its hydrogen projects and plans. For example, PGNiG, a natural gas and crude oil producer, also intends to develop blue hydrogen production and storage capacities (PGNiG, 2020, 2022). Gaz-System, the Polish natural gas transmission system operator, has reached an agreement (Gaz-System, 2022a) with gas TSOs from Slovakia (Eustream), Hungary (FGSZ) and Romania (Transgaz), which includes plans to explore the possibility of international hydrogen transmission. Gaz-System has also recently submitted applications for financing two

---

[7] 50 percent of Orlen's hydrogen in 2030 is planned to be low-carbon, which includes hydrogen generated from natural gas with CCS, biomethane, municipal waste and electrolysis (Orlen, 2022).

[8] Orlen's role in these collaborations is to provide hydrogen fuel and its corresponding necessary infrastructure.

[9] Orlen's hydrogen academy is a step towards building a skilled staff for the future hydrogen economy by means of training university students and recent graduates (Polish Ministry of Climate and Environment, 2023e).

international hydrogen transmission network projects—BEMIP Hydrogen and HI East—in order for them to acquire the status of a PCI (Project of Common Interest) (WNP, 2023). On the other hand, the domestic hydrogen transmission infrastructure is expected to be developed by Gaz-System as a result of actions taken within the European Hydrogen Backbone, e.g. building a north–south hydrogen corridor, connecting hydrogen valleys or creating underground storage facilities (Gaz-System, 2022b). In fact, Gaz-System kickstarted a project "The Hydrogen Map of Poland" with the aim to assess the future hydrogen supply and demand. The end goal is to determine the level of the required development of hydrogen transmission infrastructure and present its preliminary shape (Gaz-System, 2024).

Other large industrial players have launched their initiatives as well. ZEPAK, the largest private energy group in Poland, has been actively engaged in developing the hydrogen economy—it already has mobile hydrogen storage units and a fleet of around 100 fuel cell vehicles, and in September 2023 it opened a stationary hydrogen fuel station in Warsaw, the first one in Poland (Polish Ministry of Climate & Environment, 2023d). ZEPAK has also recently finished building a hydrogen bus factory in Świdnik (Poland) (PAP Biznes, 2023). Grupa Azoty, one of the leading players in the European fertiliser and chemical markets, as part of its Green Azoty strategy (Grupa Azoty, 2021), intends to set up an alternative fuels laboratory to analyse hydrogen fuel quality (and potentially take part in its certification), produce green hydrogen and ammonia in its biggest facility in Puławy (Poland), and use the Port of Police (Poland) as a hydrogen and ammonia hub. They are also considering importing green ammonia from other countries and utilising small modular reactors (SMR) to power their operations (including hydrogen production). The JSW Group, the largest producer of high-quality hard coking coal in the EU, has started working on sourcing hydrogen from coke-oven gas (JSW, 2022a), and has plans to establish a fuel cell factory, along with a production line for hydrogen buses, at the site of the former Krupiński mine (JSW, 2022b). Ultimately, the Port of Gdynia (Poland) can become a hydrogen hub on its path towards decarbonisation (Port of Gdynia, 2022).

Not all business stakeholders come from fossil fuel backgrounds. On the other side of the spectrum, there is a notable company HYnfra, which promotes hydrogen as part of local, integrated, semi-independent energy systems based on renewables, through partnerships with the industry and local administrations (though investments are still at an early stage) (HYnfra, 2023a). One example of a multitude of projects being developed by this company is to use green hydrogen to facilitate the energy transition of the city of Sanok (Poland) (HYnfra, 2023c).[10] It is to be part of a local system based on renewables, electricity and heat storage, as well as charging and hydrogen fuel infrastructure, which is supposed to increase the energy security of the city and neighbouring region. Mainstream domestic RES industry associations are also interested in hydrogen (Polish Wind Energy Association & The Lower Silesian Institute for Energy Studies, 2021), although major progress has so far not been achieved.

---

[10] Apart from the projects mentioned here, HYnfra plans to create a Green Industrial Zone in Bucha (Ukraine) (HYnfra, 2023b) and establish a green ammonia production facility in Jordan (HYnfra, 2023d).

3   Polish Hydrogen Strategy—Key Information

The Polish Hydrogen Strategy until 2030 with an outlook until 2040 (hereinafter also referred to as: the Strategy) was approved in late 2021. The 37-page document presents a vision for the Polish hydrogen economy, proposes six key goals (see Fig. 2) and lists legislative and non-legislative actions (Polish Ministry of Climate & Environment, 2021a). Its annex, written by the Institute of Power Engineering, the Faculty of Management at the University of Warsaw and the Institute for Ecology of Industrial Areas, is the 516-page Analysis of the potential of hydrogen technologies in Poland until 2030 with an outlook until 2040 (hereinafter also referred to as: the Analysis), covering the potential for developing these technologies in Poland and providing additional insights into some of the underlying assumptions (Kupecki et al., 2021). This potential has been assessed across a whole value chain through two separate expert surveys, whose end products are sets of recommendations for facilitating the growth of the national hydrogen economy (Kupecki et al., 2021, pp. 452–454 and 463–466).

The Strategy is formulated within the context of climate and decarbonisation policy. Hydrogen is described as both a clean energy storage medium and a viable solution for industrial uses where direct electrification is not possible. There is also a significant focus on prospective economic gains related to emerging hydrogen value chains. Issues of national energy security are present but less pronounced (though it

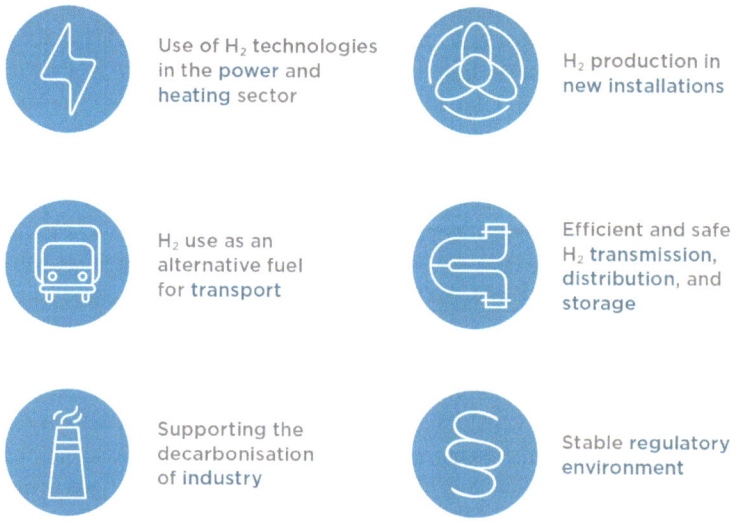

Fig. 2   Strategic objectives of the Polish Hydrogen Strategy. *Source* Authors' own elaboration based on the information from the Polish Hydrogen Strategy

must be noted that the Strategy was written before the onset of the European energy crisis).

The Strategy presents an optimistic, undiscriminating and maximalist approach to the hydrogen economy. It states the support for all low-emission methods of production, including niche or controversial technologies such as generation from coal (with CCS/CCU), although, at the same time, scaled-up production capacity envisioned for 2030 is said to consist "in particular" of electrolysers[11] (Polish Ministry of Climate & Environment, 2021a, p. 20). Even though blue hydrogen is supported, there is surprisingly little focus on adding CCS to existing grey hydrogen production[12] (and to decarbonising this process in general—the Strategy gives the impression that a new green hydrogen economy is supposed to emerge alongside, and not directly replace, the old grey one). The same approach can be seen with favoured hydrogen uses that encompass somewhat contentious positions, such as residential and district heating, light vehicles and blending into the natural gas network, in addition to various more conventional options, such as industrial processes (e.g. steelmaking), heavy-duty transport (fuel cell-powered buses, lorries and locomotives), as well as power grid flexibility enhancement.

The implementation of the Strategy is the responsibility of several different actors—primarily central and regional administrative bodies, government agencies and scientific institutes. The realisation roadmap consists of legislative (e.g. introduction of a hydrogen act) and non-legislative actions (e.g. a sector deal,[13] a hydrogen technologies centre), embedded in a general timeline extending until 2030. The Strategy will be monitored along key strategic indicators (see Fig. 3), which mostly cover production capacity (2 GW) and application in transport (1000 hydrogen buses), as well as some legislative and policy goals (Polish Ministry of Climate and Environment, 2021a, p. 35). On the other hand, there is no measurable objective for hydrogen use in the industry.

---

[11] The Strategy projects that hydrogen made by electrolysis will achieve economic attractiveness in the fastest way by powering this process from domestic offshore wind farms, which are expected no sooner than the late 2020s. The rough estimate of production costs put them at 6 EUR/kgH2, with a potential to fall to 3 EUR/kgH2 (Polish Ministry of Climate and Environment, 2021a, p. 32).

[12] Grey hydrogen, presently generated at a large scale in Poland, is seen by the government as a somewhat bridging energy carrier, necessary before green hydrogen is produced in adequate amounts (Polskie Radio 24, 2023). Independent analyses point out, however, that the pressure coming from the reduction of free CO2 allowances for the hydrogen sector might significantly decrease grey hydrogen competitiveness in favour of a green one, and subsequently remove the current Polish advantage of having an advanced (grey) hydrogen economy in place (Institute of Power Engineering et al., 2023a).

[13] This sector deal is a first-of-a-kind initiative launched in the EU. Over 250 entities are its signatories, aiming to take actions in line with five strategic goals (local content, R&D, investments, people and cooperation). The deal is an executive instrument of the Strategy, set to underpin the growth of the domestic hydrogen economy. The realisation of the goals is to be facilitated by cooperation within the Coordinating Council of Hydrogen Economy, set up for this very reason (Polish Ministry of Climate and Environment, 2022g).

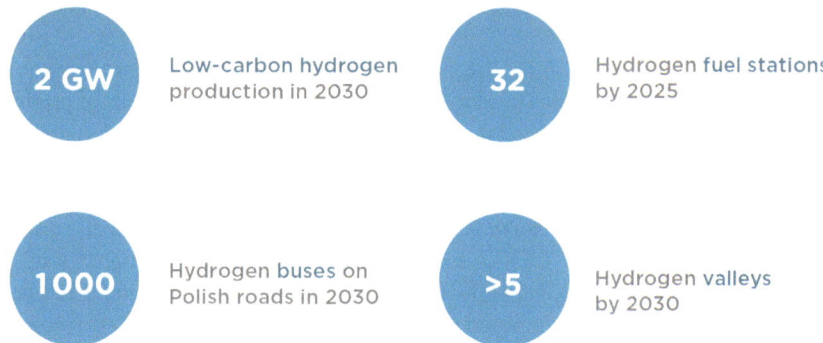

Fig. 3 Key numerical targets in the Polish Hydrogen Strategy. *Source* Authors' own elaboration based on the data from the Polish Hydrogen Strategy

Planned actions and support programmes are to be financed by the National Fund for Environmental Protection and Water Management (for infrastructure, such as fuel stations and fuel cell buses) and the National Centre for Research and Development (for R&D and innovations). The former has already launched initiatives aimed at financing the building of hydrogen fuel stations and the purchase of fuel cell buses with several industry stakeholders and local administrations, respectively. The cost of achieving key indicators concerning hydrogen mobility and production capacity, is estimated at roughly 2.1 billion EUR[14] (Polish Ministry of Climate & Environment, 2021a, p. 31), while the first round of support programmes listed in the Strategy have a value of at least 410 million EUR[15] (Polish Ministry of Climate & Environment, 2021a, p. 31). However, access to a key source of hydrogen funding—the EU's Recovery and Resilience Facility (RRF)—remains partially blocked[16] as of June 2024 due to the former government's refusal to adhere to related EU requirements,

---

[14] 1 EUR = 4.3 PLN.

[15] 1 EUR = 4.3 PLN.

[16] Financing from RRF, worth 800 million EUR, is meant to be devoted to supporting the National Fund for Environmental Protection and Water Management programmes, IPCEI projects and new hydrogen production capacities. What is striking, the former government itself has indirectly admitted that the possible failure to acquire these funds would likely lead to a significant slowdown in the development of the hydrogen economy in Poland, thus the decarbonisation of industry and transport (Polish Ministry of Development Funds and Regional Policy, 2023).

which might threaten the growth of future hydrogen projects.[17] This might change in the upcoming months in the aftermath of the elections. In April 2024, Poland asked the European Commission to approve the revised version of its Recovery and Resilience Plan, which would unlock further payments (to date, only two have been made) (CIRE, 2024).

Based on the factors above, it may seem that the Strategy often loses sight of clear sustainability goals and focuses instead on promoting the development of the broad domestic hydrogen economy for its own sake—or on appeasing all stakeholders interested in possessing a share of the emerging market. Many insights from the Analysis, for example regarding the viability of different methods of production or applications, are not reflected in the strategic document itself. The most glaring problem is the contradiction between existing renewable energy targets in the Energy Policy of Poland until 2040 (PEP2040) and the scale of planned renewable hydrogen production. The Analysis calculates that 2 GW of (primarily) renewable hydrogen production capacity envisioned for 2030 by the Strategy would require 40% of the entire renewable energy capacity predicted for that year by the PEP2040 (Kupecki et al., 2021, pp. 428–429), significantly affecting the decarbonisation of the power generation. While the exact number can be disputed, the fact remains that the projected demand for renewable electricity by electrolysers is not aligned with current official supply forecasts (or policies) and that the issue is not directly addressed in the Strategy itself—this is related to the reasons discussed in the next chapter.

Ongoing government efforts with regard to the Strategy have been concentrated on drafting a so-called hydrogen constitution, which was originally supposed to be completed in 2022. This legislative package is expected to establish a regulatory framework for the operation of the national hydrogen economy. In 2023, the former Law and Justice government confirmed that the constitution was still under development by beginning the work on and publishing a two-part external analysis about financial support instruments (such as contracts for difference [CfD]) for the early hydrogen projects[18] (Polish Ministry of Climate & Environment, 2023f). Legislation

---

[17] Since the start of its office in 2015, the former Polish government had been criticised at the EU level for multiple actions that were said to threaten the rule of law and the independence of judges, among various other controversies. Moreover, that government did not introduce some of the agreed milestones necessary to access Recovery and Resilience Facility funds, although partial liberalisation of the "10H rule", which in practice blocked the development or repowering of onshore wind power plants in Poland, was finally introduced in 2023. There were public disagreements within the former ruling coalition about de-escalating the conflict to access funds and strengthen EU unity against the background of the Russian threat. However, the results of the October 2023 parliamentary elections might turn this difficult situation on its head, as the new government was formed by three opposition parties, which have a substantially more pro-European stance and could achieve the milestones. In November 2023, the European Commission confirmed that Poland will receive 5 billion EUR within the RRF, with the payment of the rest of the funds dependent on fulfilling the milestones, especially the one concerning the judiciary.

[18] Apart from contracts for difference, exempting green hydrogen from VAT (value added tax) at the early stage of hydrogen economy development is considered as a potential financial aid, but its details have not been clarified. Such exemption would have to be accepted by the EU.

regarding these instruments is expected to be published in 2024, whereas the first hydrogen auction is said to be held in 2025. However, the work on the constitution has not been finished by the Law and Justice government, which means that the new government has to take over and possibly make adjustments in its final form. As for the Analysis, it has reportedly received an update as well, yet now it is unclear when (or if) it will eventually happen. This update would revise the initial version's conclusions and include some additional content, such as an examination of the possibility of creating local hydrogen ecosystems (e.g. valleys, hubs, clusters) or the viability of particular business models in the national hydrogen economy.

Apart from the constitution, there is a focus on introducing a hydrogen act to regulate rules for conducting hydrogen-related business activities. The act is in fact a major revision of the Energy Act (Polish Government Legislative Centre, 2023b), including some fundamental definitions (like that of hydrogen or electrolytic conversion) and rules (e.g. regarding operating hydrogen storage sites and hydrogen grids) for the emerging hydrogen economy (Institute of Power Engineering, 2023a, pp. 25, 49, 69). Over 40 entities, including some industry players and the Energy Regulatory Office, have submitted their remarks in the consultation process. As a result, the act is likely to undergo significant changes in its final form since many reservations have been expressed about it, e.g. that this document overregulates and hampers the growth of the emerging hydrogen market by opening it only partially and discouraging investors from building the necessary infrastructure. As the legislative path was not completed before the end of the parliamentary term, the whole process ha d to formally start from the beginning in 2024. The hydrogen act is a milestone of Poland's Recovery and Resilience Plan, and for this reason was expected to be introduced in Q4 2023 at the latest (Business Insider, 2023). The former government's representatives have indicated that this act is ready to be implemented, although it is going to happen in 2024. The act, as of June 2024, is in the public consultation stage (Polish Ministry of Climate & Environment, 2024c).

Other parts of the legislation that were being developed include, for example, draft regulations on requirements for measurement, registration and calculation of renewable hydrogen (Polish Government Legislative Centre, 2023a). In October 2023, the ordinance on state aid for the development of hydrogen technologies (Polish Journal of Laws, 2023) was published and is officially in force. It is aimed at simplifying the rules for granting subsidies for investments required under the RRF, including for hydrogen production, transmission or storage projects. This regulation is also seen as an enabler of accomplishing hydrogen goals from the Strategy.

Despite that the Polish hydrogen legislative framework is still at an early stage, there are some other regulations already in place. They regard guidelines for testing hydrogen quality (Polish Journal of Laws, 2022a) and its sampling methods (Polish Journal of Laws, 2022b) by an accredited laboratory, hydrogen quality standards (Polish Journal of Laws, 2022c) and technical requirements for hydrogen fuel stations (Polish Journal of Laws, 2022d).

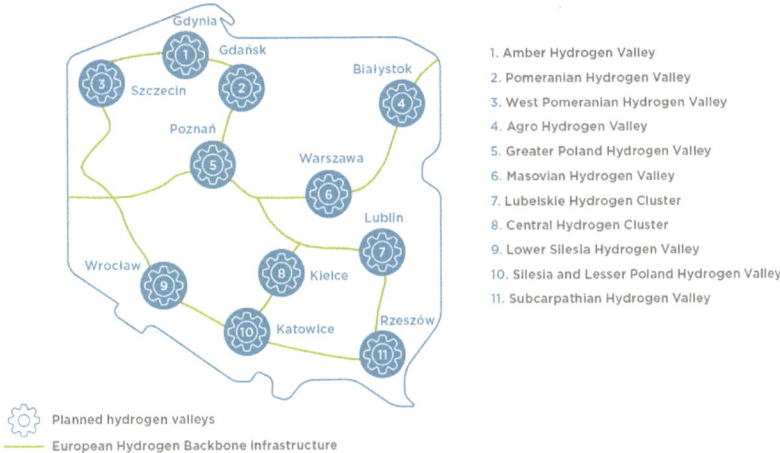

Fig. 4 Planned hydrogen valleys in Poland. *Source* Authors' own elaboration based on the data from the Industrial Development Agency JSC

4   International Dimensions of the Polish Hydrogen Strategy

The main focus of the Polish Hydrogen Strategy is on the domestic hydrogen value chain. The Strategy's stated goals are to create a Polish hydrogen industry, contribute to reaching climate neutrality, and help the Polish economy preserve its competitiveness on the path to net zero (Polish Ministry of Climate & Environment, 2021c, p. 2). The development of international hydrogen trade is generally not analysed, whereas that of hydrogen value chains—from hydrogen production to use in multiple sectors—is implicitly described as a domestic affair (initially, at the level of regional hydrogen valleys, see Fig. 4). While long-distance hydrogen transportation by pipeline is considered an option (despite the significant technical challenges associated with repurposing), partly due to the availability of new renewable energy capacity offshore, which is located away from industrial and population centres, this does not involve the notion of cross-border trade.

This does not mean that the Strategy explicitly adopts a negative approach to international trade. Instead, the topic is rather avoided, yet not entirely. The Polish document acknowledges that the EU Hydrogen Strategy aims to integrate European energy systems and incorporates an international perspective on the hydrogen market, which the Polish strategy claims to support. The Strategy also mentions the European Hydrogen Backbone,[19] although only as a way to facilitate the development of

---

[19] Within this project, several new pipelines, both new and repurposed, are expected to be built in Poland, thus constituting a link to the infrastructure of neighbouring countries, notably Germany

domestic hydrogen gas networks (Polish Ministry of Climate & Environment, 2021a, p. 22). There is also some mention of participation in knowledge sharing and joint R&D initiatives at the EU level. The Strategy, however, makes no explicit reference to hydrogen import or export. This is linked, to a certain extent, to the limitations described in the previous chapter; since the scale of domestic production potential is not critically examined, neither the hydrogen supply gap nor the resulting domestic hydrogen cost is apparent. The issue of the possible relocation of industry sectors to countries with abundant renewable energy resources is also not considered.

This contrasts with the findings of the Analysis, which provides a basis for the abovementioned considerations and addresses a host of relevant issues in this regard: factors affecting the national capacity and competitiveness of renewable hydrogen production, the role of different countries in international hydrogen trade and possible hydrogen transport modes. These questions are examined with a significant degree of detail, albeit mostly in the form of a review of numerous national strategies and international reports. There is even an original scenario analysis of the Polish hydrogen export potential, although this would only exist if production outpaced domestic consumption, which is not likely considering the substantial current industrial demand.[20] The potential need for Polish hydrogen imports, however, has been avoided entirely, which appears to be a deliberate decision. By contrast, the potential to replace fossil fuel imports with domestic hydrogen production has been assessed in detail. Nevertheless, from a 2030 perspective, the impact of this is considered to be very limited, amounting to no more than replacing just under 3% of liquid fuels, energy raw materials and energy imports (Kupecki et al., 2021, p. 334).

Why then does the Polish Hydrogen Strategy omit the issue of international hydrogen trade? And why does the scientific review in the annex avoid the topic of imports when applied to Poland, despite acknowledging its global significance and prominent role in other national strategies? Various answers have been provided by experts and stakeholders, including the following:

- Rapid obsolescence of the document: As the Strategy was implemented in the context of support for the post-COVID recovery of the Polish industry, before the energy crisis, the focus on international energy policy is for the most part missing.
- A wait-and-see approach: A large-scale global hydrogen market does not yet exist. With many uncertainties about technologies (especially the viability of hydrogen shipping), costs and environmental impacts, less advanced countries may be tempted to avoid the issue for the time being, and instead choose to follow in the footsteps of frontrunners at a later point in time and implement already proven solutions. Poland is likely to be such a country, taking the approach of waiting for the global hydrogen economy to materialise and only then defining its specific role in it.

---

and Lithuania (European Hydrogen Backbone, 2023). Nevertheless, the Strategy does not explore how these pipelines may be utilised for transnational trade.

[20] This is surprisingly not fully acknowledged: both the Strategy and the Analysis, in part, treat the low-carbon hydrogen industry as somehow separate from existing grey hydrogen value chains.

- Avoiding hard truths: The Polish Hydrogen Strategy aims to promote the potential of hydrogen technologies and build the goodwill necessary for legislative changes and pilot-stage implementations. Despite concessions to fossil fuel-based industries (see next point), it generally advocates for the development of renewables and introduces a vision of the decarbonised industry, transport and heating. Renewable energy is promoted in Poland as a way to achieve energy independence and limit fossil fuel imports. Hydrogen imports complicate this narrative, raising (not unfounded) concerns about new dependencies—which can harm not only public support for hydrogen but could also partly undermine support for decarbonisation as such. On the other hand, explicitly ruling out hydrogen imports would also be problematic and hard to justify given the circumstances. Hence, it is convenient for Poland to avoid the issue for now.
- Appeasing stakeholders: The interests of different hydrogen stakeholders (including the oil and gas industry, renewable energy industry associations, climate pressure groups, local administrations in coal regions and others) are not perfectly aligned. As a result, at this early stage of building a domestic hydrogen narrative, it might be better to steer clear of these contradictions. Thus, the Strategy reflects the interests and opinions of different pressure groups, e.g. through an optimistic outlook on the potential for domestic renewable and blue hydrogen production. In the absence of any voices supporting hydrogen imports, this perspective is simply omitted.

Despite such an approach, the Strategy may still be an effective tool for shaping some further actions and policy decisions. The Polish Hydrogen Strategy can be seen as a tool to promote the concept of a hydrogen economy, initiate first hydrogen partnerships along the domestic value chain, kick-start legislative action and prepare the country for taking part in European funding programmes. It does not, however, provide any kind of framework for Polish participation in the global hydrogen trade.

However, it seems that Polish stance on the issue of hydrogen trade may change as a result of the revised EU's REDIII Directive (European Commission, 2023d). Its mandatory targets of renewable hydrogen in the industry as well as road, aviation and maritime transport can be particularly difficult to achieve, given the current growth pace of RES capacity and the necessary volume of clean electricity in other sectors, such as power generation or district and residential heating. Because of that, Poland might be forced (or rather force itself) to import significant amounts of hydrogen—some estimates suggest that imports could account for well over 50% of the domestic demand in 2030 (Institute of Power Engineering et al., 2023b). For this reason, there are more and more voices calling for the development of large-scale hydrogen transmission and storage infrastructure in the country in light of the looming green hydrogen scarcity.

Even though Poland currently does not heavily contribute to the EU goals of establishing a vast hydrogen trade infrastructure, its focus on hydrogen mobility may position the country as a frontrunner in the rollout of hydrogen fuel stations. In fact, in Q3 2023 the government prepared a proposal for the locating of public stations (Polish Ministry of Climate & Environment, 2023h), which would comply with the newly announced Directive on Deployment of Alternative Fuels Infrastructure and be a valuable input into the bloc's ambitions in this area. Implementing the goals of this directive can result in establishing as much as 34 hydrogen fuel stations in Poland (Polish Ministry of Climate & Environment, 2023i).

Another important area for international cooperation is R&D and pilot implementations, although the participation of Polish stakeholders has so far been rather moderate. In the European Clean Hydrogen Partnership, there are some projects involving entities from Poland (Clean Hydrogen Partnership, 2023a); its Stakeholders Group, which is an official advisory body, features the Polish Alternative Fuels Association (Clean Hydrogen Partnership, 2023b). In the Horizon 2020 programme, Polish entities have been engaged in numerous hydrogen projects, e.g. technical development of underground hydrogen storage (CORDIS, 2023a) or industrial solid oxide fuel cell system (CORDIS, 2023b); likewise in Horizon Europe—some examples include raising public awareness and building trust for hydrogen technologies (CORDIS, 2023c) and hydrogen integration into the natural gas grids (CORDIS, 2023d).

With regard to IPCEI Hy2Tech (European Commission, 2022a) and Hy2Use (European Commission, 2022b) initiatives, the European Commission has approved financial support for one Polish project in each. In Hy2Tech Synthos, a private-owned chemical manufacturer, runs a project (USNC, 2020) with the final objective of generating hydrogen, principally for its facilities, using high-temperature electrolysis powered by so-called Micro Modular Reactors supplied by its US-based partner Ultra Safe Nuclear Corporation. In Hy2Use, the company Orlen, within its Hydrogen Eagle project (Orlen, 2021), aims to create over 100 hydrogen fuel stations and install a total of roughly 250 MW of electrolyser capacity in Poland, Slovakia, and Czechia, thereby supplying hydrogen to industrial and transportation customers.

Another major hydrogen project with 158 million EUR worth of support from the EU is Green H2 by Lotos, a subsidiary of Orlen. It involves the deployment of a 100 MW electrolyser, along with a 50 MW PV power plant and a 20 MW battery energy storage. Green hydrogen, generated within this project, is to be utilised in Orlen's refinery in Gdańsk (Poland) (Polish Ministry of Climate & Environment, 2023g).

The European Clean Hydrogen Alliance incorporates over 840 undertakings, over 50 of which are located in or include Poland (European Commission, 2022c); they are related to, for example, hydrogen storage in underground salt caverns or different hydrogen generation projects. Within the Innovation Fund, ZEPAK has been awarded a grant for its 5 MW green hydrogen production facility (with PEM electrolysers), with a view to generating hydrogen for the domestic public bus transport sector (European Commission, 2023b); financial aid for the abovementioned Lotos' Green H2 (strictly speaking, for its initial phase featuring the deployment of a 1 MW electrolyser) has also been approved (European Commission, 2023c). As for other international undertakings, Mission Innovation (Mission Innovation, 2023) and Breakthrough Energy Catalyst Europe (Breakthrough Energy, 2023) do not feature Poland or Polish businesses whatsoever.

There are also some other examples of R&D cooperation. The privately owned energy company Polenergia, for example, is involved in an international consortium for implementing pure hydrogen combustion in gas turbines and is also seeking to utilise green hydrogen to produce sustainable aviation fuel (Polenergia, 2022). ZEPAK, apart from the projects mentioned in the previous paragraph, has developed a 0.5 MW alkaline electrolyser, the first Polish-made electrolyser, through its company Exion Hydrogen, which has a manufacturing plant in Poland and an R&D facility in Belgium (Exion Hydrogen, 2023); its production and commercial availability began in 2023. In addition to this, ZEPAK, in collaboration with the Italian design studio Torino Design, has developed a hydrogen bus that has already been tested on Polish streets (NesoBus, 2022).

Regarding hydrogen sustainability and standards issues, Poland has not been particularly active in shaping them through certification norms either at the EU or international level. There are no Polish entities in the CertifHy programme (CertifHy, 2023), and Poland does not participate directly in processes within the International Partnership for Hydrogen and Fuel Cells in the Economy (IPHE, 2023) and has not visibly influenced the EU's delegated acts on hydrogen. A lack of action may be explained by the fact that solving these issues is still in its infancy globally (IEA, 2023). As a result, Poland is, at least for now, not deeply engaged, instead taking more of a rather secondary role in the green hydrogen economy development, which explains the omission of an international approach in the Strategy. Furthermore, significantly more focus is being given to establishing a regulatory framework for the domestic hydrogen economy, as such a framework is currently virtually non-existent.

As for general intergovernmental cooperation, in December 2022 Poland announced a hydrogen partnership with Iceland (Polish Ministry of Climate & Environment, 2022e) with a focus on clean hydrogen and hydrogen-derived fuel production by using Iceland's geothermal power. This cooperation is expected to have a positive effect on Polish energy security and independence. In addition, in May 2023, Poland signed a memorandum of cooperation on hydrogen with Japan (Polish Ministry of Climate & Environment, 2023a). Its objectives include the deepening of Polish-Japanese collaboration in the area of low-carbon hydrogen production and the development of sustainable and cost-effective hydrogen value chains in the power, transport, heating and industry sectors. Also, Poland is keen on cooperating with South Korea on hydrogen, with a particular emphasis on green hydrogen (Polish Ministry of Climate & Environment, 2024a).

5 Hydrogen in the Context of a Broader Polish Approach to International Energy Security

Given the Strategy's failure to address the question of hydrogen imports, Polish interests related to an emerging international hydrogen economy may be better understood by considering the broader national energy strategy, as well as policies and initiatives undertaken by public and private stakeholders.

The Energy Policy of Poland until 2040 is a key strategic document for the future development of the Polish energy system. Its 2021 version focused on three pillars: "Just transition", "Zero-emission energy system" and "Good air quality" (Polish

Ministry of Climate & Environment, 2021b, p. 13). Energy security and independence, however, were already important parts of the rationale behind various objectives, including the use of domestically mined coal, the diversification of fuel imports, the development of transnational energy connections and the expansion of alternative power generation technologies such as nuclear and renewables (mainly offshore wind). In March 2022, the government announced that a fourth pillar would be added in the upcoming PEP2040 update, with an explicit focus on energy security and independence (Polish Ministry of Climate & Environment, 2022a). However, the other related objectives are mainly a restatement of those previously covered, without any substantial shift in the strategic outlook or policies.

In early 2023, the Ministry of Climate and Environment announced that PEP2040 would be soon updated with a new scenario for the power sector (Polish Ministry of Climate & Environment, 2023b). According to both official communication and a leaked draft, the new scenario envisions a much quicker development of renewable energy (expected to cover up to 47% of the total electricity demand already in 2030, up from 32% envisioned in the 2021 version), dynamic additions of nuclear power in the 2030s, and a lower reliance on imported natural gas. The new draft scenario is more coherent with the Polish Hydrogen Strategy in that it is expected to take into account the additional electricity demand stemming from domestic hydrogen production. Hydrogen is mentioned among the possible solutions to the issue of electricity generation curtailment, which is expected to significantly affect the RES in the late 2030s due to renewable energy saturation and the competition with the always-on nuclear power. The announcement was generally welcomed by the stakeholders as a step in the right direction, though there were discussions about both the level of ambitions as well as about the possibility of achieving even the described targets with the current policies and investment levels (Kubiczek & Smoleń, 2023).

However, despite the Ministry's communication about the prompt amendment, the change has not been in fact officially accepted by the Council of Ministers as of June 2024. The reason for that was the brewing contest inside the formerly ruling right-wing coalition ahead of the autumn 2023 elections. A minor coalition partner, Sovereign Poland (called Solidarity Poland until 2023), was trying to raise its profile by positioning itself in direct opposition to the EU's climate policies, publicly criticising the Prime Minister for his allegedly too lenient approach. The party called for further development of domestic coal mining and leaving the EU ETS system— in fact, abandoning the decarbonisation goal altogether. As a result, the PEP2040 update was not introduced before the elections, which also negatively affected the Polish National Energy and Climate Plan update. In fact, the draft of the updated NECP should have been published in June 2023. Instead, initially only preliminary consultations were launched, and a new scenario for the power sector was published only as "inspirational material" for participants of the consultations. In March 2024, the Polish goverment sent merely a preliminary version of the updated NECP to the European Commission, as the final one was yet to be completed (Polish Ministry of Climate & Environment, 2024b). Though strategic documents do not necessarily have an immediate impact on the decarbonisation process (which is currently proceeding at a much faster rate than that envisioned by the current strategies due to the proliferation of solar power), the uncertainty especially affects the early-state solutions that are reliant on public subsidies and declarations of support, such as green hydrogen.[21]

In the near future, the Polish energy landscape may be more favourable to the faster rollout of renewables and corresponding decarbonisation efforts due to the results of the October 2023 parliamentary elections. Even though they were won by the previously ruling party—Law and Justice—the new government was formed by three committees that jointly have a secure majority in both chambers of the Parliament and appointed Donald Tusk as a prime minister. These parties voice support for more ambitious climate policies. Their stated policy goals include fully unlocking the RRF funds, bulk deployment of RES (mainly PV, onshore and offshore wind, biogas) and continued development of nuclear energy as well as some other solutions such as energy efficiency and decentralised energy communities (Teraz Środowisko, 2023). Civic Platform, in particular, has announced a goal of 68% share of RES in electricity generation already in 2030 (Biznes Alert, 2023), above both 32% from current official strategic documents and less than 50% stated in the unofficial scenario published by the former government in June 2023. These actions, if implemented successfully and on a broad scale, are likely to be beneficial to the growth of the domestic (green) hydrogen economy, although their actual and visible effects are to be felt no sooner than in a few years.

In addition to the energy transition goals, the hydrogen economy can be seen in the broader context of energy independence. Since at least 2015, the government has interpreted it primarily in the sense of reducing the country's reliance on Russian fossil fuel imports. Even before the war in Ukraine, the last contract for Russian natural gas was meant to expire at the end of 2022, which was made possible by investments in the LNG terminals in Świnoujście (Poland) and Klaipeda (Lithuania), as well as the new pipelines connecting Poland with Denmark (i.e. the Baltic Pipe) and Lithuania (Polish Ministry of Climate & Environment, 2022c). Poland also planned to continue importing Russian-sourced fossil gas from Germany, although that is of course no longer feasible. Divestment from Russian oil and coal was more gradual before the energy crisis,[22] as these ties were thought to be less strategically risky due to the existence of possible alternative sources (this, however, was shown to be not entirely true in 2022, as the import of the types of coal suited for household consumption proved to be a significant challenge).

At the risk of oversimplification, it is possible to identify the following key characteristics of the former Polish Law and Justice government's approach to international energy policy, which may also clarify the current and future potential role of hydrogen within it:

---

[21] The roadblocks in the way of PEP2040 and NECP updates are also preventing the much-needed revision of the Polish Hydrogen Strategy, which has to be aligned with the broader Polish energy objectives. This revision is expected in 2024, although much will depend on the timeline for other domestic decarbonisation efforts.

[22] Russian coal imports grew between 2015 and 2018 due to a combination of low prices, high quality and the decreasing output of Polish mines.

1. Poland wants to fully phase out its dependence on Russian fossil fuel imports, which are considered both a national security risk and a source of income for a hostile power (Kupecki et al., 2021, pp. 318–322).
2. Poland is open to importing energy carriers that it cannot produce domestically from countries other than Russia, especially from Europe and the U.S., but also from elsewhere, given an appropriate level of diversification (Polish Ministry of Climate & Environment, 2021b, pp. 36 & 41). Before the war in Ukraine, Poland planned to increase natural gas imports and continue oil imports for the foreseeable future (Polish Ministry of Climate & Environment, 2021d, pp. 16–17). Orlen has a presence in Norway's gas mining, as Poland wants to play an active part in the gas value chain there. Polish view on these issues is pragmatic and driven by economic and security interests, with little focus on the more distant issues of international relations or environmental and social sustainability.
3. Poland, in principle, supports technologies that can replace fossil fuels. In practice, however, the government opposes any measures taken exclusively for climate protection unless they are accompanied by economic benefits. Furthermore, it is wary of measures that could harm established stakeholders, such as state-owned enterprises from the energy and industrial sectors (for example the coal, fertiliser and automotive sectors). On the other hand, it sees the emergence of new value chains as an opportunity for industrial development.
4. Poland would like to expand transnational energy connections to increase the resilience of its energy system against external shocks (Polish Ministry of State Assets, 2019, pp. 40–48). For example, although additional grid connections are to a certain extent seen (or even opposed) as unwelcome competition to domestic producers, their role in emergencies is highly valued. Gas connectors with neighbouring friendly countries are perceived as factors that improve energy security in the region, while at the same time strengthening the international position of Poland as an aspiring CEE (Central Eastern Europe) leader.
5. The Law and Justice government vocally opposed the leading role of Germany in the EU. While this position was partly a narrative tool that served domestic purposes, among key decision-makers there was a genuine distrust of German motives and a sense of judgement. When applied to international energy policy, this negative outlook is now (after years of dealing with the issue of the Nord Stream gas pipeline, but also the nuclear phase-out) to a degree shared by the currently ruling liberal parties as well, which affects how German initiatives regarding international trade in energy carriers will be received.

Under the PEP2040 and, to some extent, the Polish Hydrogen Strategy, the international potential of hydrogen is viewed mostly in the context of points 1. and 3.—i.e. as a domestically produced resource which can potentially be utilised to reduce imports of Russian fossil fuels, in addition to the secondary goals of protecting the business models of "national champions" (state-owned enterprises in the oil and gas and chemical sectors). External decarbonisation pressures and possible opportunities for manufacturing industries (automotive, machinery) operating in Poland do not figure prominently here. Moreover, views on German support for hydrogen imports are likely to be affected by the issues described in point 5.

Polish documents do not cover the potential role of hydrogen in the context of points 2. (diversified import of energy carriers) and 4. (CEE energy cooperation). The import and export of hydrogen have not been thoroughly explored, although they are not excluded either. Poland is not dogmatically opposed to importing energy carriers, provided that some diversification of supply is ensured, especially if they serve as a basis for a further generation of added value. What is more, Poland welcomes initiatives that increase the resilience of the CEE energy system.

Despite strong narrative and ideological differences, the approach of the current liberal government to the issues of energy security can be largely similar. There are, however, hopes for a much better atmosphere for cooperation on the European level. Polish hydrogen economy outlook could also improve with increased RES goals and stronger energy transition commitment. But the hydrogen economy itself, nor its international dimension, was not on the elections' campaign agenda and parties have not presented detailed positions on the issue.

The analysis of the Polish approach to international energy policy shows that the country could be supportive of international hydrogen trade if it is proven in practice and presented in the right framework. This includes conceptualising renewable hydrogen as a basic commodity that could be feasibly and cheaply imported and used domestically in further stages of value creation, including generating jobs, added value and taxable incomes. An analogous role is currently played by imports of oil and gas, which are used in multiple Polish industries. Successful implementation of the international hydrogen trade by early adopters, if it materialises, could persuade Poland that it is economically feasible. Polish energy security concerns could be addressed by measures such as the diversification of hydrogen sources, the development of flexible international markets, and the creation of a hydrogen infrastructure that is resilient against acts of sabotage, at a level similar to or better than LNG. Furthermore, if developed as a new sphere for cooperation between Czechia, Slovakia, the Baltic states, Romania, Hungary and possibly Ukraine,[23] hydrogen trade could become much more attractive than the current image of the international hydrogen market as a "German" plan that is mainly tailored to Germany's needs and strategic objectives (though those concerns will be less relevant under new administration). This regional cooperation is already emerging—in May 2023, hydrogen clusters and associations from the Visegrad Group, the Baltic states, Ukraine and Slovenia initiated the 3 Seas Hydrogen Council project to support business cooperation, strengthen joint policy impact and share experience (CIRE, 2023). This new partnership has received support from the former Polish government, and such initiatives should still be supported by the new one.

---

[23] The issue of prospective hydrogen import from Ukraine to the EU in the post-war future has not been brought up in the Strategy, and Poland has not formulated a clear position on it. As of mid-2024, the main challenge that will need to be addressed in Ukraine, which requires considerable external support, will be the rebuilding or replacement of lost power generation for domestic consumption alone; Poland may play a role in that (Polish Ministry of Climate and Environment, 2022d).

Another tool that can be leveraged to improve strategic hydrogen planning in EU countries is European funding for R&D or infrastructure. At this point, one of the reasons why Poland and other EU Member States such as France (French Ministry of Ecological Transition, 2020) or Hungary (Hungarian Ministry of State for Energy and Climate Policy, 2021) are either avoiding or opposing international hydrogen trade is the lack of a fair assessment of their domestic hydrogen production capacity. If such a quantitative assessment would be an obligatory part of national hydrogen strategies, possibly as a prerequisite for receiving hydrogen-related EU funding, it might encourage the countries in question to either pursue a more ambitious development of renewable energy capacities to meet this demand and/or to look more openly at the possible necessity of hydrogen imports. Such an approach would also support a better prioritisation of hydrogen applications, especially the decarbonisation of industrial processes that are heavily reliant on grey hydrogen.

6   Conclusions

The question of a Polish position in the European and global hydrogen value chain is a complex one. In light of modest hydrogen production capacity deployment plans in comparison with the European frontrunners (e.g. Germany and France), it will be difficult for Poland to maintain its current position in the hydrogen economy. In the long run, smaller ambitions could potentially jeopardise one of the pillars of the Strategy, namely energy security, as they would make domestic hydrogen production less competitive, thus increasing the attractiveness of imports. Despite the former Law and Justice government's hesitancy, the country is under pressure from both European policymakers and international investors to pursue decarbonisation. This requires mitigating substantial GHG emissions in the existing grey hydrogen production and, in the longer term, using hydrogen to cut emissions in industry, some segments of the transport sector, and possibly energy storage and power generation. At the same time, domestic production of low-carbon hydrogen will be limited by renewable energy capacity and natural gas supply. However, a new government post-2023 elections could introduce changes to further facilitate the deployment of RES.

The issue of importing hydrogen, which so far has been largely avoided in official documents and has notably even been absent from the discussion following the release of the REPowerEU plan, will thus have to be addressed sooner or later. When (or if) the viability of the global hydrogen trade is demonstrated, Poland is likely to follow these trends, as it will not want to fully miss out on the resulting economic opportunities. At that point, the country could potentially become a supporter of European hydrogen cooperation, while at the same time improving bilateral relations with non-EU prospective green hydrogen exporters and implementing its ambition to become an energy and resource hub for Central Eastern Europe.

Acknowledgements Research for this chapter was financially supported by the German Federal Foreign Office within the framework of the project *"Geopolitics of the Energy Transformation— Implications of an International Hydrogen Economy"* (GET Hydrogen), funding reference number AA4521G125.

References

Biznes Alert. (2023). Retrieved October 23, 2023, from https://biznesalert.pl/koalicja-obywatelska-program-oze-audyt-atom-polska/

Breakthrough Energy. (2023). Retrieved October 4, 2023, from https://breakthroughenergy.org/our-work/europe/

Business Insider. (2023). Retrieved October 3, 2023, from https://businessinsider.com.pl/prawo/firma/ustawa-wodorowa-nawet-sluzby-mialy-uwagi-resort-klimatu-wprowadza-zmiany/qpt pwpw

CertifHy. (2023). Retrieved October 4, 2023, from https://www.certifhy.eu/wp-content/uploads/2023/02/CertifHy-Stakeholder-Platform-Members-Q1_2023.pdf

CIRE. (2023). Retrieved October 5, 2023, from https://www.cire.pl/artykuly/serwis-informacyjny-cire-24/z-inicjatywy-polski-powstala-rada-wodorowa-europejskich-panstw-

CIRE. (2024). *Jak wygląda stan wykorzystania środków z KPO?* https://www.cire.pl/artykuly/serwis-informacyjny-cire-24/jakwyglada-stan-wykorzystania-srodkow-z-kpo

Clean Hydrogen Partnership. (2023a). Retrieved October 4, 2023, from https://www.clean-hydrogen.europa.eu/projects-repository_en?f%5B0%5D=oe_project_title%3Apoland&page=0

Clean Hydrogen Partnership. (2023b). Retrieved October 4, 2023, from https://www.clean-hydrogen.europa.eu/about-us/organisation/stakeholders-group_en

CORDIS. (2023a). Retrieved October 4, 2023, from https://cordis.europa.eu/project/id/101007176

CORDIS. (2023b). Retrieved October 4, 2023, from https://cordis.europa.eu/project/id/875148

CORDIS. (2023c). Retrieved October 4, 2023, from https://cordis.europa.eu/project/id/101111933

CORDIS. (2023d). Retrieved October 4, 2023, from https://cordis.europa.eu/project/id/101111888

Czyżak, P., Sikorski, M., & Wrona, A. (2021). Wiatr wżagle. Zasada 10H a potencjał lądowej energetyki wiatrowej w Polsce. https://instrat.pl/wp-content/uploads/2021/05/Instrat-Wiatr-w-z%CC%87agle.pdf

Energy Regulatory Office. (2023). *Sprawozdanie z działalności Prezesa Urzędu Regulacji Energetyki 2022.* https://www.ure.gov.pl/download/9/13743/Sprawozdanie2022.pdf

European Commission. (2022a). Retrieved October 4, 2023, from https://ec.europa.eu/commission/presscorner/detail/en/ip_22_4544

European Commission. (2022b). Retrieved October 4, 2023, from https://ec.europa.eu/commission/presscorner/detail/en/ip_22_5676

European Commission. (2022c). Retrieved October 4, 2023, from https://single-market-economy.ec.europa.eu/industry/strategy/industrial-alliances/european-clean-hydrogen-alliance/project-pipeline_en

European Commission. (2023a). Retrieved October 3, 2023, from https://ec.europa.eu/assets/rtd/eis/2023/ec_rtd_eis-country-profile-pl.pdf

European Commission. (2023b). Retrieved October 4, 2023, from https://ec.europa.eu/assets/cinea/project_fiches/innovation_fund/101038982.pdf

European Commission. (2023c). Retrieved October 4, 2023, from https://ec.europa.eu/assets/cinea/project_fiches/innovation_fund/101102990.pdf

European Commission. (2023d). Retrieved November 24, 2023, from https://energy.ec.europa.eu/topics/renewable-energy/renewable-energy-directive-targets-and-rules/renewable-energy-directive_en

European Hydrogen Backbone. (2023). Retrieved October 5, 2023, from https://ehb.eu/page/eur opean-hydrogen-backbone-maps

European Hydrogen Observatory. (2023a). Retrieved November 24, 2023, from https://observ atory.clean-hydrogen.europa.eu/hydrogen-landscape/production-trade-and-cost/hydrogen-pro duction

European Hydrogen Observatory. (2023b). Retrieved November 24, 2023, from https://observatory. clean-hydrogen.europa.eu/hydrogen-landscape/production-trade-and-cost/hydrogen-trade

Exion Hydrogen. (2023). Retrieved October 4, 2023, from https://exionhydrogen.com/

French Ministry of Ecological Transition. (2020). Stratégie nationale pour le développement de l'hydrogène décarboné en France. https://www.entreprises.gouv.fr/files/files/secteurs-d-activite/industrie/decarbonation/dp_strategie_nationale_pour_le_developpement_de_l_hydr ogene_decarbone_en_france.pdf

Gaz-System. (2022a). Transmission System Operators from Poland, Romania, Slovakia and Hungary cooperate on the development of a regional hydrogen grid. Retrieved October 3, 2023, from https://www.gaz-system.pl/en/for-media/press-releases/2022/april/07-04-2022-transm ission-system-operators-from-poland-romania-slovakia-and-hungary-cooperate-on-the-develo pment-of-a-regional-hydrogen-grid.html

Gaz-System. (2022b). Retrieved October 3, 2023, from https://www.gaz-system.pl/dam/jcr:2b8 a40ac-ea03-4f0d-af90-ace06dc97688/13_Wyzwania%20w%20nadchodzącym%20roku%20g azowym%202022%202023.pdf

Gaz-System. (2024, April 18). *GAZ-SYSTEM is developing the Hydrogen Map of Poland.* https://www.gaz-system.pl/en/formedia/press-releases/2024/april/18-04-2024-gaz-sys tem-is-developing-the-hydrogen-map-of-poland.html

German Federal Ministry for Economic Affairs and Energy. (2023). The National Hydrogen Strategy Update. https://www.bmwk.de/Redaktion/EN/Hydrogen/Downloads/nat ional-hydrogen-strategy-update.pdf?__blob=publicationFile&v=4

Global CCS Institute. (2023). Retrieved October 3, 2023, from https://co2re.co/FacilityData

Grupa Azoty. (2021). Retrieved October 3, 2023, from https://grupaazoty.com/upload/1/files/2021/ strategia2021-2030/29.10.2021/Grupa_azoty_strategia_2030_prezentacja_28_10_2021_EN. pdf

Hungarian Ministry of State for Energy and Climate Policy. (2021). Hungary's National Hydrogen Strategy. https://cdn.kormany.hu/uploads/document/a/a2/a2b/a2b2b7ed5179b17694659b8f050 ba9648e75a0bf.pdf

HYNFRA. (2023a). Retrieved October 3, 2023, from https://www.hynfra.pl/?lang=en

HYNFRA. (2023b). Retrieved October 3, 2023, from https://www.hynfra.pl/post/hynfra-to-build-the-green-industrial-zone-in-bucha?lang=en

HYNFRA. (2023c). Retrieved October 3, 2023, from https://www.hynfra.pl/post/sanok-to-become-poland-s-first-hydrogen-city?lang=en

HYNFRA. (2023d). Retrieved October 3, 2023, from https://www.hynfra.pl/post/hynfra-sets-up-a-company-in-jordan-where-a-green-ammonia-factory-is-to-be-built?lang=en

IEA. (2023). Retrieved October 4, 2023, from https://www.iea.org/energy-system/low-emission-fuels/hydrogen

Industrial Development Agency JSC. (2023). Retrieved October 3, 2023, from https://arp.pl/en/ what-we-do/hydrogen-valleys/

Institute of Power Engineering, Faculty of Management at the University of Warsaw, Energy Club. (2023a). Łańcuch wartości gospodarki wodorowej w Polsce. https://2023.hydrogenconf erence.pl/

Institute of Power Engineering, Faculty of Management at the University of Warsaw, Energy Club. (2023b). *Prognoza zapotrzebowania na wodór odnawialny RFNBO.* https://hydrogenconfere nce.pl/

Instrat. (2023). Retrieved October 3, 2023, from https://energy.instrat.pl/en/electrical-system-gen eration-capacity-are/

IPHE. (2023). Retrieved October 4, 2023, from https://www.iphe.net/partners

IRENA. (2022). Global hydrogen trade to meet the 1.5°C climate goal: Part I—Trade outlook for 2050 and way forward. https://www.irena.org/-/media/Files/IRENA/Agency/Publication/2022/Jul/IRENA_Global_hydrogen_trade_part_1_2022_.pdf?rev=f70cfbdcf3d34b40bc256383f54 dbe73

JSW. (2022a). Retrieved October 3, 2023, from https://www.jsw.pl/en/responsible-business/we-res pect-natural-resources/jsws-environmental-projects/production-of-hydrogen

JSW. (2022b). Retrieved October 3, 2023, from https://www.jsw.pl/en/responsible-business/we-res pect-natural-resources/jsws-environmental-projects/hydrogen-cells-factory-electric-and-hyd rogen-vehicles

Kopeć, J. (2023). *Poland risks losing half of previously greenlighted onshore wind land potential due to recent parliamentary amendment.* https://instrat.pl/en/500m-vs-700m/

Kubiczek, P., & Smoleń, M. (2023). Poland cannot afford medium ambitions. Savings driven by fast deployment of renewables by 2030. https://instrat.pl/wp-content/uploads/2023/04/Instrat_Policy-Paper-03-2023-Poland-Cannot-Afford-Medium-Ambitions.pdf

Kupecki, J. et al. (2021). *Analiza potencjału technologii wodorowych w Polsce do roku 2030 z per-spektywą do 2040 roku.* https://www.gov.pl/attachment/1b590d54-fa1e-49fe-9096-b2d0c6 a4fe59

Maj, M., & Szpor, A. (2020). Gospodarka wodorowa w Polsce. Obserwacje na podstawie ram badaw-czych technologicznego Systemu Innowacji. https://pie.net.pl/wp-content/uploads/2021/01/PIE-PP_Wodor.pdf

Mission Innovation. (2023). Retrieved October 4, 2023, from http://mission-innovation.net/our-members/

Municipal Office in Konin. (2022). Retrieved October 3, 2023, from https://www.konin.pl/index.php/jeden-news-1432/wodorowy-solaris-wozi-pasazerow-w-koninie.html

NesoBus. (2022). Retrieved October 4, 2023, from https://www.nesobus.pl/en/

Orlen. (2021). Retrieved October 4, 2023, from https://www.orlen.pl/content/dam/internet/orlen/pl/en/about-company/media/press-releases/2021/obrazy/Orlen_prezentacja_Hydrogen_final.pdf.coredownload.pdf

Orlen. (2022). ORLEN Group Hydrogen Strategy 2030. https://www.orlen.pl/content/dam/int ernet/orlen/pl/en/sustainable-development/orlen_group_hydrogen_strategy/ORLEN-Group-Hydrogen-Strategy.pdf

Orlen. (2023). Retrieved October 3, 2023, from https://www.orlen.pl/pl/o-firmie/media/komuni katy-prasowe-kategorie/wszystkie

PAP Biznes. (2023). Retrieved November 24, 2023, from https://biznes.pap.pl/pl/news/pap/info/3510030,rusza-produkcja-autobusow-wodorowych-w-swidnickiej-fabryce-grupy-polsat-plus-i-ze-pak

PGNiG. (2020). Retrieved October 3, 2023, from https://en.pgnig.pl/news/-/news-list/id/pgnig-lau nches-new-hydrogen-program/newsGroupId/1910852

PGNiG. (2022). Retrieved October 3, 2023, from https://pgnig.pl/aktualnosci/-/news-list/id/gk-pgnig-stawia-na-paliwa-przyszlosci/newsGroupId/10184

Polenergia. (2022). Retrieved October 4, 2023, from https://www.polenergia.pl/en/polenergia-lau nches-hydrogen-rd-project/

Polish Government Legislative Centre. (2023a). Retrieved October 5, 2023, from https://legislacja.rcl.gov.pl/projekt/12375050

Polish Government Legislative Centre. (2023b). Retrieved October 5, 2023, from https://legislacja.rcl.gov.pl/projekt/12365500

Polish Journal of Laws. (2022a). Retrieved October 5, 2023, from https://dziennikustaw.gov.pl/DU/rok/2022/pozycja/2828

Polish Journal of Laws. (2022b). Retrieved October 5, 2023, from https://dziennikustaw.gov.pl/DU/rok/2022/pozycja/2824

Polish Journal of Laws. (2022c). Retrieved October 5, 2023, from https://dziennikustaw.gov.pl/DU/rok/2022/pozycja/2793

Polish Journal of Laws. (2022d). Retrieved October 5, 2023, from https://dziennikustaw.gov.pl/DU/rok/2022/pozycja/2158

Polish Journal of Laws. (2023). Retrieved October 23, 2023, from https://dziennikustaw.gov.pl/DU/rok/2023/pozycja/2189

Polish Ministry of Climate and Environment. (2021a). *Polska Strategia Wodorowa do roku 2030 z perspektywą do roku 2040.* https://www.gov.pl/attachment/4c261e63-f57d-48e3-b451-10b235aa2de8

Polish Ministry of Climate and Environment. (2021b). Polityka Energetyczna Polski do 2040 roku. https://www.gov.pl/attachment/3209a8bb-d621-4d41-9140-53c4692e9ed8

Polish Ministry of Climate and Environment. (2021c). Polish Hydrogen Strategy until 2030 with an outlook until 2040—Summary. https://www.gov.pl/attachment/06213bb3-64d3-4ca8-afbe-2e50dadfa2dc

Polish Ministry of Climate and Environment. (2021d). Polityka Energetyczna Polski do 2040 r. Załącznik 2 - wnioski z analiz prognostycznych dla sektora energetycznego. https://www.gov.pl/attachment/15a6e747-6231-4dc8-8aba-909a3aa0efb6

Polish Ministry of Climate and Environment. (2022a). Principles for the update of the Energy Policy of Poland until 2040 (EPP2040): strengthening energy security and independence. https://www.gov.pl/attachment/c38a14a2-a3ac-418b-b4e3-4d8fca2ebf5a

Polish Ministry of Climate and Environment. (2022b). Retrieved October 5, 2023, from https://www.gov.pl/web/klimat/polska-nie-popiera-pakietu-fit-for-55-minister-anna-moskwa-stanowczo-o-postulatach-polski-na-posiedzeniu-rady-unii-europejskiej-ds-srodowiska

Polish Ministry of Climate and Environment. (2022c). Retrieved October 5, 2023, from https://www.gov.pl/web/klimat/otwarcie-gazociagu-baltic-pipe

Polish Ministry of Climate and Environment. (2022d). Retrieved October 5, 2023, from https://www.gov.pl/web/klimat/polska-wspiera-odbudowe-ukrainskiego-systemu-energetycznego

Polish Ministry of Climate and Environment. (2022e). Retrieved October 4, 2023, from https://www.gov.pl/web/klimat/polska-i-islandia-wspolnie-o-wspolpracy-wodorowej

Polish Ministry of Climate and Environment. (2022f). Retrieved October 3, 2023, from https://www.gov.pl/web/klimat/polsko-norweskie-spotkanie-okraglego-stolu-dla-rozwoju-technologii-wychwytu-wykorzystania-oraz-skladowania-dwutlenku-wegla

Polish Ministry of Climate and Environment. (2022g). Retrieved October 3, 2023, from https://www.gov.pl/web/klimat/pierwsze-posiedzenie-rady-koordynacyjnej-ds-gospodarki-wodorowej

Polish Ministry of Climate and Environment. (2023a). Retrieved October 4, 2023, from https://www.gov.pl/web/klimat/polska-i-japonia-wzmocnily-wspolprace-w-dziedzinie-wodoru

Polish Ministry of Climate and Environment. (2023b). Retrieved October 5, 2023, from https://www.gov.pl/web/premier/zalozenia-do-aktualizacji-polityki-energetycznej-polski-do-2040-r-pep2040--wzmocnienie-bezpieczenstwa-i-niezaleznosci-energetycznej

Polish Ministry of Climate and Environment. (2023c). Retrieved October 3, 2023, from https://www.gov.pl/web/klimat/nowelizacja-prawa-geologicznego-i-gorniczego2

Polish Ministry of Climate and Environment. (2023d). Retrieved October 3, 2023, from https://www.gov.pl/web/klimat/wiceminister-ireneusz-zyska-na-otwarciu-stacji-wodorowej

Polish Ministry of Climate and Environment. (2023e). Retrieved October 3, 2023, from https://www.gov.pl/web/klimat/inauguracja-akademii-wodorowej-pkn-orlen-z-udzialem-wiceministra-ireneusza-zyski

Polish Ministry of Climate and Environment. (2023f). Retrieved October 3, 2023, from https://www.gov.pl/web/klimat/ministerstwo-klimatu-i-srodowiska-rozpoczyna-prace-nad-instrumentami-wsparcia-dla-wykorzystania-wodoru-niskoemisyjnego-w-gospodarce

Polish Ministry of Climate and Environment. (2023g). Retrieved October 4, 2023, from https://www.gov.pl/web/klimat/projekt-lotos-green-h2-zatwierdzony-przez-komisje-europejska

Polish Ministry of Climate and Environment. (2023h). Retrieved October 4, 2023, from https://www.gov.pl/web/klimat/propozycja-rozmieszczania-ogolnodostepnej-infrastruktury-ladowania

Polish Ministry of Climate and Environment. (2023i). Retrieved November 24, 2023, from https://www.gov.pl/web/gddkia/zalozenia-afir-jak-unijne-rozporzadzenie-wplynie-na-rozwoj-elektromobilnosci-w-polsce

Polish Ministry of Climate and Environment. (2024a, February 8). *Spotkanie Ministrów Miłosza Motyki i Macieja Bando z Ambasadorem Republiki Korei.* https://www.gov.pl/web/klimat/spo tkanie-ministrow-milosza-motyki-i-macieja-bando-zambasadorem-republiki-korei

Polish Ministry of Climate and Environment. (2024b, March). *Krajowy plan na rzecz energii i klimatu na lata 2021–2030.* https://www.gov.pl/web/klimat/krajowy-plan-na-rzecz-energii-i-kli matu

Polish Ministry of Climate and Environment. (2024c, May 27). *Konsultacje publiczne projektu ustawy o zmianie ustawy—Prawo energetyczne raz niektórych innych ustaw (UD36).* https://www.gov.pl/web/klimat/konsultacje-publiczne-projektu-ustawy-ozmianie-ust awy--prawo-energetyczne-raz-niektorych-innych-ustaw-ud36

Polish Ministry of Development Funds and Regional Policy. (2023). Retrieved October 3, 2023, from https://www.funduszeeuropejskie.gov.pl/media/120484/Zalacznik_nr_1_Rewizja_KPO_ czesc_dotacyjna.pdf

Polish Ministry of State Assets. (2019). Krajowy plan na rzecz energii i klimatu na lata 2021–2030. https://www.gov.pl/attachment/df8c4c37-808c-44ff-9278-676fb94add88

Polish Supreme Audit Office. (2022). Informacja o wynikach kontroli. Rozwój morskiej ener-getyki wiatrowej. https://www.nik.gov.pl/kontrole/wyniki-kontroli-nik/pobierz,NIK-P-21-065-rozwoj-morskiej-energetyki-wiatrowej-1,typ,kk.pdf

Polish Wind Energy Association. (2022). Potencjał Morskiej Energetyki Wiatrowej w Polsce. https://konferencja-offshore.pl/wp-content/uploads/2022/11/FarmyMorskie_RaportShort_ Prev.pdf

Polish Wind Energy Association, The Lower Silesian Institute for Energy Studies. (2021). Green hydrogen from RES in Poland. The use of wind and PV power for the production of green hydrogen as an opportunity to implement the assumptions of the EU Climate and Energy Policy in Poland. http://psew.pl/en/wp-content/uploads/sites/2/2021/12/Report-Green-hydrogen-from-RES-75MB.pdf

Polskie Radio 24. (2023). Retrieved October 3, 2023, from https://polskieradio24.pl/42/273/art ykul/3249784,polska-jest-potentatem-w-produkcji-wodoru-minister-moskwa-to-nie-przysz losc-ale-terazniejszosc

Port of Gdynia. (2022). Retrieved October 3, 2023, from https://www.port.gdynia.pl/hub-wod orowy-w-porcie-gdynia/

Statistics Poland. (2023). Energy 2023. https://stat.gov.pl/download/gfx/portalinformacyjny/en/def aultaktualnosci/3304/1/11/1/energy_2023.pdf

Teraz Środowisko. (2023). Retrieved October 23, 2023, from https://www.teraz-srodowisko.pl/akt ualnosci/programy-wyborcze-energetyka-14015.html

USNC. (2020). Retrieved October 4, 2023, from https://www.usnc.com/ultra-safe-nuclear-corpor ation-partners-with-synthos-green-energy-to-develop-micro-reactor-based-systems-to-decarb onize-company-chemical-facilities/

WNP. (2023). Retrieved October 3, 2023, from https://www.wnp.pl/energetyka/gaz-system-zlozyl-wnioski-o-finansowania-dwoch-projektow-przesylania-wodoru,679560.html

Michał Smoleń is the Head of the Energy and Climate Programme at the Instrat Foundation, a Warsaw-based think-tank. He leads policy work on Poland's and the EU's climate and energy policy. He previously worked as a senior consultant at PwC Poland and holds a Sociology degree from the University of Warsaw.

Wojciech Żelisko is an analyst at the Energy and Climate Programme at the Instrat Foundation. His main area of research is the hydrogen economy. Previously, he was a junior consultant at Audytel, a Warsaw-based energy sector consultancy. He is a Power Engineering graduate of the Faculty of Power and Aeronautical Engineering at the Warsaw University of Technology.

Hydrogen Affairs in Hungary's Politically Confined Ambition

John Szabo

Abstract Hydrogen is a much-discussed facet of Hungary's energy transition that has seen little progress in practice but offers an important tool to extend the government's foreign and energy policy. Policy-making in Hungary is highly centralised and government ambitions have prioritised the continued role of nuclear power, natural gas, and solar photovoltaics. These closely trace foreign policy priorities are well, given that the former two entrench relations with Russia and the latter enhances energy autonomy and allows the country to meet EU climate targets. A hydrogen economy supports such ambitions, while the government has also welcomed EU funds and foreign investment into novel (green) technologies that increase the value added in the economy. Domestic demand offers a secondary, but nonetheless important, push for the uptake of hydrogen, as it is envisioned to play a role in industry and transportation. Hungary's case shows how pre-existing political economic confines shape the uptake of hydrogen, as governments and other key actors take action while disrupting pre-existing practices to the least extent possible.

1 Introduction

Hungary has been quick to articulate hydrogen ambitions. The government and a number of companies were hasty to jump on the "hydrogen train", with the former taking the lead by publishing the National Hydrogen Strategy (henceforth, simply Strategy) in May 2021 (Government of Hungary, 2021). This followed the European Commission's (2020) 'A hydrogen strategy for a climate-neutral Europe' and other strategies relatively quickly (World Energy Council, 2021), which may be

Research for this chapter was supported by the British Academy Visiting Fellowships Programme 2023 (ref.: VF2\100827).

J. Szabo (✉)
Research Fellow, Institute of World Economics, Centre for Economic and Regional Studies, HUN-REN, Tóth Kálmán street 4, Budapest, Hungary
e-mail: szabo.john@krtk.hu

British Academy Visiting Fellow, Brunel University London, London, UK

© Helmholtz-Zentrum Potsdam, Deutsches GeoForschungsZentrum GFZ 2024
R. Quitzow and Y. Zabanova (eds.), *The Geopolitics of Hydrogen*, Studies in Energy,
Resource and Environmental Economics, https://doi.org/10.1007/978-3-031-59515-8_6

surprising given the size of Hungary's hydrogen market and relatively moderate climate and renewable ambitions. This chapter explores the roots of Hungary's interest in hydrogen and pieces together the role international forces play in shaping this energy carrier's and related technology's role in the country's energy system and, more broadly, the economy.

Hydrogen's uptake largely responds to external forces and how a relatively small economy takes decisions within pre-existing confines. It shows how the European Union's (EU) climate goals and the funds made available for the development and diffusion of new technologies is a powerful force emanating from Western Europe and EU institutions. Closely linked to this is the government's ambition to attract companies that are rolling out novel technologies. Meanwhile, energy relations with Russia are a mainstay of energy and foreign policy, confining the extent and form of change in the energy sector. At the intersection of these forces is a deeply natural gas and nuclear power dependent country with little appetite to change the structure of its energy demand. Albeit, the recent shift in its industrial policy to welcome energy-hungry multinational corporations is forcing action. This chapter explores the autonomy and structural constraints of small countries and how they may approach the unfolding energy transition.

Following this introduction, section two discusses demand- and supply-side drivers, actors, research and innovation, as well as hydrogen's link to the automotive sector. Section three then moves on to explore the international facet of hydrogen affairs by assessing strategy and the interlinkages between domestic and international policy. The final, fourth section, draws conclusions.

2   Domestic Drivers: Enchantment with Nuclear, Natural Gas, and Solar PV

National strategies and their demand-side drivers

Hungarian government policy first engaged with hydrogen in the early-2010s which then took on renewed momentum in the early-2020s, but tangible results have been slow to follow. The Ministry for National Development's (2012) 'Energy Strategy 2030' already discussed hydrogen's potential role in the transportation sector and as a form of energy storage in the early-2010s. Ambitions did not materialise, as initial enthusiasm toward hydrogen fizzled out in Europe and the USA with the decline in oil and natural gas prices (Szabo, 2020). Hungarian interest towards hydrogen rekindled with rising climate ambitions and the general discussion on the matter throughout the EU following the publication of the 'Clean Energy for All Europeans' (European Commission, 2016). As EU-level decarbonisation became a top-priority and experts pointed out the limitations to electrification (Eurelectric, 2018), policy-makers in Hungary began to explore the potential of hydrogen, yet again. Support for the energy carrier was foreshadowed in the 2019 National Energy Climate and Energy

Plan (NECP) (Government of Hungary, 2019a), which projected that Hungary may consume as much as 51 ktoe of hydrogen by 2030.

The NECP indicated that Hungary's energy system will rely on hydrogen to decarbonise transportation and natural gas consumption, store energy, as well as balance the electricity grid. Albeit, the Plan did not indicate specific volumes. Enthusiasm towards hydrogen emerged in a setting where Hungary's single aging coal plant was bound to be retired and decarbonisation policy would have to tackle transportation as well as natural gas consumption in the industrial and household heating sectors. What is more, there was a Europe-wide "hype" around hydrogen backed by policy-makers and natural gas interests, since the former saw it as a "silver bullet" in the transition while the latter as a tool for the survival of their industry (Szabo, 2020). The Government of Hungary (2021) was relatively quick to publish its 'National Hydrogen Strategy', in which it refined its ambitions included in the NECP and ideas for hydrogen's applications became more circumspect. It still discusses hydrogen as the "Swiss army knife" of the energy transition, but it shifted emphasis to meeting demand in the industrial sector alongside transportation and balancing the electricity grid.

Hydrogen already plays an important role in Hungary's industrial sector. The country consumed 160,000 t of hydrogen in 2020, which was driven by fertiliser production (115,000 t), oil refining (25,000 t), and the chemical industry (20,000 t). These industrial actors currently consume emitting methane-based hydrogen at the fertiliser plant in Pétfürdő, the Danube Refinery in Százhalombatta, and the chemical plant in Kazinbarcika. The Strategy proposes that this demand should be met via a combination of low-carbon hydrogen,[1] carbon free,[2] or green hydrogen[3] by forming two "hydrogen valleys": a "Northeast hydrogen valley" and the "Transdanubia hydrogen ecosystem". The former is based on existing heavy-, petrochemical-, and chemical industry, while the latter builds on existing demand from fertiliser produc-tion and oil refining as well as potential demand from an iron mill (Dunaújváros) and cement production facilities (Beremend and Királyegyháza). These sources of demand could be paired with hydrogen supply based on the electricity generated by the Paks nuclear power station and green electricity, enabling the formation of a complete ecosystem.

The hydrogen strategy indicates that only a portion of future hydrogen demand will be met via renewable and carbon free hydrogen (16,000 t) or low carbon hydrogen (20,000 t) by 2030 (Government of Hungary, 2021). It suggests that 24,000 t of green, low carbon, or carbon free hydrogen per annum will substitute emitting (grey) hydrogen in the industrial sector. Moreover, the transportation sector's demand will reach 10,000 t by 2030, which will grow to 65,000 t by 2040 and 212,000 t by 2050. Heavy-duty transportation is the driver of the latter's growth as the diesel engines of trucks, buses, and garbage trucks are phased out. The Strategy envisions a need

---

[1] Low-carbon hydrogen in this case is understood as blue (natural gas-based and paired with carbon capture and storage) or turquois (natural gas-based using pyrolysis) hydrogen.

[2] Carbon free hydrogen refers to pink hydrogen (nuclear-based).

[3] Green hydrogen refers to renewable-based hydrogen.

for twenty charging stations by 2030, which would optimally be distributed along highways near larger cities (Szabó et al., 2022). It also envisions that hydrogen be blended into the natural gas grid to decarbonise natural gas, which plays an important role in meeting household and commercial energy needs—73% of households used natural gas directly throughout the country (KSH, 2019). Hydrogen is also presented as a tool that could support balancing the grid (it indicates 60 MW of installed capacity) by helping meet peak electricity demand and store energy when necessary.

The Strategy identified four pillars for hydrogen demand—(1) industry, (2) transportation, (3) balancing the electricity grid, and (4) decarbonising gas consumption by blending—reflected in the draft of the 2023 NECP Revision as well (Government of Hungary, 2023). Most demand originates from the decarbonisation of natural gas (heating, industry, electricity generation) and petroleum products (transportation), but the scale of this is relatively modest when considering the size of the total energy sector. The 2023 NECP 'with existing measures' (WEM) scenarios indicate a negligible role for renewable hydrogen in most sectors that the government modelled: 0.4% of total energy demand by 2030. Albeit, this is much higher in the 'with added measures' (WAM) scenario. These figures should be taken with a grain of salt, as they are indicative of government goals, but specific figures and trajectories will only be final with the finalised 2024 NECP revision. And, what materialises is a whole other question.

A force that is set to alter the Strategy is the government's recent move to support battery manufacturing. Industrial policy has welcomed the development of battery-related megaprojects undertaken by a host of multinational companies, including South Korean Samsung and Chinese CATL. Battery production is a tremendously energy-intensive activity and current estimates suggest that projects in the pipeline will increase domestic electricity demand by 3.5 TWh, which equates to approximately 8–9% of 2021 total demand (Bucsky, 2023). Government policy has sought to facilitate this expansion, and state-owned energy enterprises will invest in capacities to substantially increase electricity generation. Measures that can help dampen the mismatch between supply and demand include three natural gas combined cycle gas turbines with a total installed capacity of 1650 MW. These can both help meet peak demand and the baseload demand of battery manufacturing facilities from the mid-2020s onwards. They are also planned to be hydrogen-ready, further increasing hydrogen's potential role. The output of battery plants may also help dampen electricity balancing needs of Hungary (including hydrogen), as the produced batteries need to be tested and there is thinking about linking these to the electricity network. Feasibility related questions linger, but the sheer scale of battery projects in the pipeline are poised to rewrite energy needs and courses of action in Hungary.

Supply-side drivers

Hydrogen's uptake is also supported by supply-side factors, given the infrastructure disposition and energy politics of Hungary. The country's long-term hydrogen ambitions emphasise a gradual shift to green hydrogen while relying on low carbon and carbon free hydrogen in the interim. Green ambitions are premised on the continued boom of solar photovoltaics. The installed capacity of solar PV rapidly rose over the course of the past 6–7 years, from effectively nil to nearly 5 GW by mid-2023 (MTI, 2023). This is especially substantial when considering the size of the electricity market, where peak demand highs were just over 7 GW in 2022 (MAVIR, 2023). The solar boom led a lopsided development, as government policy effectively banned the construction of wind installations that would have been able to balance some of solar PV's intermittency (Antal, 2019). Meanwhile, the grid has become saturated and pilot projects that connect batteries to the grid are only being unrolled. The intermittency stemming from solar output and limitations for the grid to absorb generation underpins the case for project developers to explore hydrogen production, but the economic case is not there just yet.

The Paks nuclear power station is a further driver of zero carbon hydrogen. The four reactors that began operations in the 1980s have a total installed capacity of 2 GW and provide approximately half of Hungary's domestically generated electricity. They will remain operational into the 2030s and, in parallel, the government began to explore the option to build a nuclear power station—Paks 2—in the early-2010s that would substitute the existing one. Initial plans indicated that the two power stations' operations would overlap, leading a number of experts to propose that surplus electricity be used to produce hydrogen. Delays in the construction of Paks 2 make overlapping operations increasingly unlikely, but nuclear is still framed as a mainstay in Hungarian energy affairs and has also underpinned the government's pro-nuclear politics in European affairs. The power station remains the basis of plans to construct an electrolyser that would harness abundant baseload electricity from the plant to produce hydrogen (Balogh, 2021). This introduces a feedback loop whereby nuclear-based electricity underpins a case for hydrogen and interest in hydrogen underpins the case for nuclear generation.

The third supply-side driver for hydrogen is the extensive role natural gas plays in Hungary's energy system. Hungary has an extensive gas grid, large natural gas storage facilities, and natural gas infrastructure developed by end-users to meet energy needs (e.g. household heating or industrial demand). This enables the uptake of low carbon hydrogen in a number of ways. The gas grid could be converted to support the blending of hydrogen which can either be domestically produced or imported. If domestically produced, Hungary can continue to import natural gas, which it can convert into hydrogen via steam methane reforming paired with CCS. Hungary's geological potential for CCS is held to be good, but projects have not yet been implemented (CCS4CEE 2023). MOL—a regionally important oil and gas company—has experience with CO_2 injection for enhanced oil recovery (EOR) and Hungarian stakeholders have participated in three large EU projects on CCS (EU GeoCapacity, CGS Europe, and CASTOR). Despite limited experience, stakeholder involvement in projects alongside common discourse and objectives in the Strategy indicate that there is an interest to produce blue hydrogen and utilise the stakeholders' experience with natural gas.

Private and public stakeholders

Decision-making in Hungary's energy system is centralised in the hands of the government and the largest stakeholders align their decisions with state objectives

(Szabo et al., 2020). The government follows a top-down approach to decision-making, whereby political directions are set at the highest levels of government—generally by Prime Minister Viktor Orbán—which then provide the basis for the policy ministries develop. Ministries play a role in identifying newly emerging issues and objectives that can be translated into policy, if these align with high politics. High politics' approach to hydrogen largely derives from foreign and energy policy—the two are closely linked—and a bid to boost innovation in the domestic economy. The government's foreign policy orientation includes maintaining close ties with Russia based on energy trade (e.g. natural gas and nuclear technology) and a general interest to explore whether emergent technologies can help increase the added value and technological sophistication of the country, especially if that links to the automotive or manufacturing sectors and is preferably financed by external actors (e.g. the EU).

Hydrogen's recent emergence on the EU-level coincided with the centralisation of most energy as well as innovation and development-related topics in the Ministry of Innovation and Technology between 2018–2022. This ministry explored hydrogen's role in the context of a generally ambitious research and development agenda and it was also able to include it into energy planning—the NECP and the National Hydrogen Strategy offer a testament to this. Following the 2022 elections and with the emergence a global energy crisis, the government dissolved this Ministry and created a stand-alone Ministry of Energy, while moving roles pertinent to the development of a hydrogen ecosystem to the Ministry for Culture and Innovation, the Ministry for Construction and Transport, as well as the Ministry for Economic Development. Hydrogen-related issues that rely on energy, innovation, education, infrastructure, transport, economic, and development policy are now splintered between these ministries, ultimately, slowing the adoption of the technology.

Supply-side actors involved with hydrogen affairs tend to be linked to one of two energy champions: the Hungarian Public Utility (*Magyar Villamos Művek* i.e. MVM) or MOL. MVM is an entirely state-owned public utility holding that has a broad portfolio ranging from power generation to natural gas storage. Hungarian Gas Storage (*Magyar Földgáztároló*, i.e. MFGT) is the MVM subsidiary responsible for the latter and plays a key role in exploring hydrogen storage at scale. Its strategy is typically subjugated or closely guided by government policy, including natural gas storage level targets or exploring hydrogen storage to enhance future energy security. MOL is a publicly traded company and therefore enjoys somewhat more autonomy. However, shareholders include state-aligned foundations that were directly owned by the state until relatively recently and personal ties have underpinned a close working relation between the government and the company. Both companies are key players in the hydrogen ecosystem that respond to European trends but take action in close coordination with the state.

Together with MOL, Nitrogénművek and Wanhua–BorsodChem consume the overwhelming majority of hydrogen making them key players in shaping the domestic market. They have also all begun to explore options for low carbon hydrogen. Nitrogénművek is a privately owned company that is the largest domestic producer of various fertilisers, for which it produces substantial amounts of hydrogen-intensive Calcium Ammonium Nitrate (CAN). Its owner is among the wealthiest individuals in Hungary, but one that falls outside of PM Orbán-aligned businesspersons and he has conflicted with the government on numerous occasions. The company's leadership's involvement in policy is inhibited by personal relations, but it has to be taken into account thoroughly given that it is the largest consumer of hydrogen. Wanhua–BorsodChem is a producer of isocyanates, PVC, and other vinyl products. It is owned by a Chinese holding company and fortifies China-Hungary business and political ties, but its impact on policy is limited to the government considering that they will need to cooperate in the future to meet hydrogen demand. MOL tends to coordinate with the government, but it being an international oil company has engaged in piloting green hydrogen alone. All three of these companies began to explore non-emitting hydrogen, but have done so largely independently; albeit, while coordinating with the government on strategy.

Hydrogen associations, discussion platforms, and organisations have been active in supporting the diffusion of hydrogen. However, their role in policy is muted, as foreign and energy policy alongside investment tends to drive policy in Hungary. Nonetheless, the Hungarian Hydrogen and Fuel Cell Association has been active since 2011, while the Hungarian Hydrogen Technology Association was established in 2020 and became a member of Hydrogen Europe—the European hydrogen advocacy organization—and Global Hydrogen Industrial Association Alliances. Through Hydrogen Europe, Hungary's hydrogen industry is also represented at UNIDO's International Hydrogen Energy Centre. The country is a member of the International Energy Agency (IEA), but is only represented in the IEA's Hydrogen Technology Collaboration Programme through the European Commission. Finally, the second Budapest Hydrogen Summit was organised in 2023, with high-level participants from the public and private sector indicating the general interest towards the matter. These offer important platforms for discussion that can then be incorporated into policy, but their impact is limited in comparison to the effect high politics has on decisions.

Research and innovation on hydrogen

Hungarian stakeholders launched a number of innovative projects that support the development of hydrogen. A grand undertaking and one deemed strategically important by the state is MFGT's endeavour to explore large-scale hydrogen storage. The company's preliminary study found that "[h]ydrogen is a long-term solution and a pathfinder for the decarbonisation objective" (MFGT, 2019, p. 64) leading the company to develop the 'Aquamarine' pilot project. This explores how it can store the hydrogen produced from a 2.5 MW electrolysis system in the underground storage facility at Kardoskút—a 280 million m³ natural gas storage site. The project is financed by the Ministry of Innovation and Technology, the National Research, Development and Innovation Office, and MFGT. After delays, the company completed on-site construction in March 2023 and began to run test operations prior to the pilot. Tests have potentially regional implications, since Hungary's gas storage capacities are substantial in a regional context and especially when compared to demand. In principle, it could offer sites for storing hydrogen, as it does for natural gas, if this becomes technologically feasible, establishing itself a regionally strategic role.

FGSZ—the natural gas transmission system operator (TSO) and a subsidiary of MOL—is exploring the transit of hydrogen in its pipeline infrastructure system. It joined the European Hydrogen Backbone (EHB) initiative, a cooperation between thirty-two energy infrastructure operators (at the time of writing), most of which are natural gas grid operators. FGSZ (2022) indicated in its 2022 ten-year development plan that it will facilitate the uptake of hydrogen by adapting pipelines and compressor stations during the 2026–2032 period. Hungary's location at the intersection of a number of natural gas markets entails that ambitions to develop hydrogen compatible and dedicated hydrogen infrastructure are at the top of the company's agenda, as FGSZ can position Hungary as an important hydrogen transit country in the future—something the draft 2023 NECP revision also aims to achieve.

Large-scale demand-side developments are also on the way in the Hungarian hydrogen sector. Nitrogénművek has teamed up with the European Bank for Reconstruction and Development (EBRD) to explore emission reduction opportunities (Nitrogénművek, 2023). The project aims to enhance energy efficiency at Nitrogénművek's Pétfürdő fertiliser plant and use a renewable energy-powered electrolyser to produce ammonia. Meanwhile, MOL announced in 2022 that it will invest €22 million to develop a 10 MW electrolyser in Százhalombatta near the Danube Refinery (MOL, 2022). This will produce 1600 t of green hydrogen per annum, but this is only a fraction of the company's demand. The electrolyser is not directly paired with the development of a renewable energy plant, but will connect to the grid and the company will procure green electricity by purchasing green certificates. The need to comply with the EU's definition of renewable hydrogen would lead it to sign power purchase agreements (PPA) with green electricity generators and prove geographic and temporal correlation between supply and demand, but they can rely on existing installations if signed by 2028. This would thus draw green electricity from a not fully renewable mix, prioritising hydrogen production over consuming the green electricity where possible. Nitrogénművek's and MOL's endeavours signal interest and some action to green their activities, but these remain a marginal fraction of their overall portfolio.

Basic research has also been ongoing on hydrogen in Hungary. A consortium of nine university and research institutes won a €17 million project to establish a 'National Laboratory for Renewable Energies', which will prioritise hydrogen technologies and carbon capture and utilisation (CCU) (ELKH TTK, 2023). The endeavour is financed by government and EU funds, and also seeks to facilitate that it support future international research collaborations and applications for funding (e.g. Horizon). Hungary also joined the Clean Hydrogen Partnership, but has not yet participated in any projects. Its involvement may, however, have taken a turn for the worse, as the majority of Hungarian universities and research institutes that adapted the so-called 'foundation model'[4] cannot participate in Horizon projects due

[4] Most of the Hungarian higher education system was reorganised beginning in 2019. What were publicly funded universities were brought under the control of asset management foundations that

to rule of law breaches in the country. The government has sought to fill the void with some funding, but given the poor shape of the central budget this is unlikely to fill the gap. Further resources for research, development, and innovation will be made available through various programmes (e.g. GINOP+, IKOP+, KEHOP+), the EU's Modernisation Fund, and national budget programmes in the forthcoming years. These continue to heavily draw on EU funds with a general focus on developing the domestic hydrogen scene (Kaderják, 2022).

Domestic actors were involved with their European counterparts in developing a number of so-called Important Projects of Common European Interest (IPCEI) (Government of Hungary, 2021, p. 10). These include: Blue Danube @ Green Hydrogen, Black Horse, and Silver Frog. Blue Danube @ Green Hydrogen is a cooperation between Austria, Germany, Romania, Netherlands, EU and targets the production of large-scale green hydrogen in Southeastern Europe and its subsequent transport to industrial actors along the Danube River (Verbund, 2020). Black Horse tackles the diffusion of hydrogen-fuelled heavy-duty transport and the development of requisite fuelling infrastructure through a cooperation between Visegrád countries (REKK, 2021). Finally, Silver Frog aims to develop a 2 GW per annum solar module factory, which will help the deployment of 10 GW of solar PV capacity to produce green hydrogen. The consortium is composed of Hungarian heterojunction PV module manufacturer Ecosolifer, Swiss mechanical engineering company Meyer Burger, the Belgian-based unit of Canadian hydrogen business Hydrogenics, and Danish renewables developer European Energy (Enkhardt, 2019). Hungarian companies and research entities are thus involved in international research throughout the entire supply-chain, but results of these projects are yet to materialise.

Hydrogen cells and the transportation sector

The automotive sector has a long-standing tradition in Hungary. The country was an important producer of especially heavy-duty vehicles and buses during the communist era, which, paired with low wages, offered an attractive location for original equipment manufacturers primarily from Western Europe to establish operations. Audi, Mercedes Benz, Stellantis, Suzuki have major production capacities in the country and BMW is constructing a plant as well. In addition, Credobus, ITE Bus & Truck, BYD, Volvo, and Ikarus manufacture buses and other heavy-duty vehicles. Electrification is at the top of the most companies' agendas active in Hungary, but some have suggested that they will produce vehicles propelled by hydrogen cells as well. Passenger vehicle manufacturers Audi (h-tron), Mercedes Benz (GLC F-CELL), and BMW (iX5) piloted hydrogen cell-propelled vehicles, but there is no indication that these will be manufactured in Hungary. Bus producers also tend to focus on electrification, as Credobus was the only one with plans to launch the Econell 12 Next with fuel cells in 2023–2024 (Credobus, 2022). Meanwhile, a handful of

---

were governed by a board of trustees. This not only made universities reliant on returns from investments, but the trustees were frequently government officials, Members of Parliament, or other individuals with close ties to the ruling Fidesz party. The European Commission has launched inquiries into how this has impeded the autonomy of universities.

hydrogen fuel cell propelled buses were added to the Budapest and Paks fleets (e-Cars, 2023; Zomborácz, 2022). Thus, the technology has only begun to penetrate the transportation market and vehicle manufacturing in Hungary. The government—drawing on EU funds—has launched programmes to support the uptake of buses with alternative propulsion systems (HUMDA, 2023), including hydrogen, but domestic automotive manufacturing is largely shifting to the production of electric vehicles.

3   International Approach

Strategic objectives and energy relations

Hungary's overarching approach to international affairs balances the country politically and economically between the "West" and the "East". Links to the "West" were fast-tracked with the collapse of the Eastern bloc, which would set the country on a path to join the North Atlantic Treaty Organization (NATO) and the EU. Political integration was paired with deepening economic ties, as Western capital flowed into the country and into sectors such as the automotive or electronics industries. The second Orbán government took office in 2010 and shortly launched its "Eastern Opening", as a part of which it began to strengthen ties with Eastern powers. This included the extension of hydrocarbon and nuclear-based relations with Russia and attracting investment from Far Eastern partners (e.g. China or South Korea) (Bernek, 2014). Subsequent Orbán governments have maneuvered between these two poles, seeking to form a bridge, while reaping the economic and political benefits of respective diplomacy.

Foreign relations in both directions are closely linked to the energy sector. Hungary has been reliant on energy imported from the Soviet Union since the middle of the twentieth century, as demand far outpaced domestic energy produced from a limited hydrocarbon and lignite resource base. The government developed a reliance on Soviet imports, by first importing crude oil, then electricity, followed by natural gas, and, finally, nuclear technology (Szabo & Deák, 2020). The order of these and their availability was dictated by leaders in Moscow; essentially, Hungary's energy mix was externally imposed. The dissolution of the Soviet Union did not dissipate reliance on imports, as Hungary continues to import the overwhelming majority of its crude oil and natural gas from Russia in addition to having commissioned the Paks 2 nuclear power station by Russia's Rosatom. It integrated its natural gas and electricity grid into the European system upon joining the EU, but sustained most of its reliance on Russian resources. This was essential to politics that ensured low energy prices for domestic consumers, which was a key pillar of the government's domestic policy and a force that helped PM Orbán's Fidesz secure electoral wins (Weiner & Szép, 2022).

The Hungarian government has indicated little propensity to substantially change the structure of its energy mix. If it does change, this tends to be driven by the lowest cost option or that which is subsidised by external actors. Historically, the inexpensive resources offered by the Soviet Union drove change, while the EU now plays this role by mandating renewable energy targets and offering funding to support their diffusion. The 2019 NECP, the 'National Energy Strategy 2030, with an Outlook to 2040' (Government of Hungary, 2019b), and the 2023 NECP offer glimpses of future plans; all of which emphasise the need to substitute natural gas and electricity imports. The government plans to tackle the former through energy efficiency measures, while the latter is based on nuclear ambitions and the continuance of the solar PV boom. The direction of change—energy efficiency and renewables—draw on EU support, be that directly through EU funds (e.g. energy efficiency programmes) or by developing essential auxiliary technologies (e.g. the grid, electricity storage, etc.) that enable the renewable energy's expansion. With this, the government continues the balancing act by leaving room for EU funds and Western investment in renewables, while maintaining close energy-based ties with Russia through cooperation on nuclear power as well as sustained natural gas and oil imports.

The link between international and domestic strategy

Government strategy that prioritises Russian nuclear power- and natural gas-based cooperation in addition to attracting EU funds for renewable-related innovation prefigure hydrogen ambitions. Simply put, there are two dominant geopolitical forces driving a hydrogen economy: Russia and the EU. Natural gas trade with Russia enables low carbon hydrogen and Rosatom's role in Paks 2 that of its nuclear-based variant. The EU plays a role in driving Hungary's transition to renewables and promoting innovation in the energy sector, which increasingly includes support for green hydrogen as well. The government transposes these external drivers—international relations and funding—into domestic policy, which is met by demand from a handful of industrial actors, alongside prospective consumption from transportation, heating, and electricity generation.

A fixture in Hungarian energy affairs is nuclear power. The government and key stakeholders have discussed this as a source of electricity for zero carbon hydrogen production. The Paks power station is of Soviet design, tying Hungary to Russia with the latter providing fuel rods and storing nuclear waste. The sanctions imposed by Western powers due to Russia's war in Ukraine have not yet targeted nuclear power, but leaders have discussed such measures and the government is exploring alternative fuel rod supplies. Westinghouse may offer a compatible solution, which could be essential in the long-term as the government announced that it will investigate whether the reactor's operational lifetime can be extended into the 2040s. In parallel, the government of Hungary is pursuing Paks 2 for which MVM and Rosatom signed an agreement under the auspices of their respective governments, renewing the two countries' nuclear-based cooperation.

Initial nuclear plans suggested that the operations of the two nuclear power stations would overlap, but delays have forced the latter's planned completion into the late-2020s or early-2030s. The government continues to argue that the operation of these projects will overlap—even if this seems increasingly unlikely—and the surplus electricity it generates can either be exported or used to run an electrolyser. This underpins the economic rationale of the project and allows for Hungary to develop hydrogen production capacities that can be based on Paks 2 generation and renewable electricity once MVM retires Paks. Thus, the supply-side of Transdanubia Hydrogen Ecosystem would be solved. However, many raised concerns over the need for a new nuclear power station (Sáfián, 2015) and the opacity of the contract (Erdélyi, 2021). Moreover, the war in Ukraine and related sanctions undermine the viability of Paks 2 as key suppliers, such as Siemens or General Electric, are forced to reconsider their involvement (Szabó, 2023). Moreover, there is substantial political pressure from EU institutions to consider imposing sanctions on Russian nuclear projects, obstructing the completion of Paks 2. Despite risks of completion and potential further delays, government support has not wavered as the project is a prominent component of Hungary-Russia cooperation and neither party is willing to outright abandon it.

The other mainstay of Hungarian energy affairs is the government's reluctance to shift away from Russian natural gas. Hungary—like most countries in the region— diversified natural gas import routes since joining the EU and especially following the shock of the 2009 supply crisis. However, it continues to largely rely on Russian supplies. Government rhetoric emphasises that this is the cheapest source of energy which enable the "utility price reduction" programmes for households. Meanwhile, these programmes have been essential in securing electoral victories. In practice, contracted volumes are essential in meeting Hungary's relatively large natural gas demand, at a typically lower price point to that of imported LNG—not only do LNG deliveries in the Mediterranean tend to come at a premium, but Croatian transit is also costly. Nonetheless, these flows are far from cheap, as they are linked to natural gas hub prices.

A high natural gas price environment undermines Hungary's blue hydrogen ambitions, but the government's commitment to Russian natural gas has not wavered. It re-committed when state-owned utility MVM inked a fifteen-year contract in September 2021 (Euronews, 2021), which it expanded—against EU ambitions to move away from Russian resource reliance amidst the war—in 2023 (VG, 2023). The Kremlin may have halted most supplies to Europe, but the Orbán government discusses Russian natural gas as a pillar of energy security, which it presumes will sustain given the stability of imports through the Balkan Stream pipeline. Strategic cooperation based on this energy carrier's trade is still said to remain stable and there is no indication that the government plans to substantially reduce the country's dependence through 2035. Blending low carbon hydrogen extends demand further into the future, but the costs of this may be quite high.

The other leg of hydrogen developments is the green transition led by the EU and a number of Western member states. The recent solar PV boom in Hungary emerged as a combination of EU climate and energy policy, market forces, as well as domestic drivers. The EU's climate objectives and renewable energy targets provided support through incentives, such as those allocated from EU cohesion policy or income from selling carbon quotas. These also stimulated government to take action, as a part of which it rolled out a generous feed-in tariff system until relatively recently and a net-metering that was simply accessible for household consumers, hence the solar PV boom. Alongside the diffusion of solar PV, new frontiers of green technologies pave the way for the state and a handful of enterprises to pursue "technological opportunism" (Chen & Lien, 2013). These overwhelmingly draw on EU funds which underpins further decarbonisation and enhance the competitiveness of domestic firms, while also allocating substantial wealth to national entities. However, these have not yet materialised in the field of hydrogen as most projects are still in their infancy.

A key vehicle in Hungary's green transition plans is based on the EU's Recovery and Resilience Facility (RRF). This would provide the country with just over €7 billion in grants and nearly €10 billion in loans. However, issues with the rule of law in Hungary have led the European Commission to withhold funds, which impedes investment into the energy transition. Grants have been made available and the energy chapter has indicated that investment will be channeled into electricity. Loans have not been approved and details are unclear as is the status of negotiations between the government and the European Commission, but there is a stand-off around the issue that is intertwined with other political issues ranging from the rule of law in Hungary to PM Orbán vetoing aid to Ukraine and voting against EU legislation (e.g. Energy Efficiency Directive or the Renewable Energy Directive III). Nonetheless, these are essential to kick-starting investment in the energy sector, including CCS projects, the construction of electrolysers, industrial application of hydrogen, and purchasing hydrogen-propelled buses and other large vehicles (Brückner, 2023). Without RRF financing, Hungary's energy transition is bound to gradually grind to a halt.

4 Conclusion

Hungary's approach to hydrogen reflects its general engagement with new green technologies: it conveys enthusiasm and seeks to attract as much foreign funding as possible, but it adapts its actions to pre-existing political confines. At the policy-level hydrogen has been oft-discussed and included into various planning documents, but action is yet to materialise. Only the largest industrial consumers and natural gas infrastructure actors have begun to plan and execute smaller projects. MOL's refinery, Nitrogénművek's fertiliser plant, the FGSZ-operated pipeline system, and MFGT's natural gas storage facilities are subject to hydrogen pilots. There is some pressure to explore alternative fuel supplies due to Hungary's deep reliance on natural gas and the extensive grid that can become an asset in a low carbon setting. Action is usually limited to areas where foreign grants and financing is available, with actors involved typically large consumers that need to convey a convincing plan on how they will sustain their businesses. Thus, hydrogen endeavours tend to reflect corporations seeking to convince investors and clients that they are preparing for a low carbon future, while state-led action has been largely limited to planning.

The supply-side of the hydrogen story tends to be more important in the government's hydrogen ambitions. Producing the energy carrier aligns closely with its energy politics. Natural gas and nuclear power are bedrocks of the government's energy plans, and with these leaders sustain the country's close relation to Russia. Pivoting away is a non-option, as it is deemed costly and the government claims that Russian energy offers stable and inexpensive resources that allows for the continuance of its low utility price policy. Low carbon hydrogen extend the country's dependence on overwhelmingly Russian natural gas and offer a relatively low cost, low CO_2 emission energy resource in difficult to decarbonise sectors. In similar vein, nuclear-based hydrogen lends rationale to pursuing the Paks 2 project, which is a strategic endeavour for both parties that neither wants to forfeit due to the high economic and political costs. Hydrogen sustains nuclear ambitions, and nuclear ambitions underpin hydrogen, offering a loop that favours the country's pro-Russia foreign policy. Diplomacy is thus important in Hungary's hydrogen affairs, but it tends to be indirect as the basis of discussion are the energy carriers that underpin the current energy system and hydrogen is a hypothetical discussed to extend their role into the future.

There is also ample room for green hydrogen in Hungary's energy system, especially given the prospects EU funds offer. Hydrogen can dampen the lopsided focus on a solar boom since 2016 that has left experts in the energy sector scurrying to solve the issue of storage and grid-related bottlenecks. It can help store surplus electricity generated by renewables, while depleted hydrocarbon fields could play a role in storing this renewable hydrogen. All-in-all, Hungary's hydrogen ambitions—covering production, storage, CCS, etc.—allow for the state to continue manoeuvring between the Western Europe/European institutions and Russia, by maintaining ties with the latter through nuclear and natural gas while meeting EU goals, attracting funds and investment, and facilitating hydrogen-related cooperation with Western counterparts.

Acknowledgements Research for this chapter was supported by the British Academy Visiting Fellowships Programme 2023 (ref.: VF2\100827).

References

Antal, M. (2019). How the regime hampered a transition to renewable electricity in Hungary. *Environmental Innovation and Societal Transitions, 33*, 162–182. https://www.sciencedirect.com/science/article/pii/S2210422418300029

Balogh, Á. (2021, 28 April). Atomerőmű-kapacitások és a karbonsemleges hidrogénelőállítás lehetőségei Magyarországon. *Green Policy Center.* https://www.greenpolicycenter.com/2021/04/28/atomeromu-kapacitasok-es-a-karbon-semleges-hidrogeneloallitas-lehetosegei-magyarorszagon/

Bernek, Á. (2014). Hazánk keleti nyitás politikája és a 21. századi geopolitikai stratégiák összefüggései. *Külügyi Szemle, 17(2)*, 122–144. https://kki.hu/assets/upload/06_Bernek_Agnes.pdf

BMWI. (2020). *The national hydrogen strategy.* BMWI. Retrieved November 8, 2023, from https://www.bmwk.de/Redaktion/EN/Publikationen/Energie/the-national-hydrogen-strategy.pdf?__blob=publicationFile&v=6

Brückner, G. (2023, 6 February). Lantos Csaba és Navracsics Tibor soha nem látott, közel 6000 milliárd forintos energiatervet készít. *Telex.* https://telex.hu/gazdasag/2023/02/06/lantos-csaba-es-navracsics-tibor-soha-nem-latott-kozel-6000-milliard-forintos-energiatervet-keszit

Bucsky, P. (2023, 15 May). A Magyar cégek is megfizetik az új akkumulátorgyárak árát. *G7.* https://g7.hu/vallalat/20230515/a-magyar-cegek-is-megfizetik-az-uj-akkumulatorgyarak-arat/

Chen, C.-W., & Lien, N.-H. (2013). Technological opportunism and firm performance: Moderating contexts. *Journal of Business Research, 66*(11), 2218–2225. Available at: https://www.sciencedirect.com/science/article/abs/pii/S0148296312000331

Credobus. (2022). *Megérkezett az econell 12 next.* Retrieved November 8, 2023, from https://credobus.hu/media/2022/08/31/megerkezett-az-econell-12-next/

e-Cars. (2023, 18 January). Hidrogén-üzemanyagcellás busz közlekedik Pakson. *e-cars.* https://e-cars.hu/2023/01/18/hidrogen-uzemanyagcellas-busz-kozlekedik-pakson/

ELKH TTK. (2023). *Nemzeti Laborok.* Eötvös Loránd Research Network Research Centre for Natural Sciences. Retrieved November 8, 2023, from http://www.ttk.hu/nemzeti-laborok/mel

Enkhardt, S. (2019, 11 October). Operation Silver Frog: Innovative 2 GW solar production plan for green hydrogen in Europe. *PV Magazine.* https://www.pv-magazine.com/2019/10/11/operation-silver-frog-innovative-2-gw-solar-production-plan-for-green-hydrogen-in-europe/

Erdélyi, K. (2021, 25 January). Az Alkotmánybíróság szerint rendben van, hogy a kormány 30 évre titkosította a Paks2-projektet. *Átlátszó.* https://atlatszo.hu/kozugy/2021/01/25/az-alkotmanybirosag-szerint-rendben-van-hogy-a-kormany-30-evre-titkositotta-a-paks2-projektet/

Eurelectric. (2018). *Decarbonisation pathways.* Eurelectric. Retrieved November 8, 2023, from https://cdn.eurelectric.org/media/3457/decarbonisation-pathways-h-5A25D8D1.pdf

Euronews. (2021, September 27). 15 évre kötött gázvásárlási szerződést a Gazprommal Magyarország. *Euronews.* https://hu.euronews.com/2021/09/27/15-evre-kotott-gazvasarlasi-szerzodest-a-gazprommal-magyarorszag

European Commission. (2016). *Clean energy for all Europeans.* European Commission. Retrieved November 8, 2023, from https://eur-lex.europa.eu/legal-content/EN/TXT/?qid=1582103368596&uri=CELEX:52016DC0860

European Commission. (2020). *A hydrogen strategy for a climate-neutral Europe.* European Commission. Retrieved November 8, 2023, from https://eur-lex.europa.eu/legal-content/EN/TXT/?uri=CELEX:52020DC0301

FGSZ. (2022). *10 Éves Fejlesztési Terv.* FGSZ Natural Gas Transmission. Retrieved November 8, 2023, from https://fgsz.hu/file/documents/2/2308/2022_10_eves_fejlesztesi_terv.pdf

Government of Hungary (2019a) *National energy and climate plan.* Government of Hungary. Retrieved November 8, 2023, from https://energy.ec.europa.eu/system/files/2020-01/hu_final_necp_main_hu_0.pdf

Government of Hungary. (2019b). *Nemzeti Energiastratégia 2030.* Government of Hungary. Retrieved November 8, 2023, from https://www.enhat.mekh.hu/strategiak

Government of Hungary. (2021). *Magyarország Nemzeti Hirdogénstratégiája.* Government of Hungary. Retrieved November 8, 2023, from https://cdn.kormany.hu/uploads/document/6/61/61a/61aa5f835ccf3e726fb5795f766f3768f7f829c1.pdf

Government of Hungary. (2023). *Nemzeti Energia- és Klímaterv: 2023. évi felülvizsgált változat.* Government of Hungary. Retrieved November 8, 2023, from https://commission.europa.eu/system/files/2023-09/HUNGARY%20-%20DRAFT%20UPDATED%20NECP%202021-2030%20_HU.pdf

HUMDA. (2023). *Zöld Busz Program.* Hungarian Mobility Development Agency. Retrieved November 8, 2023, from https://humda.hu/green/zold-busz-program

Kaderják, P. (2022). Magyarország hidrogén stratégiája, különös tekintettel a távhőszektorra. Presentation at XXI. Távhőszolgáltatási Konferenciát és Szakmai Kiállítás in Eger, Hungary. https://tavho.org/uploads/rendezvenyeink/Konferencia%202022/Magyarorsz%C3%A1g%20hidrog%C3%A9n%20strat%C3%A9gi%C3%A1ja%2C%20k%C3%BCl%C3%B6n%C3%B6s%20tekintettel%20a%20t%C3%A1vh%C5%91szektorra%20-%20Dr.%20Kaderj%C3%A1k%20P%C3%A9ter.pdf

KSH. (2019). *A települések infrastrukturális ellátottsága.* Central Statistical Office. Retrieved November 8, 2023, from https://www.ksh.hu/docs/hun/xftp/stattukor/telepinfra/2019/index. html

MAVIR. (2023). *Rendszerterhelés.* MAVIR Hungarian Electricity Transmission System Operator. Retrieved November 8, 2023, from https://www.mavir.hu/web/mavir/rendszerterheles

MFGT. (2019). *Energy storage Danube region study.* MFGT. Retrieved November 8, 2023, from https://mfgt.hu/-/media/MFGT/P2G/MFGT_EnergyStorageDanubeRegionStudy_ 2019.pdf?la=hu-HU

Ministry for National Development. (2012). *National energy strategy.* Government of Hungary. Retrieved November 8, 2023, from https://2010-2014.kormany.hu/download/7/d7/70000/Hun garian%20Energy%20Strategy%202030.pdf

MOL. (2022). *Újabb Lépés az Energiafüggetlenség Felé: Zöld Hidrogén Gyártásába Kezd a Mol.* Retrieved November 8, 2023, from https://molgroup.info/hu/media-kozpont/sajtokozlemenyek/ ujabb-lepes-az-energiafuggetlenseg-fele-zold-hidrogen-gyartasaba-kezd-a-mol

MTI. (2023, June). Mérföldkőnél a hazai naperőművek kapacitása. *VG.hu.* https://www.vg.hu/ene rgia-vgplus/2023/06/merfoldkonel-a-hazai-naperomuvek-kapacitasa

Nitrogénművek. (2023). *EBRD and Nitrogenmuvek to start study on energy improvement and green-ammonia development in Hungary.* Nitrogénművek. Retrieved November 8, 2023, from https://www.nitrogen.hu/en/news/328-ebrd-and-nitrogenmuvek-to-start-study-on-energy- improvement-and-green-ammonia-development-in-hunagry

REKK. (2021). *V4 hydrogen cooperation opportunities.* Regional Centre for Energy Policy Research. Retrieved November 8, 2023, from https://rekk.hu/downloads/events/V4_hydrogen_ cooperation_opportunities.pdf

Sáfián, F. (2015, January). Paks Nélkül a Világ. *Energiaklub.* https://energiaklub.hu/files/study/pak sii_nelkul_a_vilag_web.pdf

Szabó, Á., Orosz, L., Borsi, Z., Telekesi, T., Főglein, K., Faragó, G., & Schváb, Z. (2022). Hidrogén Tüzelőanyagcellás Nehézgépjárművek Hidrogéntöltő Állomásainak Optimális Hely- meghatározása—*Külügyi Műhely, 2022*(1). http://real.mtak.hu/143666/1/04_KulugyiMuhelySz aboAdam.pdf

Szabó, D. (2023, 10 February). Szankciós kereszttűzbe került Paks II. *Portfolio.* https://www.por tfolio.hu/gazdasag/20230210/szankcios-kereszttuzbe-kerult-paks-ii-595504

Szabo, J., Weiner, C., & Deák, A. (2020). Energy Governance in Hungary. In M. Knodt & J. Kemmerzell (Eds.), *Handbook of energy governance in Europe.* https://link.springer.com/refere nceworkentry/https://doi.org/10.1007/978-3-319-73526-9_13-1

Szabo, J. (2021). Fossil capitalism's lock-ins: The natural gas-hydrogen nexus. *Capitalism Nature Socialism, 32*(4). https://www.tandfonline.com/doi/full/https://doi.org/10.1080/ 10455752.2020.1843186

Verbund. (2020). *Green hydrogen blue danube.* Verbund. Retrieved November 8, 2023, from https://www.verbund.com/en-de/about-verbund/news-press/press-releases/2020/11/17/greenh ydrogenbluedanube

VG. (2023, 11 April). Szijjártó Péter Moszkvából üzeni: biztosítva van Magyarország földgáz- és kőolajellátása, megújult a paksi beruházásról szóló szerződés. *Világgazdaság.* https://www.vg. hu/vilaggazdasag-magyar-gazdasag/2023/04/szijjarto-peter-moszkvaban-biztositott-magyar orszag-foldgaz-es-koolaj-ellatasa

Weiner, C., & Szép, T. (2022). The Hungarian utility cost reduction programme: An impact assess- ment. *Energy Strategy Reviews, 40.* https://www.sciencedirect.com/science/article/pii/S22114 67X22000177?via%3Dihub

World Energy Council. (2021). *Working paper: National hydrogen strategies.* World Energy Council. Retrieved November 8, 2023, from https://www.worldenergy.org/assets/downloads/ Working_Paper_-_National_Hydrogen_Strategies_-_September_2021.pdf

Zomborácz, I. (2022, 11 February). Ilyen a budapesti hidrogénes busz. *Totalcar.* https://totalcar.hu/ magazin/kozelet/2022/02/11/ilyen-a-budapesti-hidrogenes-busz/

John Szabo is a Research Fellow at the Centre for Economic and Regional Studies, Budapest as well as a British Academy Visiting Fellow at Brunel University London. His interests focus on the society-environment nexus in energy affairs.

Spain's Hydrogen Ambition: Between Reindustrialisation and Export-Led Energy Integration with the EU

Ignacio Urbasos and Gonzalo Escribano

Abstract The Spanish approach to renewable hydrogen has evolved significantly driven by economic and geopolitical factors. Initially framed as a tool for domestic industrial development during the COVID-19 crisis, the 2020 Spanish Hydrogen Roadmap emphasised creating hydrogen clusters for production and consumption. However, the Russian invasion of Ukraine and the European Commission's call for increased ambition in renewable hydrogen in REPowerEU prompted a strategic shift. Spain now focuses on exports and infrastructure development, balancing the drive for domestic green reindustrialisation with an export-oriented model to integrate the Iberian Peninsula's energy with Europe. Despite concerns about historical obstacles to interconnections, Spain prioritises hydrogen diplomacy with key European allies. Looking ahead, Spain envisions hydrogen as a vector for Euro-Mediterranean energy integration, energy cooperation, and business collaboration in Latin America. The chapter concludes that Spain needs to build a national strategy that aligns the domestic and international dimensions of hydrogen development, sending a coherent message to civil society, the private sector and institutions.

1 Introduction

In Spain, renewable hydrogen has been enthusiastically embraced by the Government, devolved administrations, the private sector and, to some extent, civil society. When the Spanish Hydrogen Roadmap was published in 2020, in the midst of the economic and social crisis caused by COVID-19, renewable hydrogen was seen as a tool for industrial development and economic diversification. The strategy focused on the creation of hydrogen valleys or clusters that could concentrate hydrogen production and consumption, attracting economic activity associated with the molecule.

I. Urbasos (✉) · G. Escribano
Energy and Climate Program, Elcano Royal Institute, Madrid, Spain
e-mail: iurbasos@rielcano.org

G. Escribano
Department of Applied Economics, Universidad Nacional de Educación a Distancia-UNED, Madrid, Spain

© Helmholtz-Zentrum Potsdam, Deutsches GeoForschungsZentrum GFZ 2024   131
R. Quitzow and Y. Zabanova (eds.), *The Geopolitics of Hydrogen*, Studies in Energy,
Resource and Environmental Economics, https://doi.org/10.1007/978-3-031-59515-8_7

For this reason, the external dimension of Spain's hydrogen strategy was relatively modest in its initial design.

The Russian invasion of Ukraine and the European Commission's call in the REPowerEU to increase the level of ambition for renewable hydrogen led to a change in the Spanish hydrogen policy, reinforcing its external dimension with a new focus on exports and infrastructure development. This strategic shift exposes a trade-off between hydrogen production for domestic consumption as a driver of green reindustrialisation, versus an export model that enhances the Iberian Peninsula's energy integration with Europe.

Both the Spanish Roadmap and the private sector have shown a clear preference for renewable hydrogen over other low-carbon alternatives, which has endowed its domestic development with a certain strategic continuity compared to other European players with more ambiguous technological preferences.

Spain is concerned that the obstacles to Iberian gas and electricity interconnections with European markets historically posed by France will be replicated for hydrogen, even more so given the path of self-sufficiency based on nuclear-derived hydrogen presented by the Elysée. Until now, Spain has focused its hydrogen diplomacy on key European allies (Portugal, France, Italy, Germany and the Netherlands). In the long run, hydrogen is a vector for integration and energy cooperation in the Euro-Mediterranean energy space, as well as for business and investment cooperation in Latin America.

The main documents that define the Spanish hydrogen strategy are the Hydrogen Roadmap published in 2020, the National Integrated Energy and Climate Plan (PNIEC) 2023–2030, the Climate Change and Energy Transition Law approved in 2021, the 2050 Long-Term Decarbonisation Strategy, the Just Transition Strategy and the Energy Storage Strategy.

2   Domestic Development of Hydrogen

Spain has focused its strategy on renewable hydrogen, prioritising it over other forms of decarbonised hydrogen production in the 2020 Hydrogen Roadmap. With the exception of a few specific projects associated with blue hydrogen production in refineries, private sector plans are aligned with this preference. Its lower environmental credentials (Howarth & Jacobson, 2021), and the absence of oil and gas extraction activities in Spain has left blue hydrogen (with carbon capture) out of the equation, while the nuclear phase-out programme planned for 2027–2035 excludes its use for hydrogen production. According to the Spanish Hydrogen Strategy, initial development will focus on existing industrial uses around hydrogen clusters or valleys. It is expected that the sectors that already use hydrogen as a raw material, mainly refining and fertilisers, will be the first movers for its production and consumption, progressively adding new sectors such as methanol, synthetic fuels or steel. Currently, Spain consumes approximately 500,000 tons of fossil hydrogen

Table 1 Installed capacity of the electricity sector in Spain in 2020 and the interim 2023–2030 PNIEC targets in MW (PNIEC de España, 2023)

	2020	2030 PNIEC target
Solar PV	11.004	76.387
Solar CSP	2.300	4.800
Onshore and offshore wind	26.754	62.044
Hydropower	14.011	14.511
Other renewables	616	1.929
Coal	10.159	0
Natural Gas	32.058	30.392
Nuclear	7.399	3.181
Other fossil and waste	4.260	2.174
Hydrogen electrolysis	0	11.000

Note MW means megawatt of installed capacity. Source based on the data from PNIEC de España, 2023

per year, almost exclusively for industrial uses (70% in refineries and 25% in chemical industries) with captive production of hydrogen in situ and limited trade of the molecule.

In the 2023 revision of the PNIEC, Spain has significantly elevated its renewable energy objectives. Specifically, the country has revised its electrolyser capacity target for 2030, raising it from 4 to 11 GW (Table 1). However, no hydrogen production targets are set, and no estimates of electrolyser load factors are presented. The PNIEC does not envisage hydrogen blending in the gas network, but rather proposes an initial production of hydrogen close to consumption centers and a subsequent development of a backbone network dedicated exclusively to hydrogen and developed by the national gas TSO Enagás. The new PNIEC also increases the renewable hydrogen quota in conventional industrial uses from 25 to 74% by 2030.

In Spain, the development of relatively cheap green hydrogen has been seen as a unique opportunity to attract investment in the decarbonisation of the metallurgical, chemical and petrochemical sectors, where hydrogen feedstock represents an important part of the final costs. Renewable hydrogen is also seen as a backup tool for the Iberian electricity system in the absence of new interconnections, allowing better management of surpluses and possible grid bottlenecks. As the penetration of renewables progresses, generation surpluses are expected during autumn and spring as opposed to deficits on many winter and summer days. Although still at an early stage, renewable hydrogen is identified as one of the key technologies for seasonal storage in decarbonised systems (Guerra et al., 2020).

Finally, hydrogen is also an element of the Just Transition Strategy (MITECO, 2022) and is expected to serve as a lever for industrial development to provide territorial cohesion. Renewable hydrogen has been added to the Just Transition Agreements and is playing a leading role in the auctions for grid access points available after the closure of coal-fired power plants (and in the future, nuclear energy), integrating elements of local employment, reindustrialisation and territorial cohesion (Cuesta et al., 2022). There is growing optimism about the role of hydrogen in the reindustrialisation of regions in economic and demographic decline after several decades of loss of competitiveness due in part to high energy costs.

For these reasons, hydrogen has enjoyed a strong drive from the public sector, in particular with the PERTE (Strategic Projects for Economic Recovery and Transformation), funded by the NextGenerationEU Funds (Lázaro et al, 2022). The total budget allocated to the hydrogen PERTE in the Recovery Plan was initially endowed with 1.555 million euros, further increased to 3.000 million euros with the addendum made after the publication of the REpowerEU. Other support lines funded by the NextGenerationEU Funds include the allocation of 450 million euros for the transformation of ArcelorMittal's Avilés steel mill and the renewable energy and storage PERTE support initiatives with indirect connections to hydrogen. In another public initiative, the public company Navantia, specialised in the construction of military vessels and submarines, has announced its intention to develop a renewable energy business line. This would include the manufacture of electrolysers with know-how from the assembly of state-of-the-art submarines in collaboration with Repsol.

2.1 A Decentralised Hydrogen Strategy

In Spain, hydrogen development is also attracting the interest of devolved administrations. One of the key elements of this regional development is the formation of trans-regional alliances and partnerships for the creation of hydrogen corridors. Such initiatives are essential to avoid regulatory fragmentation and encourage the development of a national hydrogen market. They can also boost regional development by focusing on areas of greatest competitive advantage. This is the case of the Basque Country, which seeks to strengthen and revitalise its local industry with a clear technological development component, while Castilla y León perceives hydrogen as a way to accelerate investment in renewables and their associated value chain. Furthermore, the involvement of the Autonomous Communities in the development of renewable hydrogen is key to guaranteeing the simplification of administrative processes.

This regional development seems to respond to the growing concern in Spain about the phenomenon of rural depopulation (Junquera, 2021). Recent analyses suggest that renewable energies, especially onshore wind, are not creating long-term jobs at the local level (Fabra et al., 2022) and renewable hydrogen is expected to enable greater wealth creation in the territory. However, such initiatives involving the harnessing (directly or indirectly) of renewable resources in these regions for its consumption elsewhere have already generated scepticism among regional political forces (Sánchez, 2021).

Understanding this phenomenon and knowing how to address it will be a key element in ensuring social support for green hydrogen development in Spain, particularly if there is a long-term export ambition. Hydrogen development will have to be carried out under strict environmental and social sustainability criteria. Large projects and infrastructures such as H2Med pipeline are already facing opposition from relevant civil society actors who doubt their environmental, economic and social benefits (Fundación Renovables, 2023).

Water availability can be a limiting factor for green hydrogen production in Spain. The irregular rainfall pattern of the Mediterranean climate means that water consumption for hydrogen projects can be in direct competition with agricultural consumption in certain regions of Spain during specific dry seasons, which are expected to intensify under a + 2 °C global warming scenario (Roudier et al., 2016). Many of the regions with the best solar resources are also producers of intensive irrigated agriculture, sometimes with unsustainable water consumption patterns. In coastal areas, water desalination may be a viable solution, given its low impact on hydrogen production costs: $0.02 hydrogen per kg (Khan et al., 2021). Inland, the scarcity of water resources may become a barrier to the installation of electrolysers in the absence of a nearby wastewater treatment plant (Simoes et al., 2021). The persistent drought in southern Spain during 2020–2023 has increased media attention on the water consumption of unregulated agriculture and its impact on protected natural areas.

2.2 Leading Sectors: Refining, Fertilisers, Steel and Synthetic Fuels

The Spanish refining sector has positioned itself as a world pioneer in the production and use of renewable hydrogen. By 2030, all operators in the sector (Repsol, BP and Cepsa) expect to have decarbonised all their consumption of hydrogen. Spain has one of the most modern refining fleets in Europe following counter-cyclical investments of €6.5 billion made by the main players in the sector between 2012 and 2015 at a critical time for the sector (MINCOTUR, 2018). Indeed, Spain is a net exporter of refined products, which in 2022 reached a combined value of €29 billion, equivalent to 7.5% of total exports (Comex, 2023). The adoption of green hydrogen would enable a significant reduction in scope 1 carbon emissions in the short term, a priority for the three companies involved in the refining activity in Spain, all with ambitious decarbonisation strategies.

The use of green hydrogen would also be favoured by high natural gas prices, the end of free EU ETS allowances in 2034 as the Carbon Border Adjustment Mechanism (CBAM) is introduced, the proven renewable development capabilities of Spanish refining operators and the excellent financial results between 2021 and 2023, with refining margins at record highs. This use of renewable hydrogen would initially focus on its traditional uses in refining (hydrogenation and hydrocracking) and subsequently on the production of synthetic fuels. Renewable hydrogen is therefore presented as a transition vector between the traditional activities of refineries and their subsequent reconversion to the production of decarbonised fuels.

In the fertiliser sector, Fertiberia, a national leader with a dominant market position in Spain and Portugal, is making rapid progress in the integration of green hydrogen for ammonia production with a project portfolio of 860 MW of electrolytic capacity. Together with Iberdrola, Fertiberia is operating one of Europe's first large-scale green-hydrogen to ammonia plant with a 20 MW electrolyser at its Puertollano plant and plans to build another in Huelva of 200 MW together with Cepsa by 2026. Again, the alliance between Fertiberia, Cepsa and Iberdrola comes at a key moment for the fertiliser sector, with the phasing out of free EU ETS emission allowances, the entry into force of the CBAM and raw material prices at record highs. In addition, Fertiberia divested its Algerian subsidiary (Fertial) under political pressure from Algiers in 2019, losing a key cost-competitive ammonium supplier. Consequently, investment in renewable hydrogen also presents itself as a hedging strategy against geopolitical risks. Fertiberia has announced plans to achieve carbon neutrality by 2035, hoping to become a leading company in low-carbon fertiliser production and compete in the EU market (Fertiberia, 2019).

Although there was no domestic methanol production in Spain, projects have emerged for the production of low-carbon methanol associated with renewable hydrogen involving chemical companies (traditionally importers) and energy companies. While obtaining a sustainable source of CO_2 is becoming a challenge for the development of these projects, there is still no discussion about a potential pipeline network for CO_2 transport.

In recent decades, the Spanish steel industry has undergone a profound restructuring to electric arc furnace (EAF) steel production, leaving only ArcelorMittal as an integrated blast furnace steel producer at its plant in Gijón (Asturias). This ArcelorMittal plant, with an output of around five million tonnes of primary steel, is the core of HyDeal, one of the world's largest integrated hydrogen and renewables projects. Although its initial ambitions have been substantially scaled down from 7.4 to 3.3 GW by 2031, this project aims to produce electricity in Castilla y León, taking advantage of renewable resources and land availability, to produce green hydrogen that would be used for direct reduction of iron (DRI). Doubts remain as to whether the resizing of the project is a sign of rationalisation or evidence of the difficulties in implementing it. In February 2023, the European Commission approved a €460 million grant for Spain's ArcelorMittal's HyDeal project, paving the way for the development of the project, which has not yet reached a final investment decision.

Spain holds the third position in the EU for high-temperature industrial consumption, utilizing 33 TWh of natural gas across sectors like ceramics, metallurgy, glass, and cement (Agora Energiewende, 2021). Notably, these industries are tackling technical hurdles by initiating programs like Orange. Bat, focused on incorporating low-carbon hydrogen in ceramics production. In a similar vein, Hydro, a Norwegian aluminium company, achieved in Spain a world-first by producing recycled aluminium using green hydrogen instead of natural gas in the extrusion process.

2.3 Technological Development, R+D+I and Professional Skills

Spain lacked the know-how associated with renewable hydrogen when the Hydrogen Roadmap was published, having to develop its entire industrial and innovation complex. This technological situation contrasts with the initiative shown by the private sector to start up green hydrogen projects, which has been forced to look abroad for electrolytic technology to develop the first pilot projects. The main projects announced use technology from Nel, John Cockerill, Siemens and Cummins. The preference for electrolysers from Western manufacturers is explained by greater reliability and ease of obtaining financing. However, Chinese-manufactured alkaline electrolysers have managed to significantly reduce their costs (Ansari et al., 2022), sparking growing interest among Spanish project developers (Urbasos, 2023). In the field of power electronics, a considerable part of the hardware investment costs for large-scale electrolysis plants, leading Spanish companies are successfully extending their product lines and services associated with renewable hydrogen.

In the future, Spain is expected to develop a significant local electrolyser assembly industry to supply the domestic market without necessarily relying on innovation by Spanish companies. This could replicate the situation in the automotive or onshore wind industry, where Spain is one of the European leaders in terms of industrial output without a major national champion. The first large project is the Cummins plant in Guadalajara, which will have the capacity to manufacture 500 MW of electrolysers per year by 2024, scalable in the future to 1 GW. The foreseeable development of green hydrogen in Morocco, Tunisia and Portugal would offer a potential market for the world's leading Spanish companies in technical project development and infrastructure construction.

Spain's hydrogen-related innovation strategy has also followed a process of increasing internationalisation. Spain joined the Mission Innovation 2.0 project in September 2022 and four Spanish projects were selected in the Commission's €5.4 billion Hydrogen IPCEI (Important Projects of Common European Interest) call to support the first deployment and innovation in the hydrogen technology value chain in July 2022. In the second edition of the IPCEI, held in September 2022 and focused on the development of industrial and infrastructure development projects, Spanish companies obtained 20% of the projects selected.

Spain is experiencing two bottlenecks associated with the lack of skilled human capital. On the one hand, many companies, in particular SMEs, are unable to benefit from the generous funds of the Recovery and Resilience Mechanism due to a lack of qualified personnel in the public and private sectors to carry out administrative procedures (Olcese, 2022). This lack of skilled personnel not only affects project development in the short term, but may also slow down project development in the future, when the demand for skilled labour will be higher. For this reason, the Recovery Fund hydrogen package included vocational training and re-skilling as one of its five strategic objectives.

3   External Dimension of Hydrogen in Spain

The external dimension of hydrogen in Spain has evolved since the publication of the Hydrogen Roadmap in 2020, moving from a focus on domestic industrial development to one of greater integration through exports in response to the Russian invasion of Ukraine. Spain took on the mandate set out in the REPowerEU to design hydrogen strategies not only in terms of decarbonisation but also in terms of energy security. So far, Spain has signed bilateral hydrogen cooperation agreements with Italy in 2020 and Portugal in 2021, highlighting the need to share information, explore private investment opportunities and integrate supply chains. With Germany there is a broad understanding on energy transition and climate ambition that extends to hydrogen. However, there are some elements of antagonism, in particular on hydrogen imports from outside the EU and potential conflicts over state aid and industrial policy. With France, the announcement of the construction of the H2Med pipeline in December 2022 marked the beginning of a new phase of cooperation, although the rapprochement between Madrid and Paris on energy integration appears fragile. For Spain, it is a priority to secure France's commitment to the construction of new infrastructures that integrate Spain into the EU energy market, especially electricity but also hydrogen. The construction of the H2Med pipeline, which will connect the Iberian Peninsula with Europe via France, will be crucial in determining whether Spain's hydrogen strategy is geared towards domestic consumption or also targets exports (Table 2).

Spain has remained on the sidelines in the development of bilateral and multilateral agreements with non-European players in the hydrogen sector. The absence of a cooperation agreement with Morocco is noteworthy, bearing in mind the importance that the North African country is giving to the external dimension of green hydrogen with the signing of agreements with Germany and Portugal. The Spanish private sector is actively participating in the development of the external dimension of hydrogen in Spain. In October 2022, Cepsa and the Port of Rotterdam announced the establishment of Europe's first green (ammonia-based) hydrogen shipping corridor from the Port of Algeciras. The private energy sector has also shown interest in investing in markets with great potential for hydrogen development, such as Chile and Brazil, constituting an incipient vector of business diplomacy and decarbonised

Table 2   Matrix of the external dimension of hydrogen in Spain based on domestic consumption levels and infrastructure development

| | | H_2 domestic consumption | |
		High	Low
International H_2 pipelines	Development	H_2 transit and industrial hub	H_2 pipeline exports
	Absence	H_2-based industrial development	Seaborne exports of H_2 derivatives

Source The authors

cooperation in Latin America. Spain's 2023 rotating Presidency of the EU Council served to position Spain as a key interlocutor with Latin America, with hydrogen enjoying a relevant position in the EU-Celac Summit negotiations.

3.1 The Role of Exports and the Iberian Peninsula as a Hub for Hydrogen Trade

Spain and Portugal aspire to contribute to the development of a European hydrogen market that would result in a hydrogen corridor between the Iberian Peninsula and Central Europe, potentially later extended to include North Africa. France's historical reluctance to help Spain's energy integration beyond the Pyrenees raises fears that the current lack of natural gas and electricity interconnections could be replicated: Spain has excess LNG regasification and electricity generation capacity (both gas and renewables) that it cannot export due to this lack of infrastructure (Escribano, 2021). In the absence of an integrated hydrogen pipeline network in Western Europe, Spain's exports would risk losing the competitive advantage generated by its abundant renewable resources, its geographical proximity to European import markets and its institutional and geopolitical stability (Núñez-Liménez and De Blasio, 2022). Hydrogen pipelines are considered the most efficient option for transporting hydrogen in large quantities and over distances of less than 3.000 km, especially if there is fossil infrastructure that can be retrofitted (IEA, 2022).

Another key element of the external dimension of green hydrogen is the possibility of integrating the North African and European markets into the Mediterranean import corridor already identified in the REPowerEU package. Morocco's ambitions to become a green hydrogen exporter would be coupled with Algeria's pressing need to replace its hydrocarbon exports and the existence of gas pipelines. Both countries, with excellent geographical conditions for the development of green hydrogen (and blue in the case of Algeria) would offer the possibility of establishing a trans-Mediterranean corridor through Spain. In addition, Spain and Morocco are experiencing growing energy integration in the areas of electricity, natural gas and trade in refined products that could be extended to the green hydrogen sector (Urbasos, 2023). Although the costs of adapting hydrogen pipelines would be significant, the Maghreb-Europe gas pipeline (GME) connecting Algeria and Spain via Morocco would have favourable characteristics for trans-Mediterranean hydrogen transport with an underwater route of only 45 km, which would considerably reduce the compression problems that these infrastructures have in their conversion (Amore-Domenech et al., 2023). The Algeria-Morocco section of the GME ceased operating in October 2021 in a context of growing rivalry between Algiers and Rabat, exposing a key barrier to future Mediterranean energy integration: the increasing geopolitical polarisation of North Africa.

Hydrogen exports from North Africa would face the same obstacles as those from the Iberian Peninsula to reach North European markets, unless alternative transit pipelines, such as the proposed H2Med, are built or transport by sea becomes competitive, in this case avoiding the peninsula as a transit hub (Escribano, 2021). Retrofitting LNG terminals could be a long-term solution if a hydrogen pipeline network is not fully developed in the EU. The main LNG plant operator in Spain, Enagás, is prioritising the development of hydrogen pipeline infrastructure in the framework of the European Hydrogen Backbone, leaving LNG plants operational for natural gas and biomethane, as well as for bunkering in the long run (Enagás, 2022). Given that all new LNG regasification plants in Germany must be hydrogen-ready, whether in the form of ammonia, biomethane or liquid hydrogen, these infrastructures could be decisive in the flow of renewable hydrogen from the Iberian Peninsula if their conversion is eventually techno-economically feasible.

In a scenario of isolation from the rest of Europe, Spain could alternatively focus on generating a low-carbon hydrogen ecosystem in the Western Mediterranean based on local industrial consumption. Industrial activities in Mediterranean countries would benefit from relatively lower hydrogen production costs than in the rest of the EU, which could be a driver for the competitiveness of associated industry. This industrial development could have a greater or lesser level of complexity and added value: from the simple production of ammonia, methanol or iron with DRI to the production of green steel or more advanced petrochemicals. The supply of green hydrogen from North Africa would allow for increased scale and industrial competitiveness. It would also enable a higher degree of economic integration in the Western Mediterranean at a time of rapid transformations, such as the entry into force of the CBAM and the Net Zero Industry Act, offering attractive transition prospects and greater interconnections with the Southern Neighbourhood.

3.2 Infrastructure: Hydrogen as a Driver for Greater Energy Integration

Intensifying interconnections with the rest of Europe has been a traditional priority of Spanish energy policy since its accession to the EU (Escribano et al., 2019). Given Spain's ambitions to be a key country in hydrogen transit, a strategic element is the development of infrastructures that enable the achievement of a green corridor from the Iberian Peninsula to Central Europe. So far, hydrogen infrastructures are limited due to the industrially concentrated and poorly distributed use of hydrogen in Spain. For this reason, interconnections must be developed simultaneously at the national and international level to connect renewable resources, hydrogen production and industrial centres.

Spain and Portugal have progressively integrated the electricity (MIBEL) and, to a lesser extent, the gas (MIBGAS) markets. This has provided a well-developed market structure for energy trading that would facilitate further coordination between the electricity and gas sectors in the future creation of an Iberian-wide hydrogen value chain. Portugal's Hydrogen Strategy aims to install 2 GW of electrolysis capacity by

2030 for export markets. Spain's role as a final destination for Portuguese hydrogen will depend on Iberian industrial consumption and its level of integration in the European hydrogen market (Table 2). The signing between Spain and Portugal in December 2022 of an agreement for the construction of an interconnection between the cities of Zelorico and Zamora is the first major bilateral project in this area.

The proposed IPCEI candidate pipeline, with a budget of 350 million euros, is slated to commence construction in 2025, contingent upon securing 50% of the project's financing from EU funds. The pipeline is designed to supply hydrogen produced in Portugal to the new green steel industrial cluster in Asturias (see Fig. 1), functioning as an element of Iberian integration and industrialisation, rather than as an export-oriented project.

In the case of Italy, Spain signed a cooperation agreement aimed at exchanging technology and taking advantage of Italy's vast diplomatic-business networks in the Mediterranean region and the African continent (Giulli, 2022). This agreement came in a context of high-level bilateral energy understanding with the creation of a virtual LNG pipeline between Barcelona and Livorno, and shared interests in pressing France on new energy interconnections. After the victory of Giorgia Meloni, and the change of government in Italy, bilateral energy initiatives, including those related to hydrogen, have been substantially reduced.

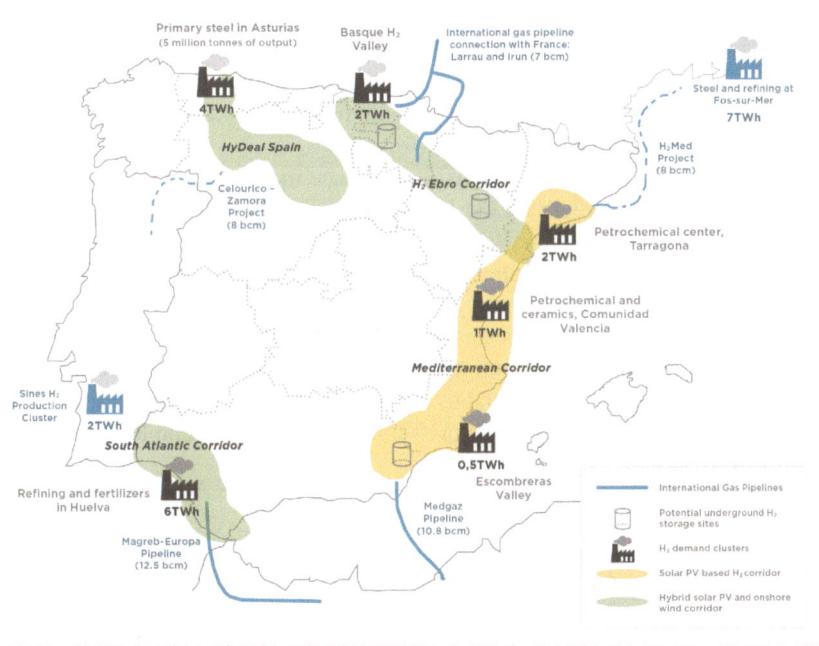

Fig. 1 Regional development of hydrogen in Spain and its international implications. *Source* the authors

France represents a somewhat paradoxical case. Although there are clear bilateral synergies in energy, and in particular hydrogen, the lack of collaboration on infrastructure development in recent decades has generated some suspicion south of the Pyrenees. The construction of the H2Med offshore pipeline would provide a greater incentive for cooperation between Portugal, Spain and France. H2Med entails major technical complexities, such as the need for a 140 MW compressor station or the irregularities of the seabed in the Gulf of Leon. Nevertheless, H2Med would link the Ebro-Mediterranean hydrogen corridors with a potential exchange capacity of two million tonnes, equivalent to 20% of REPowerEU domestic production target for 2030 (Urbasos, 2022). Enagás, the Spanish gas TSO, has already launched a first non-binding Call for Interest and a binding Open Season would be launched in 2025 to obtain access to the infrastructure.

However, France has been inconsistent in its position on the future hydrogen pipeline. First, French national preferences for an offshore route make the project substantially more expensive and technically more difficult. Secondly, France has shown a transactional attitude by pressuring Spain on H2Med in order to get Madrid's support in European negotiations on the inclusion of nuclear hydrogen as renewable or low-carbon. Finally, the French government has been ambiguous when it comes to defining certain technical details of the project, such as the location of the compressor stations. Enagás plan is to build a single station in Barcelona to provide a one-way Spain to France flow, while Paris is considering possible exports of nuclear hydrogen to Spain, advocating for a second station in Marseille. Moreover, France's credibility in the development of infrastructure has been damaged with the Spanish-French electricity interconnection in the Bay of Biscay accumulating delays due to cost overruns caused by inflation in materials and the design of the route (also offshore under French imposition).

While electricity interconnections with France are progressing slowly and are unlikely to meet the targets set out in the Spanish PNIEC to reach 8.000 MW of interconnection by 2030, renewable hydrogen could be a complementary option to connect the Iberian Peninsula renewable resources with the rest of continental Europe. This dual electricity and hydrogen connection would increase the robustness and resilience of the European and Iberian decarbonised energy systems, while providing Spain with the opportunity to capitalise on its renewable resources and land availability. However, it is advisable to rethink the compatibility of H2Med with the reindustrialisation and territorial structuring narrative employed in the 2020 Hydrogen Roadmap.

3.3   Regulatory Diplomacy

Spain aims to contribute to the development of a common European hydrogen market with a preference for hydrogen produced with renewable electricity in the EU. The

Spanish government's regulatory efforts are aimed at ensuring that strict sustainability standards are in place to guarantee low emissions from hydrogen domestically produced and imported. The Spanish government has been a strong supporter of separating nuclear hydrogen, labelled as low-carbon, from renewable hydrogen. Despite pressure from Paris to change its position, Spain has made clear its preference for renewable hydrogen at both national and European level over other low-carbon alternatives.

Regarding the Hydrogen Delegated Acts (DA) of February 2023, the Spanish government welcomed its main outcomes. Spain supports a transitional period in which time correlation and additionality are extended to a monthly period instead of an immediate application of the stricter hourly correlation. This flexibility is expected to allow the development of the first projects at a competitive cost, enabling their integration into industrial processes, fertilisers and refineries, which need a continuous supply of hydrogen. Although the first reactions from the private sector pointed out that the DAs excessively favoured projects associated with renewable sources with high-capacity factors (hydropower and offshore wind), the Spanish government expressed its satisfaction with the final result.

In May 2022, Spain launched a system of Guarantees of Origin for renewable gases, including biogas, biomethane and green hydrogen. This is expected to help increase their production and allow Spain to reach other EU markets. As a transitional measure, and due to the lack of human and technical resources in the public administration, the Spanish gas TSO, Enagás, will oversee the implementation and monitoring of this mechanism, signalling the administrative and technical challenges that green hydrogen could generate. In addition, Spanish private companies have advocated the creation of premium low-carbon products and the implementation of Europe-wide guarantees of origin for industrial products related to green hydrogen. These mechanisms would enable the implementation of low-carbon fertiliser, steel, aluminium and ceramics projects being studied in Spain. Spain defended the creation of the European Hydrogen Bank, seeing it as an opportunity to deploy a common European policy of subsidies for hydrogen production. However, it has favoured a strategy of preferentially subsidising European production. The success of Spain and Portugal in the first auction of the European Hydrogen Bank has demonstrated the relative competitiveness of Iberian projects in the European market. More than 80% of the €720 million tendered in the auction were awarded to three Spanish and two Portuguese projects out of a total of seven.

3.4 Allies and Competitors in the Hydrogen Economy

As Eicke and De Blasio (2022) argue, new tensions may arise between industrialised countries with hydrogen import needs and potential producers with a renewable resource and skilled labour force. Potential exporters, such as Spain, would be interested in attracting green hydrogen-intensive industrial activities rather than exporting hydrogen directly. Locating industrial facilities close to renewable hydrogen production would create a higher level of added value in the territory, fostering hydrogen-related industrial policies that leverage this competitive advantage. These competing interests may lead to conflicts; EU importers would protect domestic markets through state aid and subsidised hydrogen imports, as in the case of H2Global or Germany's proposed carbon contract for difference support scheme, while hydrogen producers would try to attract these hydrogen-related industries.

This could lead to tensions within the EU at a time when industrial policies, especially those associated with the key technologies of the energy transition, have regained their geostrategic value (Maihold, 2022). The relaxation of state aid allocation criteria in the EU in response to the US Inflation Reduction Act (IRA) could lead to a subsidy race that would benefit countries with the deepest fiscal pockets (Meester, 2022). In parallel, renewables appear to be increasingly integrated into narratives of natural resource nationalism traditionally limited to fossil commodities, such as hydropower exports from Scandinavian countries (Hansen & Moe, 2022), with possible ramifications for other low-carbon energy exporters. This process of diverging interests between EU hydrogen producers and importers must be taken into account when analysing the future of Spain's hydrogen strategy and its potential competitors and partners.

4   Conclusions

Renewable hydrogen represents a dual opportunity for Spain. At the domestic level, it is presented as a vector for decarbonised reindustrialisation, while at the international level, the construction of new interconnections has the potential to reconfigure the peripheral and relatively isolated position of the Iberian Peninsula within the European energy space.

In addition to contributing to the Energy Union with a diversified supply of LNG and electricity, Spain can provide a significant amount of renewable hydrogen obtained in an economically, socially, politically and environmentally sustainable manner. To this end, Spain's regulatory efforts in Brussels have been directed towards establishing a regulatory framework that values renewable hydrogen by applying strict sustainability criteria against other low-carbon options with lesser environmental, energy security and socio-economic credentials.

The H2Med offshore pipeline between Barcelona and Marseille is a strategic project to increase energy interconnections between France and the Iberian Peninsula. Its successful development requires the collaboration of other European partners such as Portugal, Germany or the Netherlands to extend its route by incorporating producers and offtakers to build a true European hydrogen market in which Spain can play a leading role. In the long term, this market could subsequently incorporate future producers of hydrogen and its derivatives from North Africa, serving as a renovating element of the Euro-Mediterranean energy space. Hydrogen also presents itself as another lever of business diplomacy for the Spanish private sector, with particular potential as a key investor in Latin America.

At the domestic level, the Spanish strategy has focused on the development of industrial clusters and valleys based on a narrative of industrialisation, just transition and territorial cohesion that has generated great expectations. This strategy involves dedicating a substantial part of national and imported hydrogen production to consumption in these new industries. Therefore, the domestic dimension, designed in the 2020 Hydrogen Roadmap, and the external dimension, updated after the Russian invasion of Ukraine in 2022, must be reconciled, assuming that renewable hydrogen will be a limited resource, at least in the medium term. Spain has yet to define the development of renewable hydrogen between two models: an interconnected one with a preference for exports versus another with a leaning towards domestic industrial development. It is urgent to build a national strategy that aligns the domestic and international dimensions of hydrogen development in Spain, sending a coherent message to civil society, the private sector and institutions.

Acknowledgements Research for this chapter was financially supported by the German Federal Foreign Office within the framework of the project *"Geopolitics of the Energy Transformation— Implications of an International Hydrogen Economy"* (GET Hydrogen), funding reference number AA4521G125.

References

Agora Energiewende. (2021). *No-regret hydrogen: charting early steps for H_2 infrastructure in Europe.* https://static.agora-energiewende.de/fileadmin/Projekte/2021/2021_02_EU_H 2Grid/A-EW_203_No-regret-hydrogen_WEB.pdf

Amore-Domenech, R., Meca, V. L., Pollet, B. G., & Leo, T. J. (2023). On the bulk transport of green hydrogen at sea: Comparison between submarine pipeline and compressed and liquefied transport by ship. *Energy, 267,* 126621.

Ansari, D., Grinschgl, J., & Pepe, J. M. (2022). *Electrolysers for the hydrogen revolution: Challenges, dependencies, and solutions.* Comment, 57, SWP. https://www.swp-berlin.org/publicati ons/products/comments/2022C57_Electrolysers_HydrogenRevolution.pdf

Comex. (2023). *Informe mensual de Comercio Exterior. Diciembre 2022.* Informe mensual, Ministerio de Industria Comercio y Turismo, December release. https://comercio.gob.es/Imp ortacionExportacion/Informes_Estadisticas/Historico_Informes/Mensuales/2022/2022-12_Inf orme_Mensual_Comercio_Exterior.pdf

Cuesta, H. Á. (2022). Transición energética y políticas de empleo verde. *LABOS Revista De Derecho Del Trabajo y Protección, 3,* 154–179.

Enagás. (2022). *Financial results 2021.* https://www.enagas.es/content/dam/enagas/es/ficheros/ sala-de-comunicacion/publicaciones/informe-anual/historico/INFORME%20ANUAL%202 021_ENAGAS.pdf

Escribano, G. (2021). *H2 Med: hydrogen's geo-economic and geopolitical drivers and barriers in the Mediterranean. Policy Paper,* 4/2021, Elcano Royal Institute. https://media.realinstitutoel cano.org/wp-content/uploads/2021/09/policy-paper-escribano-h2-med-impulsores-barreras-geopoliticas-geoeconomicas-para-hidrogeno-en-mediterrane.pdf

Escribano, G., Lázaro, L., & Lledó, E. (2019). *La influencia de España en el ecosistema europeo de energía y clima. ARI,* 87/2019, Real Instituto Elcano. https://media.realinstitutoelcano.org/wp-content/uploads/2021/11/ari87-2019-escribano-lazaro-lledo-influencia-espana-ecosistema-eur opeo-energia-clima.pdf

Eicke, L., & De Blasio, N. (2022). *The future of green hydrogen value chains: Geopolitical and market implications in the industrial sector*. Belfer Center for Science and International Affairs, Harvard Kennedy School. https://www.belfercenter.org/sites/default/files/files/public ation/Paper_MappingHydrogen_Final.pdf

Fabra, N., Gutiérrez, E., Lacuesta, A., & Ramos, R. (2022). Do renewables create local jobs? *Documentos De Trabajo Del Banco De España, 7*, 1.

Fertiberia Group. (2019). *Annual Report.*

Fundación Renovables. (2023). *Dismantling hydrogen H2med excuse for a false energy transition*. Documento de análisis. https://fundacionrenovables.org/wp-content/uploads/2023/03/202 30309-H2Med-hydrogen-and-the-energy-policy_EN-FINAL-DEF.pdf

Giuli, M. (2022). *The geopolitics of clean hydrogen—Opportunities and challenges for Italy*. IAI Paper, 22/27, IAI. https://www.iai.it/sites/default/files/iaip2217.pdf

Guerra, O. J., Zhang, J., Eichman, P., Denholm, P., Kurtz, J., & Hodge, B. M. (2020). The value of seasonal energy storage technologies for the integration of wind and solar power. *Energy & Environmental Science, 13*(7), 1909–1922.

Hansen, S. T., & Moe, E. (2022). Renewable energy expansion or the preservation of national energy sovereignty? Norwegian renewable energy policy meets resource nationalism. *Political Geography, 99*, 102760.

Howarth, R. W., & Jacobson, M. Z. (2021). How green is blue hydrogen?. *Energy Science & Engineering, 9*(10), 1676–1687.

IEA. (2022). *Global Hydrogen Review 2022*. OECD Publishing. https://iea.blob.core.windows.net/assets/c5bc75b1-9e4d-460d-9056-6e8e626a11c4/GlobalHydrogenReview2022.pdf

Junquera, N., Llanera, K., & Andrino, B. (2021, November 25). How depopulation of rural areas is fueling political protest against 'emptied Spain'. *El País*. https://english.elpais.com/spain/2021-11-24/how-depopulation-of-rural-areas-is-fueling-political-protest-against-emptied-spain.html

Khan, M. A., Al-Attas, T. A., Roy, S., Rahman, M. M., Ghaffour, N., Thangadurai, V., Kibria, M. G., et al. (2021). Seawater electrolysis for hydrogen production: A solution looking for a problem? *Energy, Environment and Science, 9*, 1–16.

Lambert, M., & Schulte, S. (2021). *Contrasting European hydrogen pathways: An analysis of differing approaches in key markets*. OIES Paper, 166, NG. https://www.oxfordenergy.org/wpcms/wp-content/uploads/2021/03/Contrasting-European-hydrogen-pathways-An-analysis-of-differing-approaches-in-key-markets-NG166.pdf

Lázaro, L., Averchenkova, A., & Escribano, G. (2022). *High-impact green recovery in the EU's 'big five' (emitters): Key elements and caveats*. Policy Paper, 5/2022, Elcano Royal Institute. https://www.realinstitutoelcano.org/en/policy-paper/high-impact-green-recovery-in-the-eus-big-five-emitters-key-elements-and-caveats/

Maihold, G. (2022). *A new geopolitics of supply chains: The rise of friend-shoring*. SWP Comment, 45. https://www.swp-berlin.org/publications/products/comments/2022C45_Geopoli tics_Supply_Chains.pdf

Meester, W. (2022). *Subsidies, competition, and trade. Directorate for Financial and Enterprise Affairs Competition Committee at OECD*. https://www.oecd.org/daf/competition/subsidies-com petition-and-trade-2022.pdf

MITECO. (2022). *Spain, towards a just energy transition*. Just Transition Institute. Ministerio para la Transición Ecológica y el Reto Demográfico https://www.transicionjusta.gob.es/Documents/Noticias/common/220707_Spain_JustTransition.pdf

Núñez-Jiménez, A., & De Blasio, N. (2022). *The future of renewable hydrogen in the European Union*. Report March 2022, Belfer Center. https://www.belfercenter.org/sites/default/files/filcs/publication/Report_EU%20Hydrogen_FINAL.pdf

Olcese, A. (2022, September 29). *La falta de funcionarios cualificados atasca la ejecución de los fondos europeos*. El Mundo. https://www.elmundo.es/economia/macroeconomia/2022/08/29/630c8eeffc6c8351768b4590.html

PNIEC de España. (2023). *PNIEC, Madrid*. Ministerio para la Transición Ecológica y el Reto Demográfico https://www.miteco.gob.es/content/dam/miteco/images/es/pnieccompleto_t cm30-508410.pdf

PNIEC de España. (2020). *PNIEC, Madrid*. Ministerio para la Transición Ecológica y el Reto Demográfico. https://www.miteco.gob.es/content/dam/miteco/images/es/pnieccompleto_ tcm30-508410.pdf

Roudier, P., Andersson, J., Donnelly, C., Feyen, L., Gruell, W., & Ludwig, F. (2016). Projections of future floods and hydrological droughts in Europe under a +2°C global warming. *Climatic Change, 135*(2), 341–355.

Sánchez Herrero, G. (2021, October 14). *Del 'not in my backyard' al 'sí, pero así no'*. Agenda Pública. https://agendapublica.elpais.com/noticia/17720/not-my-backyard-al-si-pero-asi-no

Simoes, S. G., Catarino, J., Picado, A., Lopes, T. F., Di Berardino, S., Amorim, F., de Leao, T. P., et al. (2021). Water availability and water usage solutions for electrolysis in hydrogen production. *Journal of Cleaner Production, 315*, 128124.

Urbasos, I. (2022, October 21). *España aprovecha la crisis energética para redoblar su apuesta europea por el hidrógeno verde*. Blog Elcano, Real Instituto Elcano. https://www.realinstitut oelcano.org/comentarios/espana-aprovecha-la-crisis-energetica-para-redoblar-su-apuesta-eur opea-por-el-hidrogeno-verde/

Urbasos, I. (2023, March 29). *La geopolítica de la tecnología del hidrógeno: un nuevo foco de rivalidad UE-China*. Blog Elcano, Real Instituto Elcano. https://www.realinstitutoelcano.org/ blog/la-geopolitica-de-la-tecnologia-del-hidrogeno-un-nuevo-foco-de-rivalidad-ue-china/

Italian Hydrogen Policy: Drivers, Constraints and Recent Developments

Andrea Prontera

Abstract Italy presents, potentially, some important comparative advantages in the emerging clean hydrogen economy. However, unlike other large EU countries, Italy has not yet issued a comprehensive strategy on hydrogen nor has it developed a coherent hydrogen diplomacy. It was only with the National Recovery and Resilience Plan, launched in the wake of the COVID-19 crisis, that Italy upgraded its measures for promoting green hydrogen and related activities. The main players in the Italian hydrogen landscape are national industrial actors, especially state-owned energy companies. This leading role can be an important asset to overcome the problems related to an industrial system otherwise composed of small and medium companies, which can suffer competition from hydrogen frontrunners. Externally, Italy has supported all the European and multilateral initiatives on hydrogen. The main focus of the Italian international approach, however, is linked to the hydrogen hub concept, which targets the MENA region and is also supported by its national energy companies. This concept has received a further push after the beginning of the war in Ukraine. Yet, its practical realization is very problematic because of domestic and external challenges.

1   Hydrogen in Italy: In Search of a National Strategy

Unlike other large European countries, as of 2023, Italy was yet to issue a comprehensive strategy on hydrogen. At the end of 2020, the government published the first policy document—the 'Preliminary Guidelines for a National Hydrogen Strategy'—launching a consultation with stakeholders. However, the actual national hydrogen strategy, initially expected at the beginning of 2021, was never finalised. Despite this delay—which is not surprising for a country like Italy not used to adopting a forward-looking and comprehensive approach to energy and industrial policy issues—the development of hydrogen has become a topic more and more discussed in policy and business circles (e.g. Confindustria, 2020; European House-Ambrosetti, 2020,

A. Prontera (✉)
University of Macerata, Macerata, Italy
e-mail: andrea.prontera@unimc.it

© Helmholtz-Zentrum Potsdam, Deutsches GeoForschungsZentrum GFZ 2024
R. Quitzow and Y. Zabanova (eds.), *The Geopolitics of Hydrogen*, Studies in Energy,
Resource and Environmental Economics, https://doi.org/10.1007/978-3-031-59515-8_8

2022). This debate has also focused on the external dimension of the Italian hydrogen policy, as the country might play the role of a 'hub' in the emerging international hydrogen trade by connecting North Africa with the EU market. Moreover, important national industrial actors, particularly state-owned energy companies, have recently increased their involvement in several initiatives, projects and alliances on hydrogen at home and abroad. This trend has further accelerated in the wake of the Russian invasion of Ukraine and the growing role that hydrogen has gained in the EU policy response to the energy and climate crises under the REPowerEU plan.

1.1 Overview and Comparative Advantages

Italy presents, potentially, some important comparative advantages in the emerging clean hydrogen economy (European House-Ambrosetti, 2020; Franza, 2021; FCH, 2020; Giuli, 2022; SNAM, 2019). On the supply side, Italy could become the first EU market where green hydrogen would be the cheapest source of hydrogen. Italy is the third EU country for renewable power capacity (excluding hydro) installed after Germany and Spain (IRENA, 2022). According to a study of SNAM (2019)—the national gas Transmission System Operator (TSO)—given the country's access to low-cost renewable generation (particularly because of higher solar irradiation), green hydrogen will outcompete grey hydrogen by 2030, 5–10 years earlier than Germany. A study by the FCH Joint Undertaking (2020) estimates that covering Italy's 2030 clean hydrogen demand, ranging between 113,100 and 571,800 tH_2, could require between 4.1 and 21 GW of dedicated renewable capacity (as of 2022 Italian renewable capacity totalled around 60 GW). Italy has the potential to develop a similar capacity. However, the slow pace in new renewable energy installations (around 1 GW per year) since the mid-2010s—when in the wake of the economic crisis the government dismantled several supporting measures (Prontera, 2021)—is a factor of major concern for Italy's hydrogen ambitions (GSE, 2022).

On the demand side, Italy's use of hydrogen accounts for about 500,000 tH_2/yr (MITE, 2022). This amount is manly produced by using steam methane reformers (grey hydrogen) and is almost entirely used by the chemical and refining industry. The state-owned oil and gas company ENI is the major producer and consumer of hydrogen in Italy. According to the FCH-JU (2020) projection, however, the country has significant potential for expanding hydrogen demand in several sectors. First of all, hydrogen could play an important role in the decarbonisation of industrial energy demand, where natural gas accounts for about 35%. Moreover, there are several steel plants in Italy (5% of the primary steel that is produced in Europe) that could switch to hydrogen. Overall, FCH-JU sees Italian industry clean hydrogen demand oscillating between 1.13 TWh (low estimate) and 6.41 TWh (high estimate) by 2030 (FCH-JU, 2020). Further opportunity for hydrogen demand is in the heating and cooling sector, where natural gas accounts for almost 50% of the final energy demand. In these sectors the FCH-JU estimates oscillated between 1.98 and 5.4 TWh by 2030. Finally, the larger opportunity for hydrogen in Italy comes from the transport sector:

particularly road transport (trucks, buses and light commercial vehicles), as railway transport is electrified to a large extent and shipping does not play a major role in the economy. In the transport sector the FCH-JU (2020) estimates oscillate between 0.75 and 7.5 TWh by 2030. Currently, however, in Italy there are only five hydrogen stations for road transport: two for public transport (buses) and three for private vehicles (two of them realised thanks to a collaboration between ENI and Toyota).

Opportunities for hydrogen development in Italy also come from the presence in the country of industrial actors working on core hydrogen technologies, such as electrolysers and fuel cells, and related equipment (European House-Ambrosetti, 2020; Giuli, 2022). Italy is the second EU producer of electrolysis-related core technologies (25.2% of the EU total) (European House-Ambrosetti, 2020). This industrial capacity, however, is not aimed at producing clean hydrogen. The Italian electrolysis-related industry should adapt to this end if intends to exploit export opportunities among partners willing to upscale clean hydrogen production. Fuel cells production is not particularly developed in Italy: €1 million in 2018 compared to 21.8 million in Germany (European House-Ambrosetti, 2020). The picture is more promising with respect to some supporting technologies and equipment: mechanical, thermal, electric, and control systems. Italy is the leading EU manufacturer of thermal equipment such as evaporators, condensers, burners, and boilers for blue hydrogen. In 2018, the Italian market share was almost 25% of total EU production. As for mechanical technologies (valves, pumps, compressors and pressure converters) Italy ranks second in Europe behind Germany, with a 19% market share (European House-Ambrosetti, 2020). However, it is worth noting that Italian companies in all these sectors are generally small size companies. This can represent a disadvantage with respect to the key competitors such as Germany, China, Japan and Korea (Giuli, 2022).

Finally, thanks to its geographical position and its large gas infrastructure network, Italy might become a hydrogen hub. Italy could use the existing gas pipelines from North Africa (Algeria, Tunisia and Libya), where hydrogen can be generated from low-cost renewables and then transported to Italy and possibly Northern Europe. According to a study of SNAM (2019) importing North African green hydrogen, in the long-term, would cost 10–15% less than producing hydrogen in Italy. It is worth stressing that the hydrogen hub concept is strongly supported by SNAM itself (see also below). SNAM and ENI manage the gas pipelines connecting Algeria (Transmed) and Libya (Greenstream) with Italy. SNAM's stated vision, particularly, is to 'transport entirely decarbonised gas (not only hydrogen, but also biomethane), helping to strengthen Italy's role as a European hub, with a view to exporting clean energy to Northern Europe' (SNAM, 2022). The 50% of the approximately €7.4 billion of SNAM's 2020–2024 business plan is for the replacement and development of pipelines that are also compatible with hydrogen transport.

Box 1. Italy as a Hydrogen Hub? Lessons Learned from the Failure of the Gas Hub Concept

In the mid-2000s the concept of transforming Italy into a gas 'hub' in the Mediterranean gained popularity in policy and business circles. However, this goal was never achieved and the country remained an end market for gas export with very high prices. Several factors contributed to this failure, particularly:

- Instability in the North African region
- Instability of the Italian policy and regulatory framework
- Local opposition to energy infrastructures and length of the procedures for their authorisation
- Flat domestic gas demand.

1.2   Towards a Hydrogen Policy: Drivers and Key Pillars

As explained, Italy has not yet issued a national hydrogen strategy. An embryonic policy vision on hydrogen was only drafted along with the country's 2019 National Energy and Climate Plan (NECP). Before this EU-driven document, hydrogen was basically absent from the policy debate. In the 2013 National Energy Strategy (SEN, 2013), the term hydrogen was mentioned only one time with reference to the EU R&D policy, i.e. the Strategic Energy Technology Plan. Similarly, in the 2017 National Energy Strategy (a policy document of 300 pages), hydrogen was mentioned only one time in a footnote, referring to the possibility to 'imagine' a role for this source in the country's energy mix (SEN, 2017). Conversely, in the 2019 NECP the term hydrogen is mentioned 55 times (NECP, 2019). In this document, more importantly, the government explicitly recognised the role of hydrogen in the Italian climate and energy policy. The NECP called for exploring power-to-gas technologies to achieve flexibility and security of supply through sector integration. It also suggested that the Italian gas network could become the centrepiece of a 'hybrid' electric-gas energy system. Different possibilities were considered in the NECP, including blending as a transitional step to the development of two parallel infrastructures (on for gas and on for hydrogen). However, the NECP did not provide specific financial commitments nor did it set ambitious targets for hydrogen, mentioning only a small target in transportation: 1% of the RES target for transport (amounting to about 21,132 tH_2).

In 2020, government's attention to hydrogen further increased. In November, the Italian Ministry for Economic Development issued the 'Preliminary Guidelines for the Italian Hydrogen Strategy' (MISE, 2020), launching a consultation with relevant stakeholders (a consultation 'table' was also set up at the Ministry for Economic Development). The goal of the government was to finalise the national strategy on hydrogen at the beginning of 2021. However, this never happened. On January 2021 the government in charge—the so-called Conte II government—was dismissed. The newly appointed government lead by Mario Draghi (February 2021–July 2022) did not issue a national strategy on hydrogen. However, a set of new measures on hydrogen were included in the National Recovery and Resilience Plan (NRRP), transmitted by the government to the European Commission on 30 April 2021 and approved on 13 July 2021.

In the 2020 'Preliminary Guidelines for the Italian Hydrogen Strategy' the government framed hydrogen mainly as a way of matching the country's increasingly ambitious decarbonisation objectives and as a vehicle for industrial and technological development (MISE, 2020). The intention of the government was to outline a roadmap with targets to establish a hydrogen economy in Italy. Hydrogen was expected to represent 2% of Italy's total final energy consumption by 2030 (about 700,000 tH_2), targeting a 20% share by 2050. The 2030 goal would be delivered by 5 GW of electrolyser capacity, which would however cover only slightly more than half the targeted amount. This suggests a role for blue hydrogen as well (Giuli, 2022), unless significant imports of green hydrogen would be commenced before 2030 (a scenario that currently is not very likely). The government foresaw that the new targets would require up to €10 billion of investments between 2020 and 2030, to which one should add investment for dedicated renewable capacity. Out of the total amount, €5–7 billion would be dedicated to production, €2–3 billion to distribution infrastructure, and €1 billion to R&D. It was expected that up to half of these investments could be supported by ad hoc national and EU resources and funds. The preliminary guidelines focused on stimulating hydrogen demand mainly in trains, heavy duty vehicles, petrochemicals and refining. Sectoral objectives, however, were only foreseen for long haul trucks: targeting 2% by 2030 and up to 80% by 2050. Blending hydrogen in the existing gas network was also considered as a viable strategy for stimulating the hydrogen market. According to the guidelines, with these hydrogen targets, Italy should benefit from a CO_2 reduction of 8 Mton by 2030, as well as promoting the creation of 200,000 temporary jobs and 10,000 permanent jobs. The preliminary guidelines also considered the establishment of 'hydrogen valleys', as ecosystem for hydrogen innovations and technological development. Finally, the preliminary guidelines highlighted the potential role of the country as a hub for the wider European market, thanks to imports of green and blue hydrogen from North Africa.

With the 2021 NRRP the Italian hydrogen policy has been upgraded and better defined. In the NRRP the term hydrogen is mentioned 85 times, underlining a trend of growing attention to the issue by the Italian governments (Fig. 1). Decarbonization and goals to obtain industrial leadership over hydrogen technologies (along all the supply chain, from production to transport, consumption and storage) have been the main drivers of this development. Moreover, with the NRRP—which is closely aligned with the 2020 EU strategy on hydrogen—the Italian government more explicitly focuses on green hydrogen, although blue hydrogen continues to be supported by important national industrial players (see below).

According to the NRRP, the Italian hydrogen policy is based on six main pillars: (i) development of flagship projects for supporting the use of hydrogen in hard-to-abate industrial sectors, starting with the steel industry; (ii) promotion of 'hydrogen valleys' for industrial and technological upgrading by using brownfield industrial sites; (iii) promotion of hydrogen use in the heavy transport sector—also supporting refuelling stations—and in the non-electrified railway transport (4763 km of the

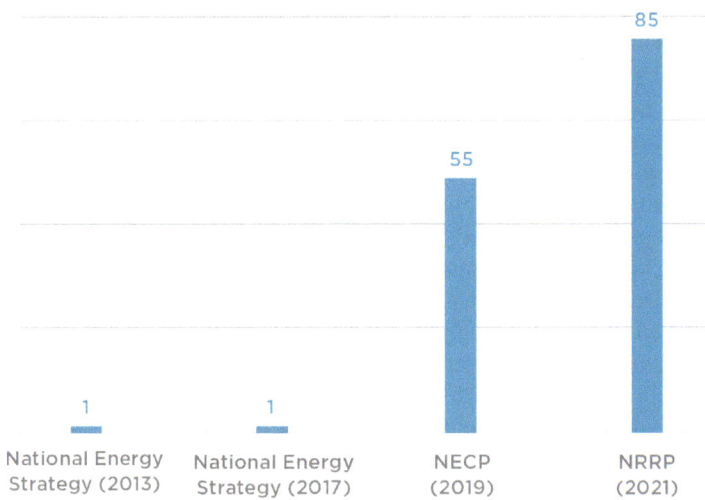

Fig. 1 Mention frequency of the term 'hydrogen' in recent Italian policy documents and plans. *Source* Author's own elaboration

Italian rail network are still served by diesel trains); (iv) creation of a gigafactory for electrolyzer production; (v) support for R&D activities, particularly in the area of green hydrogen production, transport and storage; (vi) improvement of the regulatory framework for enhancing the hydrogen economy. Overall, the Italian NRRP allocates €3.64 billion for direct investments into hydrogen to be disbursed by 2026 (Table 1).

The larger amount of resources is allocated to the decarbonization of hard-to-abate sectors, including the shift from grey to green hydrogen (€2 billion). Green hydrogen production in brownfield industrial sites also has an important role in the NRRP (€0.50 billion): the goal of the government is to establish 10 'hydrogen valleys'. As of 2023, eleven Italian regions opened a call to develop similar projects with the first €0.2 billion of NRRP financial support. For the Gigafactory, the NRRP goal is to achieve 1 GW of electrolyser production capacity by 2030, whereas for road

Table 1 Hydrogen investments in the Italian NRRP

Measure	Investments (billion €)
H_2 in hard-to-abate industrial sectors	2.00
H_2 production in brownfield industrial sites ('hydrogen valleys')	0.50
Gigafactory	0.45
H_2 for rail transport	0.30
H_2 for road transport	0.23
H_2 R&D	0.16
Total	3.64

Source PNRR (2021)

transport the target is 40 refuelling stations. In addition to these targeted measures, it is expected that the NRRP will provide indirect investments linked to the hydrogen economy in the amount of some €17 billion (MITE, 2022). However, the NRRP remains quite generic on several important aspects. For instance, the NRRP considers the opportunity to offer additional financial support for 'hydrogen production and transportation' (PNRR, 2021: 33). But it defers the details of these measures to the final publication of the national strategy on hydrogen. Such strategy, however, was not formulated even by the right-wing Meloni Government, which replaced the Draghi government in October 2022. Despite these limits, the NRRP has been instrumental in accelerating investments in hydrogen also by better linking national and EU-level strategies. For instance, in April 2023, the European Commission approved, under the State Aid Temporary Crisis and Transition Framework, the Italian scheme for hydrogen valleys (worth €450 million) (European Commission, 2023). This decision, which is in line with the RePower EU Plan and the Green Deal Investment Plan, will further help the realization the related NRRP objectives. Moreover, in its 'RePowerEU chapter'—issued on July 2023 and approved by the EU at the end of September 2023—intended to update the NRRP in the wake of the war in Ukraine, the Italian government has earmarked additional €90 million for the 'hydrogen valleys' and €140 million for R&D on hydrogen.

2 The Role of National Industrial Actors

The interest in hydrogen economy is quite recent in Italy. The country lags well behind the other large European states in terms of hydrogen projects (Fig. 2). Of the 32 Italian hydrogen projects, as of October 2022, only 5 were operational, 1 under construction, 3 in the demonstration phase and 23 in the 'concept' or 'feasibility study' phase.

Despite this gap, it is worth noting that many of these projects are promoted by important national industrial actors (i.e. state-owned energy companies), which are involved (alone or in partnership with other companies) in hydrogen development in Italy: SNAM (the Italian gas TSO); ENI (the former gas monopolist); and ENEL Green Power (ENEL is the former electricity monopolist; ENEL Green Power is the group's company that focuses on renewables). The activism of these big players in hydrogen development is in line with the Italian tradition of state-led capitalism (Schmidt, 2009), which sees national champions taking the lead in emerging industrial sectors. This development might help close the gaps of an industrial system otherwise characterised by medium and small size companies. By creating partnerships with other players, these national champions link hydrogen production with relevant offtakers (e.g. energy utilities, railway operators, refineries) and support technological upgrading (see below). However, while ENI and SNAM support both blue and green hydrogen (and carbon capture and storage, CCS, technologies), ENEL Green Power, which is a global player in the area of renewables, focuses on green hydrogen. If not properly managed, this divergence, in the long run, could undermine

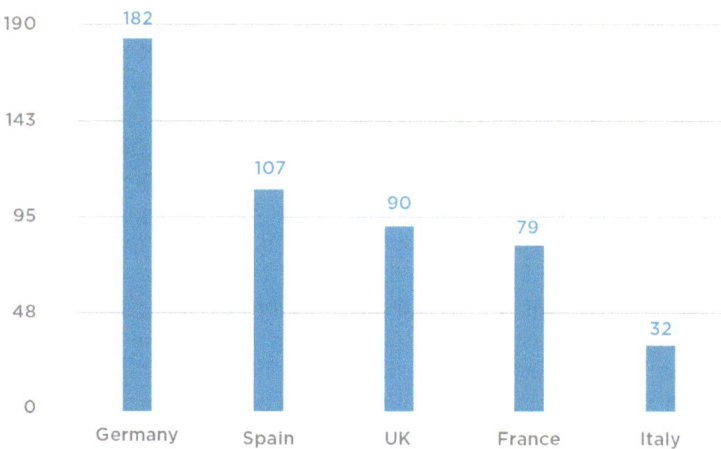

Fig. 2 Hydrogen projects in Italy and selected European countries. *Source* Author's elaboration from 'IEA Hydrogen Projects Database', available at: https://www.iea.org/data-and-statistics/data-product/hydrogen-projects-database. Accessed 10 February 2023 (data as of October 2022)

the coherence of the Italian hydrogen policy, as the energy choices of the country are widely influenced by these state-owned companies.

SNAM, along with ENI, is the main promoter of the vision of transforming Italy into a European hydrogen hub. Along with investments for adapting its gas infrastructure to this task, SNAM is also working on developing international standard for transporting hydrogen into steel pipelines, in cooperation with RINA, an Italian company leader in international certification. Moreover, SNAM has entered into partnerships with various operators for the development of the Italian hydrogen supply chain. SNAM is working with railway operators (FS Italiane and Ferrovie Nord), traction motor suppliers (Alstom) and energy utilities (ENI, A2A and Hera) to develop refuelling infrastructure for hydrogen mobility in Italy. It also started a collaboration with Wolftank Hydrogen—a company of the Austrian Wolftank-Adisa Group dedicated to hydrogen and renewable energy—to promote hydrogen mobility through the construction of refuelling stations for cars, buses and trucks. Finally, SNAM is involved in several experiments and trials for using hydrogen to decarbonise industry and power generation.

ENI is the main Italian hydrogen producer and consumer. The company's Strategic Plan for the 2022–2025 period sets a target of producing 4 million tonnes of hydrogen per year by 2050. ENI, however, supports 'a technologically neutral approach' that combines all the available low-carbon technologies. In practice, this means that ENI supports blue hydrogen and it is investing in CCS technologies and projects in Italy and abroad. In Italy, in 2022, ENI launched a cooperation with Edison (an electricity utility) and Ansaldo Energia (an Italian power engineering company) to test the use of hydrogen (both green and blue) to substitute natural gas in a power plant owned by Edison. Moreover, ENI is a member of the Hydrogen Joint Research Platform, an Italian R&D initiative launched in 2021 in cooperation with Edison, SNAM and the Polytechnic University of Milan. Finally, on September 2022, SAIPEM—an engineering company of the ENI group—entered into a partnership with Edison and Alboran Hydrogen for developing three green hydrogen projects linked to the Puglia Region 'hydrogen valley' initiative.

Unlike ENI, ENEL Green Power focuses on green hydrogen. In 2021, the company began cooperation with Saras (an Italian energy company with operations in petroleum refining, marketing, transportation and power generation) to develop a green hydrogen project in Sardinia. This project provides for the use of a 20 MW electrolyser powered by renewable energy produced on site to supply green hydrogen to be used as raw material in the Saras refinery at the industrial site of Sarroch, in the province of Cagliari. In 2022, then, ENEL Green Power launched, in cooperation with Sapio (an Italian engineering company), a green hydrogen project linked to a wind power plant located in Sicily (at Carlentini). Here, ENEL Green Power also launched the 'NextHy' initiative, a multi-stakeholder platform for promoting R&D on green hydrogen. Finally, ENEL also takes part in the development of the Italian hydrogen valleys. For the implementation of the one in Lombardy, it entered into a partnership with Trenord—a railway transport company—for suppling green hydrogen for hydrogen-powered trains.

3   The International Dimension of Italian Hydrogen Policy: The Hub Concept and Beyond

Italy has not (yet) developed a coherent hydrogen diplomacy. In 2021, the Italian Ministry of Foreign Affairs funded a research project led by the Rome-based think tank 'Istituto Affari Internazionali' on the role of Italy in the emerging geopolitics and geoconomics of hydrogen. However, no formal policy document has been issued by the government on the matter. The major part of the international dimension of the Italian hydrogen policy is carried on by its state-owned companies. This is not a surprise: this pattern reproduces the traditional national approach in the fossil fuel sector where the government 'delegated' the country's foreign energy policy to its oil and gas companies. As illustrated, the hydrogen hub concept—that can be considered the main strategic objective of the Italian international policy on hydrogen—is especially promoted by SNAM and ENI. This concept has gained further support in the wake of the war in Ukraine as the Italian government sees hydrogen import as a potential contribution (in the long term) to the country's energy security. This concept is also in line with the 'European Hydrogen Backbone' initiative. Launched in 2020 by the European gas TSOs, this initiative envisages a North-Africa Italy Corridor for importing green hydrogen to the EU (EHB, 2022). An idea that has also been included in the RePowerEU Plan (European Commission, 2022). After Russia's invasion of Ukraine, both the Draghi government and its successor, the Meloni government (in office since October 2022) have backed this project, known as 'SoutH2 Corridor' (see below). Along with the hydrogen hub strategic objective, which requires developing appropriate supply chains and infrastructure, Italy and its state-owned energy companies are also involved in several international initiatives on hydrogen both in the EU and outside the bloc. In this case as well, an important role is played by national companies, which take part in relevant European bodies.

3.1 Political Dialogue, Research and Innovation

Italy has traditionally supported all the EU policies on hydrogen. In December 2020, Italy, along with other 22 EU countries and Norway, signed the manifesto for the development of a 'European Hydrogen Technologies and Systems' value chain promoting the launch of important projects of common European interest (IPCEIs) in the hydrogen sector. In 2022, six Italian companies were selected for the IPCEI funding (totalling about €1 billion): Ansaldo, ENEL, De Nora, Fincantieri, Alstom, Iveco. Italy is also member of 'Mission Innovation', a global initiative of 24 countries and the European Commission. In this context, Italy joined the 'Renewable and Clean Hydrogen Innovation Challenge', a multinational research program, launched in June 2021, to accelerate the development of the hydrogen market. Moreover, Italy is involved in the EU R&D programmes (Horizon 2020, Horizon Europe) and supports EU public–private partnerships on hydrogen, such as the 'Fuel Cells and Hydrogen Joint Undertaking' and its successor 'Clean Hydrogen Partnership'.[1]

At multilateral level, Italy is a member of the major international organizations dealing, among other things, with hydrogen development, like the IEA and IRENA. Italy is also a member of global multilateral fora like the 'Clean Energy Ministerial-Hydrogen Initiative' and the 'International Partnership for Hydrogen and Fuel Cells in the Economy', launched by the US in 2003. As for private sector initiatives, SNAM is a member of the Hydrogen Council and, along with other members, it launched in 2020 the Green Hydrogen Catapult initiative, which aims at installing 25 GW of green hydrogen production capacity by 2026. Moreover, SNAM is among the TSOs promoter of the 'European Hydrogen Backbone' initiative. ENI is a member of the 'European Clean Hydrogen Alliance', which is a multi-stakeholder platform established in 2020 by the EU and gathering European industrial players. The company is also involved in the 'Hydrogen4EU' project, which aims to analyse the contribution of hydrogen to EU long-term decarbonization objectives. Finally, both ENI and SNAM, along with several other Italian companies and the 'Italian Hydrogen and Fuel Cell Association', are member of 'Hydrogen Europe'. 'Hydrogen Europe' is a pan-European industry association involved in shaping the EU regulatory and market framework for hydrogen. In July 2022, a representative of SNAM was nominated in the Boards of Directors of 'Hydrogen Europe'.

---

[1] The Italian National Agency for New Technologies, Energy and Sustainable Development (ENEA, Agenzia nazionale per le nuove tecnologie, l'energia e lo sviluppo economico sostenibile) is involved in several EU programmes focusing on hydrogen (e.g. H2PORTS, e-SHyIPS, CoMETHy).

3.2 Supply Chain Development and Overseas Financing

Italian state-owned companies' international engagement on hydrogen is very recent, starting around 2020–2021. However, it has further increased after the Russian invasion of Ukraine (Italy was the second buyer of Russian gas in the EU after Germany). Paralleling domestic developments, ENI and SNAM are involved in green and blue hydrogen initiatives, whereas ENEL focuses on green hydrogen. Moreover, ENI and SNAM are particularly active in the MENA region. These companies are exploiting their long-standing relations with oil and gas partners to establish new international hydrogen supply chains. This mirrors previous patterns when especially ENI exploited its relations in the oil sector to develop new partnerships in the natural gas sector. These new hydrogen-related initiatives are supported by the government as well. Particularly in North Africa, Algeria, Tunisia and Egypt are the main Italian partners in the perspective of realising the hydrogen hub concept. In the event of political stabilisation, Libya is (potentially) another country that could be involved in this vision.

In July 2021, ENI began cooperation with the Algerian national oil and gas company, Sonatrach, for exploring Algerian hydrogen potential. In December 2021, ENI and Sonatrach signed a Memorandum of Understanding for promoting cooperation in the area of renewables, hydrogen and CCS technologies. Later on, in May 2022, after the beginning of the war in Ukraine, ENI and Sonatrach signed (supported by the Italian and Algerian governments) a new Memorandum of Understanding for extending their joint gas activities in Algeria. However, on this occasion they also decided to improve their hydrogen-related cooperation; a move reiterated with another agreement signed in January 2023. A pilot project for developing green hydrogen has been also planned by the joint venture Sonatrach-ENI GSE in the Algerian desert (at Bir Rebaa North, where ENI has already built a solar power plant). In Egypt, in July 2021, ENI has signed Memorandums of Understanding with the Egyptian Electricity Holding Company and the Egyptian Natural Gas Holding Company for developing common projects on green hydrogen, blue hydrogen and CCS. In September 2021, ENI has signed a Memorandum of Understanding with Mubadala Petroleum, a UAE company, for cooperation in the area of hydrogen and CCS. SNAM as well signed, in March 2021, a Memorandum of Understanding with the Mubadala Investment company for developing common projects on hydrogen in the UAE and worldwide. Then, in March 2023, ENI has also signed a Memorandum of Understanding with the UAE's state-owned oil and gas company ADNOC for cooperation on similar activities.

After the outbreak of the war in Ukraine, the Italian government has increased energy cooperation with Tunisia (the Transmed gas pipeline connecting Algeria and Italy crosses Tunisia). In March 2022, the Italian Ministry for Foreign Affairs launched negotiations with Tunisia on a Memorandum of Understanding on developing green hydrogen projects. In May 2022, then, SNAM likewise began talks with the Tunisian government on green hydrogen cooperation. These efforts did not result in formal agreements, also because of the growing political instability in

Tunisia. However, in July 2023, the European Commission signed a 'Memorandum of Understanding on a strategic and global partnership between the European Union and Tunisia', which also includes support for the development of green hydrogen. In addition, in May 2023, the Italian, German and Austrian governments—and their respective TSOs—agreed to cooperate on the realization of the 'SoutH2 Corridor', a move that could facilitate the provision of financial and diplomatic backing by the EU under the 'Project of Common Interest' framework. According to the developers, this 3300 km pipeline system, that would comprise newly built sections and repurposed ones to transport hydrogen, could deliver from North Africa more than 40% of the REPowerEU import target of green hydrogen (South2Corridor, 2023).

At the same time, in May 2022, SNAM also signed a Memorandum of Understanding with the Spanish company ENAGAS for evaluating the possibility of building a direct pipeline linking Spain with Italy across the Mediterranean Sea from Barcelona to Livorno. This project should allow the transportation of natural gas from Spain to Italy, and possibly to the wider EU market, bypassing France. Indeed, despite the energy crisis aggravated by the war in Ukraine, the French government has continued to oppose the so-called Midcat project (a pipeline that could allow Spanish LNG capacity to reach the European continental market). Although in the short to medium term the Barcelona-Livorno pipeline would be used for natural gas, SNAM and ENAGAS discussed the possibility (in the long-term) to use it for green hydrogen. Finally, SNAM, through its French subsidiary Teréga, is involved in the so-called H2Med project, a pipeline with the capacity to transport up to 2 million tonnes per year of green hydrogen that should connect Portugal and Spain with France by 2030. This initiative received a boost in January 2023 when Germany joined project.

ENEL Green Power is involved in green hydrogen projects abroad as well. While ENI and SNAM engagements are also driven by energy security considerations, ENEL activities are driven mainly by industrial and commercial objectives and target countries outside the MENA region (see Fig. 3). In October 2020, ENEL signed an agreement with the Chilean electricity utility AME for developing a green hydrogen project at Cabo Negro (linked to a wind power park). In December 2020, then, ENEL signed a Memorandum of Understanding with NextChem for developing green hydrogen projects in the US. Finally, in February 2021, ENEL, through its subsidiary Endesa—the major Spanish electricity utility—planned investments for almost €3 billion for green hydrogen projects in Spain.

4   Conclusions

The development of hydrogen in Italy is still in its infancy. With the NRRP a more robust policy on hydrogen has emerged. However, an important gap in terms of projects exist with respect to the other major European countries. A similar gap exists also with regard to R&D expenditure, where Italy lags behind European frontrunners such as Germany, France and the UK (FCH-JU, 2020). In both cases, some

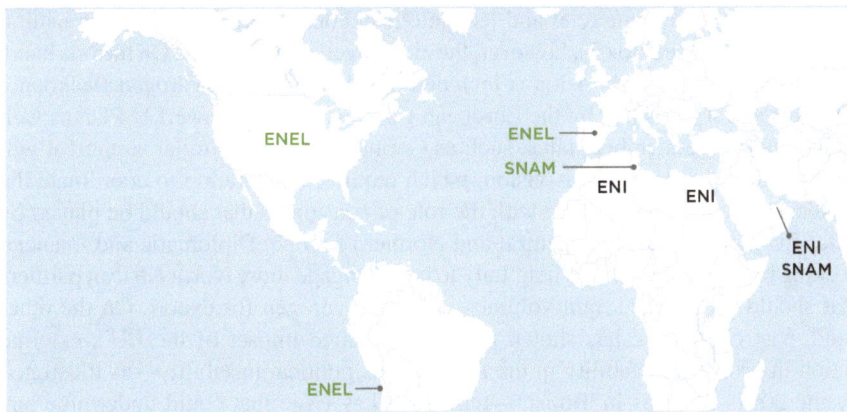

Fig. 3 Italian state-owned companies' hydrogen engagement abroad. *Source* Author's elaboration based on companies' websites. Note: green letters = green hydrogen projects; black letters = green and blue hydrogen projects

catching up is expected with the implementation of the NRRP. But it is still too soon to understand whether these measures can really accelerate hydrogen development in Italy. So far, the country has not been able to issue a comprehensive hydrogen strategy nor to launch detailed hydrogen diplomacy initiatives. The government's attention to the matter, however, has certainly increased in parallel with the EU expanded policies and targets on green hydrogen. In the drafting of the new National Energy and Climate Plan, expected for 2024, the right-wing government led by Giorgia Meloni set a goal of green hydrogen consumption for 0,25 Mton per year by 2030 (80% of which produced domestically) and of 3 GW of electrolyser capacity (NECP, 2023). In addition, as illustrated, important national industrial actors are increasing their hydrogen-related activities. Particularly, state-owned energy companies like ENI, SNAM and ENEL are taking the lead. This can be an important asset to overcome the problems related to an industrial system otherwise composed by small and medium companies, which can suffer competition from frontrunners. However, a certain divergence exists among ENI and SNAM and ENEL, with the latter focused on green hydrogen and the other two (also) on blue. Additional problems in the view of developing green hydrogen in the country are related to the difficulties in expanding renewables production, owing a fragmented and incoherent regulatory and policy framework and longstanding problems of permitting and local acceptance for wind and solar installations (e.g. Legambiente, 2023; Prontera & Lizzi, 2023).

Externally, Italy has supported all the European and multilateral initiatives on hydrogen. Italian energy companies are member of the major bodies involved in shaping hydrogen development. The main focus of the Italian international approach, however, is linked to the hydrogen hub concept. This concept has received a further push after the Russian invasion. This vision is supported especially by SNAM and ENI, which are also the main vehicle for the Italian external action in the MENA region. Conversely, ENEL's engagements abroad, focused on green hydrogen, are driven mainly by commercial and technological considerations. As for the natural gas hub concept (see Box 1), however, the challenges are numerous. On the one hand, the Italian hydrogen hub vision is included in the 'European Hydrogen Backbone' initiative and it is backed by the European Commission's RePowerEU Plan as well as by other large member states, such as Germany. Without similar support it will be difficult to implement this vision, which requires cooperation to coordinate the (potential) Italian transit's role with the role of consumers that should be played by other EU member states in Central and Northern Europe. Diplomatic and financial backing by the EU could also help Italy to better engage those North African partners that should produce relevant volumes of green hydrogen for export. On the other hand, past experience has shown the (very) limited impact of the EU's external action in promoting stability in the region. This political instability—as illustrated by the 2021–22 crisis in Tunisia—remains a key issue that could undermine any serious effort to transform North African countries into green hydrogen producers and exporters. The risk is that the investments planned for the upgrading of the South-North gas pipeline network will be more instrumental in expanding gas imports—in order to reduce dependency on Russia—rather than in opening green hydrogen corridors.

Acknowledgements Research for this chapter was financially supported by the German Federal Foreign Office within the framework of the project "Geopolitics of the Energy Transformation – Implications of an International Hydrogen Economy" (GET Hydrogen), funding reference number AA4521G125.

References

Confindustria. (2020). *Piano di azione per l'idrogeno*. Retrieved August 8, 2023, from https://www.confindustria.it/wcm/connect/552759de-3bb8-472f-a20b-07ab2aa5f21f/Position+Paper_Piano+d%27azione+per+l%27idrogeno_ott+2020_Confindustria.pdf?MOD=AJPERES&CONVERT_TO=url&CACHEID=ROOTWORKSPACE-552759de-3bb8-472f-a20b-07ab2aa5f21f-nuhfm09

EHB. (2022). *European hydrogen backbone*. Retrieved September 22, 2023 from https://ehb.eu/files/downloads/ehb-report-220428-17h00-interactive-1.pdf

European House-Ambrosetti. (2020). *Una filiera nazionale dell'idrogeno per la crescita e la decarbonizzazione dell'Italia*. Retrieved August 10, 2023, from https://www.snam.it/export/sites/snam-rp/repository/file/Media/news_eventi/2020/H2_Italy_2020_ITA.pdf

European House-Ambrosetti. (2022). *Proposal for a zero carbon technology roadmap*. Retrieved October 5, 2023, from https://www.ambrosetti.eu/en/news/proposal-for-a-zero-carbon-technology-roadmap/

European Commission. (2022). REPowerEU Plan. Brussels, 18.5.2022 COM(2022) 230 final

FCH-JU. (2020). *Italy. Opportunities for hydrogen energy technologies considering the national energy & climate plans, fuel cells and hydrogen 2 joint undertaking*. Retrieved August 22, 2023, from https://www.fch.europa.eu/sites/default/files/file_attach/Brochure%20FCH%20Italy%20%28ID%209473094%29.pdf

Franza, L. (2021). *Clean molecules across the Mediterranean. The potential for North African hydrogen imports into Italy and the EU*. Istituto Affari Internazionali.

Giuli, M. (2022). *Italy in the international hydrogen economy*. Istituto Affari Internazionali.
GSE. (2022). *Rapporto Statistico GSE-FER 2020*. Retrieved September 13, 2023, from https://www.gse.it/documenti_site/Documenti%20GSE/Rapporti%20statistici/Rapporto%20Statistico%20GSE%20-%20FER%202020.pdf
IRENA. 2022. Renewable Capacity Statistics 2022. Retrived July 23, 2023 from: https://www.irena.org/publications/2022/Apr/Renewable-Capacity-Statistics-2022
Legambiente. (2023). *Scacco matto alle rinnovabili. Gli ostacoli normative, burocratici e culturali che frenano la transizione energetica in Italia*. Retrieved October 5, 2023, from https://www.legambiente.it/comunicati-stampa/scacco-matto-alle-rinnovabili-2023/
MISE. (2020). *Strategia Nazionale Idrogeno. Linee Guida Preliminari*. Retrieved August 3, 2023, from https://www.mise.gov.it/images/stories/documenti/Strategia_Nazionale_Idrogeno_Linee_guida_preliminari_nov20.pdf
MITE. (2022). *La Situazione Energetica Nazionale nel 2021*. Retrieved August 22, 2023, from https://dgsaie.mise.gov.it/pub/sen/relazioni/relazione_annuale_situazione_energetica_nazionale_dati_2021.pdf
NCEP. (2019). *Piano Nazionale Integrato per L'energia e il Clima*. Ministero dello Sviluppo Economico, December 2019.
NCEP. (2023). *Piano Nazionale Integrato per L'energia e il Clima*. Ministro dell'Ambiente e della Sicurezza Energetica, June 2023.
PNRR. (2021). *Piano Nazionale di Ripresa e Resilienza*. Retrieved August 28, 2023, from https://www.mise.gov.it/images/stories/documenti/PNRR_Aggiornato.pdf
Prontera, A. (2021). The dismantling of renewable energy policy in Italy. *Environmental Politics, 30*(7), 1196–1216.
Prontera, A., & Lizzi, R. (2023). The necessary reorientation of Italian energy policy. *Contemporary Italian Politics, 15*(2), 252–268.
Schmidt, V. A. (2009). Putting the political back into political economy by bringing the state back in yet again. *World Politics, 61*(3), 516–546.
SEN. (2013). *Strategia Energetica Nazionale*. Ministero dello Sviluppo Economico.
SEN. (2017). *Strategia Energetica Nazionale*. Retrieved August 18, 2023, from https://www.mite.gov.it/sites/default/files/archivio/allegati/testo-integrale-sen-2017.pdf
SNAM. (2019). *The hydrogen challenge: The potential of hydrogen in Italy*. Retrieved September 2, 2023, from https://www.snam.it/it/hydrogen_challenge/repository_hy/file/The-H2-challenge-Position-Paper.pdf
SNAM. (2022). *SNAM and hydrogen*. Retrieved August 26, 2023, from https://www.snam.it/en/energy_transition/hydrogen/snam_and_hydrogen/index.html
South2Corridor. (2023). *General description of the SoutH2 Corridor*. Retrieved October 10, 2023, from: https://www.south2corridor.net/south2

Hydrogen Policy in the Netherlands: Laying the Foundations for a Scalable Hydrogen Value Chain

Roelof Stam, Coby van der Linde, and Pier Stapersma

Abstract This chapter delves into the Dutch hydrogen strategy, examining the Netherlands' starting position, its national hydrogen strategy, policy initiatives and the international approach adopted by the Dutch government in the low-carbon hydrogen economy. The Dutch Climate Agreement of 2019 identified low-carbon hydrogen as a key part of the carbon reduction strategy, especially in hard-to-abate sectors. The Netherlands is well-placed to make a substantial contribution to Europe's low-carbon hydrogen market leveraging its current role as a European energy hub, substantial chemical cluster, strategic North Sea location, offshore wind potential, and existing gas and oil infrastructure. To bolster investment security and scale up the low-carbon hydrogen market by 2030, the Dutch government has opted for a blend of obligations and subsidies. The Dutch government favours hydrogen production through electrolysis from renewable energy, while concurrently allowing for the utilization of hydrogen produced from natural gas with Carbon Capture and Storage (CCS) technology where applicable. Internationally, the Netherlands aims to position itself as the central hub for hydrogen in Northwest Europe. This involves linking Dutch domestic production at the North Sea and international exporters with users in industrial clusters across Northwest Europe.

R. Stam · C. van der Linde (✉) · P. Stapersma
Centre for International Energy Policy (CIEP), Breitnerlaan 299 2596 HA, The Hague, The Netherlands
e-mail: coby.vanderlinde@clingendaelenergy.com

R. Stam
e-mail: roelof.stam@clingendaelenergy.com

P. Stapersma
e-mail: pier.stapersma@clingendaelenergy.com

© Helmholtz-Zentrum Potsdam, Deutsches GeoForschungsZentrum GFZ 2024
R. Quitzow and Y. Zabanova (eds.), *The Geopolitics of Hydrogen*, Studies in Energy, Resource and Environmental Economics, https://doi.org/10.1007/978-3-031-59515-8_9

1   Introduction

The Dutch government has in recent years taken a more active role in the energy transition and therewith the creation of a low-carbon hydrogen[1] economy in the Netherlands. In the Dutch Climate Act of 2019, the government states that it wants to reduce CO_2 emissions by 49% by 2030, compared to 1990 levels, and aims to achieve a near 100% reduction by 2050 (Government of the Netherlands, 2019). These ambitious climate targets require drastic changes to the energy system, which is currently based largely on fossil fuels (85% of total final energy consumption) (BP, 2022). Hydrogen can potentially fulfil a 'systemic function' in the future energy system, accommodating intermittent renewable electricity (Clingendael International Energy Programme, 2019). As stated in the 'Dutch Government Strategy on Hydrogen (2020)', the government believes hydrogen will play a crucial role in the energy and feedstock transition, especially in hard-to-abate sectors (Ministry of Economic Affairs and Climate, 2020). The Netherlands prefer hydrogen produced via electrolysis from renewable energy but additionally there is also room for hydrogen produced from natural gas and/or residual gases with Carbon Capture and Storage (CCS) technology where applicable.

To achieve the Netherlands' climate targets while preventing carbon leakage and maintaining its current energy hub function, the Dutch government is working together with the private sector to realize the development of a low-carbon hydrogen market. Industry, NGOs, research institutions and government are cooperating on a large array of projects aimed at realizing a low-carbon hydrogen economy in the Netherlands. Often these projects are of an international character, connecting neighbouring countries, stakeholders in the North Sea region and global industry (Topsector energy, 2020).

In this chapter the Dutch hydrogen policy strategy will be discussed.[2] First the starting position of the Netherlands in the emerging low-carbon hydrogen economy is illustrated, after which a broad overview of the national hydrogen strategy and policy initiatives is given. The last section describes the international approach the Dutch government takes in the emerging low-carbon hydrogen economy.

---

[1] Low-carbon hydrogen is defined as all hydrogen produced with a significant carbon footprint reduction compared to unabated hydrogen production. This includes hydrogen produced via electrolysis from renewable electricity (green), hydrogen produced from natural gas or residual gases with CCS (blue) and various other 'low-carbon' hydrogen production methods. The Netherlands is an active member of IPHE (International Partnership for Hydrogen and Fuel Cells in the Economy). IPHE is involved in the certification of the carbon footprint of hydrogen produced from different sources and with different technologies to facilitate the development of international trade in low-carbon hydrogen. See their publication 'Methodology for Determining the Greenhouse Gas Emissions Associated with the Production of Hydrogen (2022) Available at: https://www.iphe.net/_files/ugd/45185a_48960ad9b26045c7a082bceb3a192bc7.pdf.

[2] The Dutch hydrogen strategy and policy approach outlined in the following chapter is predominantly based on the Dutch Climate Act (2019), the National Hydrogen Programme (2022), the National Plan Energy System (NPE) (2023) and various letters to parliament (e.g. Government Strategy on Hydrogen (2020), Development Transport System for Hydrogen (2022) and Organization and Development of the Hydrogen Market (2022)).

2 The Position of the Netherlands in the Emerging Low-Carbon Hydrogen Economy

The Netherlands is in a strong position to make a significant contribution to Europe's low-carbon hydrogen market thanks to its current role as a European energy hub, substantial chemical industry, favourable geographical location at the North Sea, offshore wind potential, and existing gas and oil infrastructure.

2.1 Energy Hub

The Netherlands is home to Europe's largest seaport, the Port of Rotterdam, located at the North Sea and in the Rhine-Meuse-Scheldt delta. With its relatively deep-water port and convenient location, connecting international waters with Northwest Europe, the port of Rotterdam functions as a global hub for international energy trade. The Port of Rotterdam plays an important part and is embedded in the Antwerp-Rotterdam-Rhine-Ruhr-Area (ARRRA), a petrochemical cluster that generates 40% of the total petrochemical output in the EU (Port of Rotterdam, retrieved on Nov 22, 2022b). Significant quantities of energy, among others in the form of crude oil, oil products, coal and gas are imported through Rotterdam daily, and transported via river barges and pipelines to industrial clusters located in Northwest Europe (van der Linde & Stapersma, 2018).

The Dutch government seeks to maintain this hub function in a future renewable energy system. As hydrogen potentially becomes a globally traded commodity and industrial demand for low-carbon hydrogen in Northwest Europe increases, the Netherlands is in a unique position to contribute to the hydrogen supply chain by both producing and importing low-carbon hydrogen and providing a gateway to Northwest Europe (Ministry of Economic Affairs and Climate, 2020; Port of Rotterdam, retrieved on Nov 22, 2022a). The offshore wind potential in the Dutch part of the North Sea is also an important incentive for the development of a low-carbon hydrogen economy.

In addition, facilities at Groningen Seaport in the north of the country are developing into a landing point for offshore wind and conversion into hydrogen. These facilities are closely located to similar German plans to develop their side of the Ems Delta, creating a potential new conversion cluster. A similar development is ongoing in the south of the country, on the banks of the river Scheldt and along the canal from Terneuzen to Ghent in Belgium.

2.2 *Industry*

The Netherlands has a relatively energy intensive economy due to its large refining and petrochemical sector and other economic activities that benefitted in the past from the availability of abundant natural gas, including greenhouse-based agriculture and horticulture, and fertilizer production. Hydrogen is already a widely used commodity in the chemical, refining and fertilizer industry. It plays a multidimensional role in these industries, as it is a by-product in some processes, while being an essential feedstock or a potential alternative energy carrier in others.

With an estimated 1.5 million tonnes per year, the Netherlands is the second largest hydrogen producer in Europe, after Germany (TNO, 2020). As such, the Netherlands has extensive experience in the safe production, transportation, storage, and consumption of hydrogen in industrial settings. Currently, most hydrogen produced in the Netherlands is created via steam methane reforming using natural gas (862,000 tonnes per year), most of the remaining of hydrogen is produced with oil and residual (refinery) gasses (574,000 tonnes per year) (TNO, 2020). Approximately 10% of the Dutch gas consumption is used to produce hydrogen, emitting significant amounts of CO_2 (Ministry of Economic Affairs and Climate, 2020).

In addition to hydrogen production via electrolysis from renewable energy, hydrogen from natural gas and residual gases with CCS provides a relatively quick solution to decrease emissions considerably and plays a role in the Dutch national carbon reduction strategy (Ministry of Economic Affairs and Climate, 2020). An example of an initiative to reduce carbon emissions in current hydrogen production is the H-vision project in Rotterdam, which aims to produce low-carbon hydrogen from natural gas and residual refinery gasses. In this project, CO_2 would be captured and either stored in empty gas fields under the North Sea or used as feedstock for basic chemicals such as methanol. The hydrogen would be used as input for the refinery process (H-vision, retrieved on Nov 22, 2022). The transportation and storage of CO_2 would be facilitated by Porthos, a joint venture of Gasunie, Port of Rotterdam and EBN.

Porthos stands for Port of Rotterdam CO_2 Transport Hub and Offshore Storage and is the flagship CCS project in the Netherlands. However, the Porthos project was delayed due to a lawsuit filed by Mobilisation for the Environment (MOB) against the state, concerning the nitrogen the project is expected to emit during its construction. The Netherlands is currently struggling with a 'nitrogen crisis'. Excessive amounts of nitrogen emissions from agriculture, industry and transportation threaten the country's biodiversity. Although these emissions have decreased significantly over the years, they continue to pose a threat to the environment. In an interim ruling issued on 2 November 2022, the court concluded that the nitrogen construction exemption used for the Porthos project does not comply with European Nature Conservation Law and may not be used for the construction of the project. Therewith, Porthos was delayed but not off track. An individual assessment of the nitrogen impact had to be made for the project (Raad van State, 2022).

On 9 December 2022, the Ministry of Climate and Energy published a letter to parliament in which it declared that it would temporarily assume liability for financial risks related to this project, pending the final verdict (Ministry of Economic Affairs and Climate, 2022d). The government considers the carbon reduction that Porthos could provide (2.5 million tonnes CO_2 per year) to be crucial to its efforts to achieve the 2030 climate goals. To prevent delays or possible cancellations, it was necessary that Porthos continued the tendering procedures already initiated. Therefore, the government aimed to guarantee financial obligations for a maximum amount of 175 million euros up until the end of 2023 (Ministry of Economic Affairs and Climate, 2022d).

The interim ruling had significant implications for the construction industry in general, as it meant that in the future the compliance of individual construction projects with the requirements of the Nature Conservation Act will need to be assessed. This will lead to considerable delays in permitting processes and to delays in the construction of homes, infrastructure, and energy projects, creating a serious obstacle to the energy transition in the Netherlands (Meijer, 2022).

On 16 August 2023, the Council of State (highest legal authority in the Netherlands) ruled that, based on the new assessment, the nitrogen deposition during the construction of Porthos does not pose a threat to neighbouring nature (Raad van State, 2023). On 18 October 2023 a final Investment decision (FID) was taken. Construction will start in 2024, and is facilitating Air Products, Air Liquide, ExxonMobil and Shell to capture and store 2.5 Mton CO_2 per year for 15 years. The CO_2 will come from their existing SMRs, which is seen as an important first step to realise low-carbon hydrogen supply. Project Aramis, an initiative of Gasunie, EBN, Total and Shell Nederland could be the next CCS project to reach FID, increasing the contribution of blue hydrogen to the energy transition targets (Netherlands Enterprise Agency, 2021c). On 22 June 2023, Belgium and the Netherlands signed an agreement on CCS collaboration and cross border CO_2 transport, facilitating also Belgian captured CO_2 for storage in depleted Dutch offshore gas fields (Aramis, 2023).

2.3 North Sea

The Netherlands' location on the North Sea is well-suited for the production of hydrogen via electrolysis from renewable energy. The Dutch part of the North Sea covers an area of about 58,000 km^2 (Government of the Netherlands, 2015). With its favourable wind conditions, relatively shallow waters, good access to ports and energy intensive industries, which are largely situated or alternatively well connected to coastal areas, the North Sea is very suitable for offshore wind power. Some of this wind energy could be used to produce hydrogen.

There are various projects underway to start producing hydrogen through electrolysis at or near the North Sea. For example, Shell has recently taken the final investment decision on Holland Hydrogen 1, a 200 MW electrolyser project in Rotterdam on the Second Maasvlakte (Shell, 2022). They intend to produce hydrogen using wind energy from the offshore wind farm Hollandse Kust Noord. Another example is H2opZee, a project in which Germany's RWE and UK-based Neptune Energy join forces to accelerate hydrogen production in the Dutch North Sea. This demonstration project will have an electrolyser capacity of 300–500 MW and will use existing pipelines to transport hydrogen produced at sea to land (RWE, retrieved on Nov 23, 2022). Another project is NortH2, an international consortium consisting of Equinor, ENECO, Gasunie, Groningen Seaports, RWE and Shell Nederland, endorsed by the Groningen provincial authority, which is currently investigating the feasibility of large-scale production, storage and transmission of hydrogen in Groningen. NortH2 aims to convert offshore wind energy into hydrogen through electrolysis in Eemshaven to supply industry with 2–4 GW of hydrogen by 2030 and upscale to more than 10 GW by 2040 (NortH2, retrieved on Nov 22, 2022).

2.4 Gas Legacy Assets

With the discovery of the Groningen gas field in 1959, the Netherlands became the number one natural gas producer in Europe. This stimulated the development of an extensive high quality natural gas network connecting the Netherlands and parts of Belgium, France, and Germany. In addition to this, it was the Dutch state that established the institutional arrangements for a European gas market, from which lessons for hydrogen can be drawn. These public–private partnerships made the development of natural gas infrastructures and natural gas markets a great European success (Correljé et al., 2003).

Although the Groningen gas field was long one of the largest onshore gas fields in the world, its supplies are finite. Therefore, in the 2000s, in order to diversify away from a predominant focus on national production, the concept of a gas hub strategy was implemented, partly shifting the focus from national production to imports. This resulted in the construction of necessary infrastructure such as import terminals and storage facilities. The gas hub strategy has proven valuable in this time of geopolitical uncertainty as it enables European countries to replace Russian flows with new liquefied gas supplies from overseas. Furthermore, the repurposing potential of this infrastructure presents a favorable starting position for the future low-carbon hydrogen economy.

On 26 September 2022, well before its anticipated depletion, Hans Vijlbrief, State Secretary for the Extractive Industries, stated that the Groningen gas field would be put on 'pilot light' as of 1 October 2022 due to the seismic risk caused by production activities (Government of the Netherlands, 2022b). This implied that a minimum amount of gas will be extracted from the field so that existing wells and infrastructure can continue to operate, ensuring that the field is still available as spare capacity in case of a severe disruption in low calorific natural gas supplies or an emergency in the Netherlands or neighbouring countries. The Groningen field has closed completely on 1 October 2023, while the decision to break down the production facilities was delayed to 2024. The latter was done to make sure that security of gas supply could be adequately organized in this period of geopolitical uncertainty and tight natural gas markets. Due to the planned closure of the Groningen field and the expected general decline of natural gas demand over the coming years, gas infrastructure will become available for repurposing to transport hydrogen production (Government of the Netherlands, retrieved on Nov 22, 2022). The Netherlands has two natural gas pipeline systems, one for high calorific gas and one for low calorific (Groningen quality) gas. With demand for low calorific gas decreasing some spare pipeline capacity is already being converted to carry hydrogen. In 2021 the government decided to refurbish gas pipelines to create the 'Dutch hydrogen backbone' connecting the Dutch industrial clusters (European Hydrogen Backbone, retrieved on Nov 22, 2022). On 27 October 2023, the first section of the hydrogen backbone, a new pipeline to connect the Maasvlakte in the Port of Rotterdam with the rest of the backbone, was officially launched. The hydrogen backbone will be further discussed later in this chapter.

3 National Hydrogen Strategy and Policy Initiatives

In the Dutch Climate Agreement of 2019 the low-carbon hydrogen economy was identified as a key part of the carbon emission reduction strategy (Government of the Netherlands, 2019). Central to the Dutch hydrogen strategy is the National Hydrogen Programme (2022), which originated from the Dutch Climate Agreement. The main task of the National Hydrogen Programme, a public–private partnership, is to investigate and stimulate the contribution of hydrogen to the realization of the energy transition. In 2022 its focus was on creating a Hydrogen Roadmap for the Netherlands together with stakeholders from the hydrogen sector. The Roadmap proposes low-carbon hydrogen targets for 2030 and describes what actions are necessary to achieve them. Figure 1 provides a schematic overview of the Hydrogen Roadmap, for more detailed information please see the Hydrogen Roadmap report (National Hydrogen Programme, 2022). The Dutch government takes an integrated approach to developing hydrogen value chains, focusing on production, import, transportation, storage, as well as the demand side, potential revenue models and on how to deal with safety and regulatory issues (National Hydrogen Programme, 2022; Netherlands Enterprise Agency, 2021a).

3.1 Cluster-Based Energy Strategy

Industry in the Netherlands is highly concentrated in regional clusters, due to economies of scale, location, cooperation opportunities and infrastructure. These clusters are identified as: Rotterdam-Moerdijk, the North Sea Cannel Area (Noordzeekanaalgebied), the Northern Netherlands (Noord Nederland), Chemelot, Zeeland/West-Brabant and 'other industries' (a sixth cluster which contains

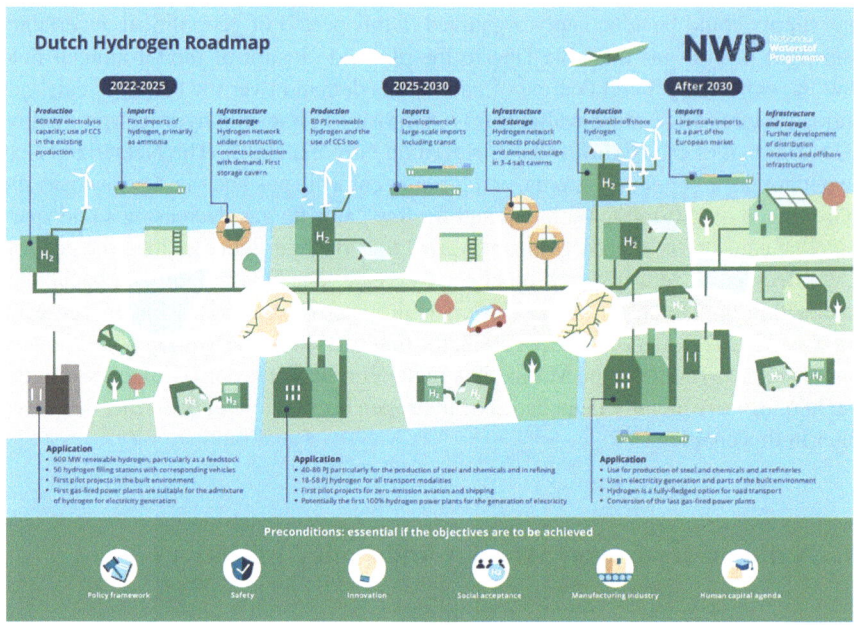

Fig. 1 Schematic overview of the Dutch Hydrogen Roadmap. *Source* National Hydrogen Programme (2022). Routekaart waterstof. Available at: https://www.nationaalwaterstofprogramma. nl/documenten/handlerdownloadfiles.ashx?idnv=2339011

remaining industries spread across the country) (Programme Sustainability Industry, retrieved on Nov 22, 2022). The government has set up Regional and Cluster-based Energy Strategies (RES & CES) to reduce carbon emissions in Dutch industry. The strategies focus on matching future supply and demand of renewable energy and on obtaining a better understanding of the necessary infrastructure. Additionally, the Dutch government created the 'Multi-year Infrastructure Energy and Climate programme' (MIEK) in 2021. MIEK describes the energy and raw material infrastructural projects that the cabinet plans to implement to accelerate the transition in the industry (Netherlands Enterprise Agency, 2021b).

Dutch industry is characterized by strong integration and cooperation between companies within these industrial clusters and a willingness to engage in open dialogue and consultation among private sector players. These unique cooperative characteristics create an environment in which innovation can flourish, making the Netherlands a good 'testing ground' for pilot projects. However, it is important to note that the investment climate has recently become less attractive due to changes in the corporate fiscal regime and nitrogen legislation and other permitting issues. National nitrogen policy, introduced to reduce nitrogen emissions and protect biodiversity, make it more difficult and in some cases impossible to get building permits for infrastructural projects in the Netherlands, including projects necessary for the energy transition, like the Porthos project.

3.2 Prioritized End-Use Sectors

The Dutch government believes that clean molecules and in particular low-carbon hydrogen will be necessary for the decarbonization of hard-to-abate industries, seasonal storage, and applications where electrification is not a viable option (Ministry of Economic Affairs and Climate, 2020). In the initial market phase, the supply of hydrogen to heavy industry and for heavy-duty transportation will therefore be prioritized.

Big industrial players are starting to prepare for a hydrogen future. For example, Tata Steel has announced plans to produce steel in the Netherlands using Direct Reduced Iron technology, a process in which iron ores are reduced using natural gas or hydrogen, instead of coal. Tata intends to commence operations before 2030 (Tata Steel, 2021). Shell aims to supply the Shell Energy and Chemicals Park Rotterdam with hydrogen produced via electrolysis in the Holland Hydrogen 1 plant using the HyTransPort gas pipeline in 2025.As the hydrogen heavy-duty truck market grows, Shell also plans to supply hydrogen as a fuel for commercial road transportation (Shell, 2022). Through the Eneco Diamond Hydrogen joint venture, Dutch utility Eneco, jointly with one of its parent companies Mitsubishi, aims to develop, an ultimately, 800 MW electrolyzer plant in the Port of Rotterdam industrial area, implying a future production of 80 kiloton of hydrogen per year (Offshore Energy Magazine, 2023). Although the project still needs to reach FID, construction is envisaged to start in 2026, and first hydrogen could be produced by 2029.

When the low-carbon hydrogen market moves into a later development phase, more sectors could be using hydrogen as an energy carrier. Besides initiatives in the industry and mobility sectors, plans for the first pilot projects in the electricity sector, built environment and agricultural sector are currently being developed; For an extensive list of 165 projects, see the Dutch National Hydrogen Programme (National Hydrogen Programme, retrieved on Nov 22, 2022d; National Hydrogen Programme, retrieved on Nov 22, 2022e).

3.3 The Role of Different Forms of Hydrogen Production

The Dutch government predominantly focuses on hydrogen produced via electrolysis from renewable electricity and hydrogen made from sustainable biogenic raw materials. Furthermore, there is also room for hydrogen produced from natural gas or residual gases with CCS, as long as this optimally contributes to the development of a broader hydrogen system, without hindering the growth of hydrogen produced via electrolysis from renewable electricity. Hydrogen from natural gas or residual gases with CCS is seen as a cost-effective way to help achieve the 2030 climate targets (National Hydrogen Programme, 2022). However, the current high price of natural gas might hinder the realisation of these plans. In the long-term the focus will be on technologies that reduce emissions to net-zero and replace fossil fuels with biotic and recycled raw material (Government of the Netherlands, 2019).

In the Dutch Climate Agreement of 2019 a target of 4 GW of electrolyser capacity for 2030 was agreed upon (Netherlands Enterprise Agency, 2021a). However, as the Netherlands is determined to reduce its dependence on Russian gas, following the invasion of Ukraine, two former coalition parties (D66 and VVD) proposed to double the already ambitious 2030 electrolyser capacity target to 8 GW (D66 & VVD, 2022). Although hydrogen will not play a role in securing energy supplies soon, it is widely viewed as an important building block for the development of domestic energy carrier production and fuel diversification to increase energy security in the future.

Besides domestic production, the Dutch government set an additional import target of 4 GW[3] of low-carbon hydrogen in 2030. The Netherlands has a rather energy intensive economy combined with a relatively small land area. As hydrogen is expected to fulfil a significant part of the total future energy demand, especially in industry and transportation, it is considered unrealistic to produce all necessary hydrogen domestically (National Hydrogen Programme, retrieved on Nov 22, 2022a). As hydrogen becomes a globally traded commodity, dependence on hydrogen comes with security of supply risks of its own. The Dutch government recognizes the importance of setting up measures to ensure the security of supply of imported hydrogen. Finding the right balance between the development of domestic production and imported flows of low-carbon hydrogen is crucial for a strong and secure hydrogen economy in the Netherlands (Clingendael International Energy Programme, 2022).

3.3.1 Policies for the Use of Low-Carbon Hydrogen in the Industrial Sector

The Dutch government has opted for a combination of obligations and subsidies to increase investment security and scale up the low-carbon hydrogen market by 2030 (Ministry of Economic Affairs and Climate, 2020; National Hydrogen Programme, retrieved on Nov 22, 2022c). Currently, with regard to obligation-based policy, the government is exploring possible options to introduce a purchase obligation for low-carbon hydrogen in industry, which would come into effect on 1 January 2026 (National Hydrogen Programme, retrieved on Nov 22, 2022c).

This is intended to ensure that the Netherlands can meet the RED III regulation, which requires that 42% of total hydrogen use for final energy and non-energy purposes in the industry by 2030 is from Renewable Fuels from Non-Biological Origin (European Parliament, 2023). Although, it is unsure whether these kind of volumes will be available in 2030, a purchase obligation would provide hydrogen

---

[3] This is a somewhat unusual unit to quantify hydrogen imports and can be interpreted in various ways. The 4 GW target could represent 1 million tonnes of hydrogen per year ((4 [GW] * 8760 [h])/33 [MWh/tonnes]), however, it is not completely clear what the Dutch government means with this target.

producers with the necessary demand security to make large scale investments. Furthermore, it would reduce the single dependence on subsidy schemes to achieve the hydrogen target in industry (Ministry of Economic Affairs and Climate, 2020).

Subsidies will provide a targeted means to adjust the market and offset part of the additional costs that come with the transition to low-carbon hydrogen as energy carrier. The financial support schemes are tailored to the various phases of the innovation development process. They are broadly subdivided into the following three categories in the Dutch Government Strategy on Hydrogen: subsidies for applied research and innovative pilot projects, scaling up projects with temporary operating cost support, and roll-out of full-scale projects via the SDE++ (Ministry of Economic Affairs and Climate, 2020).

As of 2020, the Stimulating Sustainable Energy production and Climate Transition (SDE++) scheme, one of the most important subsidy schemes for realizing large-scale renewable energy or CO_2 reduction projects, includes low-carbon hydrogen projects (Netherlands Enterprise Agency, 2023a). The SDE++ is an operating subsidy scheme. Subsidies are given to technologies that provide the most cost-effective renewable energy or carbon emission reduction. The budget of the SDE++ 2022 has recently been increased to 13 billion euros. The maximum budget for CCS in industry for the SDE++ scheme has been raised for 2022 but will gradually be phased out in the transition to a climate-neutral industry over the years.

Electrolyser projects are currently not the most cost-effective way to reduce carbon emissions, and have so far barely obtained any funding through the SDE++ scheme. Therefore, the Dutch National Growth Fund set up 'greengrowthcapacityNL' (groenvermogenNL) to scale up electrolyser projects and green chemistry ecosystems in the Netherlands. 838 million euros from the National Growth Fund was made available for the first two rounds of projects (Topsector Energy, 2022).

4 International Approach

Internationally the Netherlands seeks to position itself as the hydrogen hub of Northwest Europe, connecting Dutch domestic production at the North Sea and international exporters with users in industrial clusters in Northwest Europe. The Port of Rotterdam is currently the most important energy corridor to Northwest Europe and aims to leverage its current position to continue this role in the future hydrogen value chain (Ministry of Economic Affairs and Climate, 2020). Zeeland Port and Eemshaven are also developing hubs based on offshore wind landed in these ports. Nevertheless, Denmark and Belgium have expressed similar ambitions, while also Germany is keen to develop the Wilhelmshafen and Ems delta region for hydrogen hubs. They could be viewed as competitors, although there are also many opportunities for cooperation among these ports and countries. The regional ecosystem of various ports in Europe will help attract global suppliers, for example. Given the scale of the task, the Dutch government views cooperation as essential and is looking for partnerships along the whole value chain on a local, regional and international scale (Netherlands Enterprise Agency, 2021a).

4.1 Bilateral Partnerships

Dutch bilateral foreign policy in this field focuses predominantly on cooperation with neighbouring countries and establishing trade relationships with future exporters of low-carbon hydrogen. Together with the German and Belgian governments, the Dutch government is investigating opportunities to collaborate in the production, transportation, and usage of hydrogen. Developments in Germany are especially relevant to the Netherlands, as North Rhine-Westphalia intends to import half of its future hydrogen demand from or via the Netherlands (Eppinga, 2021). An example of current Dutch-German collaboration is the HY3 project, which recently completed a feasibility study (March 2022). They reviewed how Dutch and German offshore wind energy could be used to produce hydrogen, which would then be transported using existing Dutch gas pipelines to Dutch and German industrial clusters (HY3, retrieved on Nov 22, 2022). Bilateral cooperation was further underscored in November 2023, when Germany and the Netherlands signed Joint Declarations of Intent to strengthen cooperation in the field of H_2 infrastructure and import. This included a Joint Declaration of Intent to conduct a joint tender under the H2Global instrument (Offshore Energy Magazine, 2023, November 15). Together with the Port of Hamburg and Duisburg, the Port of Rotterdam is seen as a key delivery location for the international hydrogen flows attracted via the H2Global scheme. Participating in the initiative is important to the Netherlands as it aims to play a role in the development of the first international hydrogen value chains.

As it is of strategic importance to the Netherlands to maintain its current energy hub function in the future low-carbon hydrogen economy, the government and industry are actively pursuing potential import relationships with future prospective exporting countries. It is expected that the first import flows will come from current fossil fuel exporters, such as countries in the Middle East and North America. This is due to the relatively large potential for renewable electricity and the existing networks, infrastructure and expertise that can be utilized in these regions. Shortly thereafter, imports from European countries such as Portugal and Spain are expected to follow. The initial volumes will be small, however, in the future the Netherlands expect to import hydrogen from a growing number of countries within and outside of Europe (National Hydrogen Programme, 2022). The Netherlands was actually the first country to install a dedicated hydrogen envoy and has so far set up exploratory studies and established MoU's with numerous countries for low-carbon hydrogen trade. These countries include Namibia, Chili, South Africa, Canada, Uruguay, Oman, Morocco, Iceland, Spain, Portugal, Brazil, Denmark, Indonesia, Japan, Norway, Saudi Arabia, United Arab Emirates, United States and Australia (see Fig. 2) (National Hydrogen Programme, retrieved on Nov 22, 2022b).

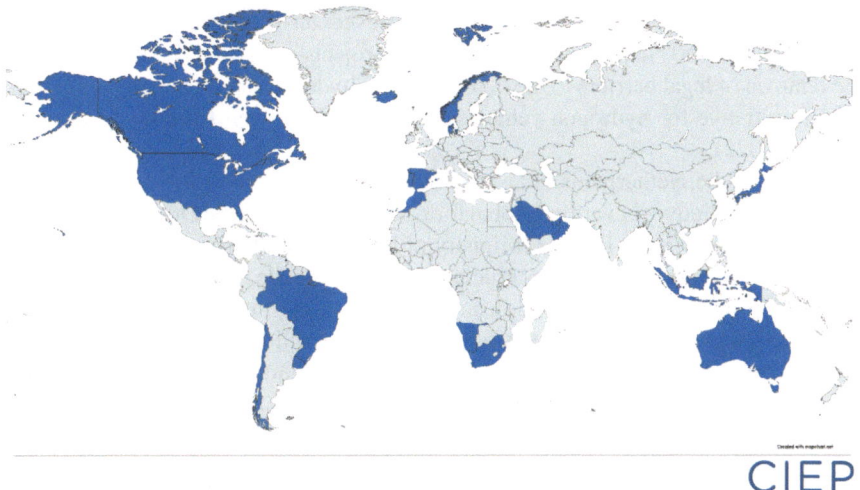

Made by CIEP, based on data from Nationaal Waterstof Programma, created with mapchart.net

Fig. 2 Dutch exploratory studies for cooperation and MoU's with potential future exporting countries. Made by CIEP, based on data from National Hydrogen Programme, created with mapchart.net

Two prominent European partnerships should be mentioned here: Cespa and the Port of Rotterdam have recently signed an MoU for a hydrogen corridor connecting Northern and Southern Europe. The corridor is expected to be operational from 2027 and will transport hydrogen derivatives produced in Spain from the Port of Algeciras to the Port of Rotterdam in the Netherlands. The project could contribute to the Port of Rotterdam's target to import 4.6 million tonnes of hydrogen annually into Northwest Europe in 2030 (Raza, 2022). Furthermore, a consortium consisting of ENGIE, Shell, Vopak and Anthony Veder, have agreed to conduct an exploratory study on the production and transportation of liquified hydrogen from Portugal to the Netherlands, with the aim to start delivering liquid hydrogen from the Port of Sines to Rotterdam no later than 2027 (Port of Rotterdam, 2022).

4.2 Multilateral Partnerships and Political Dialogue

Besides various bilateral partnerships, the Netherlands is also involved in numerous multilateral partnerships, fora, and initiatives. These include: the Clean Energy Ministerial (CEM), which the Netherlands co-leads together with Canada, the United States, Japan and the European Commission (Clean Energy Ministerial, retrieved on Nov 23, 2022); Mission Innovation (MI) and the International Partnership for Hydrogen Fuel Cells in the Economy (IPHE), in which the Netherlands is an active member and participates in R&D related initiatives (Ministry of Economic Affairs and Climate, 2020; Mission Innovation, retrieved on Nov 22, 2022); Hydrogen Europe, where the Netherlands is involved in projects such as HyLAW, focused on the removal of legal barriers to the deployment of hydrogen applications, and JIVE2, a joint initiative for hydrogen vehicles in Europe (Hydrogen Europe, retrieved on Nov 22, 2022).

Although the Netherlands is an effective participant in the global dialogue and partnership initiatives at various global fora, the main focus of the Dutch international hydrogen strategy is on Europe (Ministry of Economic Affairs and Climate, 2020). European collaboration is seen as crucial to establishing the first international supply chains, as it increases diversification and enables risk sharing of potential dependencies. The Dutch government therefore supports and will actively participate in the development of the Green Hydrogen Partnerships, Import Corridors and the global European Hydrogen Facility (Ministry of Economic Affairs and Climate, 2022c).

According to the Government Strategy on Hydrogen, some of the most important ways of engagement in political dialogue for hydrogen policymaking and cooperation in Europe—besides bilateral contact with neighbouring countries—are assumed to be:

- Continued communication with the *European Commission* about EU hydrogen policy concerning standards for safety, quality, flexible market regulations, sustainability, blending of hydrogen in existing natural gas networks, and innovation support.
- Participation in the *Pentalateral Forum,* consisting of the Benelux, Germany, Austria, Switzerland and France. Here, Austria and the Netherlands have initiated the development of common standards, market regulations, and market incentives prior to the EU discussion, calling on the European Commission to lobby for common global standards on sustainability. Efforts to establish homogenous levels of hydrogen-blending and ensure interoperability between hydrogen networks in Europe are also on the agenda (Pentalateral Energy Forum, 2020).
- Cooperation with *North Sea countries*: The North Sea wind power potential is seen as an essential source of energy for the production of hydrogen in the coming decades. The Netherlands intends to cooperate with North Sea countries through projects like the Sea Wind Power Hub, which aims to harness the North Sea's power potential (Clingendael International Energy Programme, 2021).
- Implementation of *Important Projects for Common European Interest (IPCEI)* focused on hydrogen: On 15 July 2022, the European Commission approved IPCEI Hy2Tech, the first IPCEI to support R&D in the hydrogen technology value chain. The project was initiated and prepared by the Netherlands and fourteen other member states (European Commission, 2022). The Dutch government recently made 1.385 billion euros available for IPCEI hydrogen projects (Government of the Netherlands, 2022a; Ministry of Economic Affairs and Climate, 2020).

One of the most discussed policy topics regarding hydrogen in Europe have been the Delegated acts and specifically the additionality requirements. This has also been a controversial topic in the Netherlands, where it has been met with both praise and opposition. Rob Jetten, Dutch Minister for Climate and Energy Policy, stated on 29

June 2022 that the proposed rules in the delegated acts overall seem to provide the necessary regulatory space for the Dutch hydrogen ambitions (Ministry of Economic Affairs and Climate, 2022c). He argued that the regulatory proposition provides enough flexibility for short-term projects, while ensuring that over the long-term hydrogen production will not come at the expense of extra renewable electricity on the grid. Although, the Dutch government agreed with the overall idea of the delegated acts, they provided some technical feedback to the Commission on the proposed regulations. See the document "Dutch reaction on European consultation delegated acts renewable hydrogen" for a detailed overview of the feedback (Ministry of Economic Affairs & Climate, 2022a).

Furthermore, there has been considerable pushback recently on the proposed regulations in the Dutch private sector, as companies fear that the regulations are too stringent for the initial phase of market development and are more likely to hinder than enable the low-carbon hydrogen economy from getting off the ground (Clingendael International Energy Programme, 2022). The US Inflation Reduction Act poured fuel on the fire, stirring fears that the combination of strict regulations in the EU and favourable subsidy schemes in the US might cause considerable industry displacement from the EU to the US, creating more demand uncertainty.

4.3 Shaping Hydrogen Infrastructure

In June 2021 Gasunie, the Dutch state-owned natural gas infrastructure and transportation company, received a formal mandate from the Ministry of Economic Affairs and Climate Policy to commence the development of a national hydrogen transport network (The Dutch hydrogen backbone). Gasunie currently operates and owns about 11,700 km of gas pipelines, of which approximately 8700 km are located in the Netherlands and 3000 km in Germany (European Hydrogen Backbone, 2022). The Dutch hydrogen backbone will largely be based on repurposed gas infrastructure and could be complete before 2027. The network will connect the five large industrial hubs in the Netherlands with storage facilities, overseas import, and export to Belgium and Germany (see Fig. 3). The Dutch hydrogen backbone could have a capacity of approximately 10–15 GW by the end of 2030. The development of the network will take place in multiple phases, a flexible and adaptive approach will be used, based on the needs and development of the hydrogen market (Ministry of Economic Affairs & Climate, 2022b).

Currently HyXchange is being developed, which is a trading platform for hydrogen transported through the Dutch hydrogen backbone, including global import flows and neighbouring countries (HyXchange, retrieved on Nov 22, 2022). For this hydrogen exchange to function properly, an open accessible transport infrastructure, a diverse supply of hydrogen and a dependable trading platform are needed. Next to the exchange, long-term contracts are expected to play an important role in the initial phase of market development as the number of suppliers and customers will be limited in the beginning (Clingendael International Energy Programme, 2019).

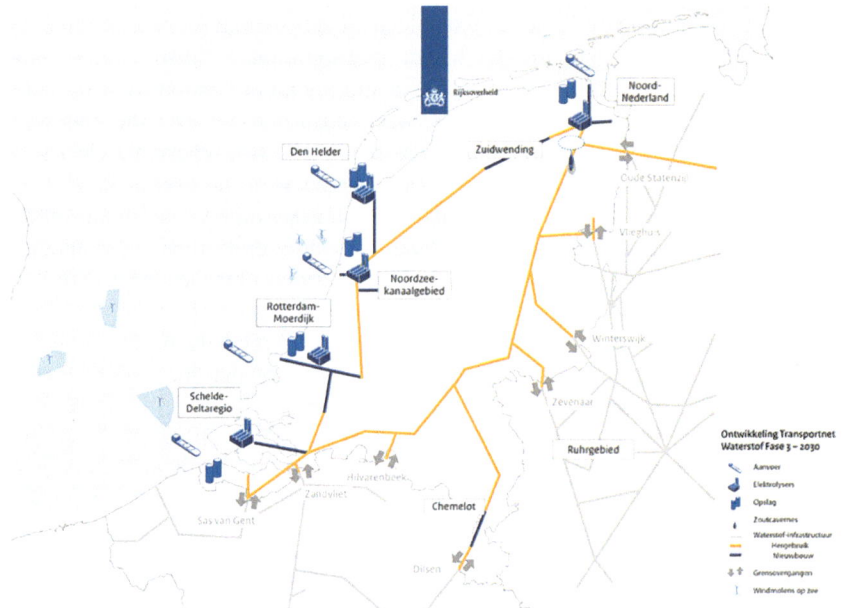

Fig. 3 The Dutch hydrogen backbone. *Source* Ministry of Economic Affairs and Climate (2022a, 2022b). Letter to parliament on the transportation network for hydrogen. Available at: https://open.overheid.nl/repository/ronl-5c57a9ba35fa907dcc805ca0da463dc33b036bb8/1/pdf/ontwikkeling-transportnet-voor-waterstof.pdf

Besides the Dutch hydrogen backbone, the Delta Corridor, a key infrastructural project for the Netherlands to secure its future position as a European energy hub, is currently being developed. After the completion of an initial feasibility study the project is now entering a second phase, to do a more detailed feasibility study, for which the necessary funding has been obtained. In this initiative the private sector has the lead but is working closely with the EU and the Dutch and German governments. The Delta Rhine Corridor Partners currently consist of Gasunie, OGE, Shell and BASF (Netherlands Enterprise Agency, 2022). The intended project consists of a bundle of pipelines between the Port of Rotterdam, Chemelot, and the German Rhineland Region.

The Final Investment Decision is expected to be taken before 2026 and the pipeline network is scheduled to enter operation by 2028 (Ministry of Economic Affairs and Climate, 2023b). It will connect large inland industrial clusters in the Netherlands and Germany with branches along the entire network, providing access to low-carbon hydrogen and CCS capacity (see Fig. 4). It will supply hydrogen to industrial hubs across, Moerdijk, Geertruidenberg, Chemelot and North Rhine-Westphalia (Gelsenkirchen, Cologne and wider areas). Additionally, in the future the Delta Corridor could also connect industrial hubs in Belgium and further into Germany. The capacity will be based on and grow with demand in Northwest Europe.

Overzichtskaart van de beoogde Delta Corridor

Fig. 4 The Delta Corridor. *Source* MIEK (2021). MIEK overview 2021 Multiyear programme Infrastructure Energy and Climate. Available at: https://www.rijksoverheid.nl/documenten/rap porten/2021/11/26/meerjarenprogramma-infrastructuur-energie-en-klimaat---overzicht-2021. Last accessed on 08.12.2022

Lastly, Gasunie is actively participating in the European Hydrogen Backbone initiative, working together with other European network operators on connecting EU member states with an intercontinental hydrogen network. The Netherlands maintains close contact with Germany and Belgium regarding the potential interconnections, development of the hydrogen network and other possibilities for cooperation.

4.4 Anticipating the Stages of Development in Organising the Low-Carbon Hydrogen Economy

The Dutch government supports the proposals made by the European Commission regarding harmonized rules for the development of a single European market for hydrogen (Ministry of Economic Affairs & Climate, 2022b). However, the regulatory framework may be a mismatch for the initial stages of hydrogen market development (Clingendael International Energy Programme, 2022). The European Union intends to regulate hydrogen in the same way as the natural gas market is regulated today, with legal separation between networks, production and distribution. This model has shown to work relatively well in mature gas markets with various suppliers and buyers operating in the market. However, legal separation might frustrate the hydrogen economy in the first phases of market development as it could hinder supply as well as demand security in a market with only few participants (Clingendael International Energy Programme, 2022).

During the development of the oil and gas industry, supply chains (production, transportation, and distribution) were built by vertically integrated oil companies in consortia and public–private partnerships. To reduce investment risks for large infrastructural projects like oil and gas exploration and extraction, security of demand is essential. Vertical integration provides the necessary demand security to significantly decrease the risk of oil and gas production investments as well as supply security for buyers. Similar investment securities are necessary for companies that intend to participate in the future low-carbon hydrogen economy (Clingendael International Energy Programme, 2022). The Dutch government is aware that it might be wise to draw lessons from the development of the world oil and gas markets and provide the regulatory space necessary for companies that are willing and able to set up the supply chains to kickstart the hydrogen economy.

5 More Developments in 2023

On 3 July 2023, the Minister of Climate and Energy Policy, Rob Jetten, submitted a letter to parliament, which accompanied the draft National Plan Energy System (NPE) (Ministry of Economic Affairs and Climate, 2023a). The NPE consists of a main document and is supported by multiple other documents dealing with various aspects of the future energy system of the Netherlands. According to the draft NPE, electricity should serve as the backbone of the energy system, promoting direct electrification where possible in combination with domestic produced and imported low-carbon hydrogen (carriers). Low-carbon hydrogen will play a crucial role in industry and international mobility, according to the plan.

The plan is focussed on meeting the 2030 and 2050 climate targets but misses to include the phases of development of new value chains. The phases of development involve an introduction, expansion, maturity, and stagnation or decline phase, stretched out over many years. The market organisation in these various phases is very different and each phase requires a different approach to stimulation and regulation (Clingendael International Energy Programme, 2022). Affordability and security of supply will also differ in these phases of development. The EU proposed draft 'gas market and hydrogen regulation' suffers from the same blindness for dynamic market developments. In the case of the EU, they try to hammer an infant low-carbon hydrogen sector into the mould of the mature gas market model, creating a mismatch in risks and benefits along the value chain in the early stages of development. In the case of the draft NPE, it would have helped if they also focused on the logic of the value chain development instead of 2030 and 2050.

In the Netherlands, they are confronted with challenges related to the pace at which offshore wind expansion can occur to realise both the direct electrification of low-temperature heating in the built environment and the powering of passenger cars, as well as fostering domestic production of hydrogen via electrolysis. In the draft NPE, direct electrification is preferred, and scarcity of solar and wind supply will come at the detriment of hydrogen production via electrolysis, increasing the investment risk due to potential low utilization rates. The increased cost of capital and materials, in addition to the risk of low utilization leaves many projects in a pre-FID phase. Towards 2030, the draft NPE foresees some government supported electrolysis hydrogen come on stream, but this is far from enough to meet the RED III requirement to replace 42% of grey hydrogen in industry with domestic supply. Dutch industry, part of the ARRRA-cluster, is relatively energy-intense, Northwest Europe is home to about 60% of EU hydrogen supply and demand (IEA & CIEP, 2021). Much will depend then on the organisation of imported flows, although also there, the increased capital and material costs may play a role in coming to FID and despite the many MoU's concluded in the past year. Relief can come from the first $7 billion tranche of the American IRA to develop H2Hubs, building on the existing hydrogen infrastructure in Texas (Petroleum Intelligence Weekly, 2023). In addition to increased oil, oil products and LNG imports from the US, imported US low-carbon hydrogen would add to a low level of diversity of supply in the years until more diverse supplies would become available on international markets.

On 17 October 2023, the report Integrale Infrastructuur-Verkenning 2030–2050 (Integral investigation into (Energy) Infrastructure 2030–2050) was published (Netbeheer Nederland, 2023). This study is a stark reminder of the huge infrastructure task to facilitate the Dutch energy transition. Already, electrification of industry and perhaps low temperature heating may be delayed for lack of electricity network capacity in the coming ten years. This applies to both the transportation and the distribution networks. The hydrogen network may offer some relief in these years, but the draft NPE does not take this sufficiently into account, perhaps creating some investment hesitation among companies on the production/import and demand side. Companies wanting to electrify will have to wait maybe a decade to receive larger volumes of power, while companies looking at hydrogen are uncertain about utilization rates and industrial demand uncertainties. The foreseen new energy system also has very large flexibility needs, which in part must come from hydrogen. In the Netherlands energy infrastructure (apart from liquids infrastructure) is run by government-owned companies, forcing the government to invest heavily, while also supporting private companies in the hydrogen value chain in the start-up phase of the value chain. Without infrastructure many of the transition routes cannot materialise. The recent report on energy infrastructure presents the situation in four scenarios, ranging from a very large challenge to an enormous undertaking for the infrastructure companies.

Moreover, in a high-over assessment, the claim on mineral commodities for electrification solutions is substantial. Competition from other countries following the same path may require re-routing, developing substitutes, and prioritizing the energy transition in a more logic step-by-step approach in line with infrastructure developments, rather than pushing everything at the same time and with government managing scarcity. In the present economic climate, with higher cost of capital and materials, there is a risk that planned project timelines are unattainable, forcing the government to re-adjust its policy support when fiscal stress increases. Instead, the government may have to reduce its interventionist stance in appointing preferred sectors of demand for electricity and hydrogen, as outlined in the draft NPE, and instead allow market forces to do this heavy lifting itself. If this scenario unfolds, markets will determine the preferred consumer of renewable power and low-carbon hydrogen, potentially altering the order of sectorial progress in CO_2 reduction. Neighbouring countries may suffer from the same fiscal stress and allow industry to take the lead, while the built environment and perhaps electrification of mobility must take an initial slower route and rely on intermediate solutions such as biofuels and insulation, until network capacities and flexibility solutions have been developed.

At the same time, the Netherlands and Germany signed a letter of intent on 14 November 2023 to launch a joint tender with Germany for renewable hydrogen through the H2Global instrument. The Netherlands reserved 300 million euros for this tender. The Netherlands sees this initiative as an important addition to its import policy. Also, discussions on connecting the Dutch hydrogen back bone to North Rhine Westphalia are ongoing and are meant to facilitate hydrogen transportation from the Netherlands to Germany (Government of the Netherlands, 2023).

6   Conclusions

The Netherlands is prepared to be among the driving forces shaping the European low-carbon hydrogen economy. This is shown by the government's hydrogen strategy and established subsidy schemes, but more so by the many hydrogen projects that have been announced and first final investment decisions that have been taken recently. Nevertheless, the recent change in the economic environment may require some adjustment to policy instruments to lure more pre-FID projects into FID in order to build out both a domestic hydrogen sector and import platforms. An international low-carbon hydrogen economy presents an opportunity for the Netherlands to continue its role as energy corridor to Northwest Europe in a future sustainable European energy system. The Netherlands can contribute to the decarbonization of the continent by connecting industrial clusters in neighbouring countries with hydrogen from international markets. However, nitrogen legislation, lack of available space, and labour market constraints are among the various issues that could yet delay or hinder the development of the low-carbon hydrogen industry in the Netherlands.

Internationally, the Netherlands focuses on cooperation, especially with neighbouring countries. Cooperation with Germany plays a particularly prominent role, especially if a significant share of German demand will have to be met via imports that enter Europe through the Netherlands. However, coordinating timing and coherence along the emerging value chain presents significant challenges. The ability to scale up production and imports will depend, among other things, on whether the necessary infrastructure, regulations, and modifications in the manufacturing industry in the wider hinterland, are in place. Intensive cooperation between all players in the chain, both nationally and internationally is required to implement a policy framework that will lead to investments in the short-term.

The Dutch government fully supports setting up harmonized rules for the development of the EU hydrogen market. However, the natural gas market model currently proposed for the low-carbon hydrogen market—legally separating production, networks and distribution—might not match with the requirements for companies in the initial stages of market development. Instead, the EU should be more open to a variety of market structure models befitting the introduction and early expansion phase of low-carbon hydrogen. The recent highly ambitious RED III targets for the industry should stimulate rather than frustrate the low-carbon hydrogen market development. It should allow for a wider variation of CO_2 emission reduction technologies in the hydrogen sector, to pave the way for no-carbon hydrogen markets in the Netherlands and its neighbours without losing sight of the CO_2-emission reduction targets.

References

Aramis (2023, June 22). *Arami Welcomes Dutch-Belgian Agreement.* Retrieved November 6, 2023 from https://www.aramis-ccs.com/news/aramis-verwelkomt-nederlands-belgisch-akk oord-over-grensoverschrijdend-transport-van-co2.

BP (2022, June 28). *bp Statistical Review of World Energy 2022 | 71st edition.* Retrieved November 23, 2022 from https://www.bp.com/content/dam/bp/business-sites/en/global/corporate/pdfs/ene rgy-economics/statistical-review/bp-stats-review-2022-full-report.pdf.

Clean Energy Ministerial. *Hydrogen.* Retrieved November 23, 2022 from https://www.cleanener gyministerial.org/initiatives-campaigns/hydrogen-initiative/

Clingendael International Energy Programme (2019, October). *From an invisible to a more visible hand? Hydrogen and electricity: towards a new energy system backbone.* Retrieved November 23, 2022 from https://www.clingendaelenergy.com/inc/upload/files/CIEP_Paper_2 019_2B_web.pdf.

Clingendael International Energy Programme (2021, April). *Hydrogen in North-Western Europe, a vision towards 2030.* Retrieved November 23, 2022 from https://www.clingendaelenergy.com/ inc/upload/files/NW-Europe-Hydrogen-Final.pdf.

Clingendael International Energy Programme (2022, July). *Managing future security of low-carbon hydrogen supply.* Retrieved November 23, 2022 from https://www.clingendaelenergy.com/inc/ upload/files/CIEP-Paper-202201-web-2.pdf.

Correljé, A., Van der Linde, C., & Westerwoudt, T. (2003). *Natural gas in the Netherlands. From cooperation to competition?.* Retrieved from https://www.clingendaelenergy.com/inc/upload/ files/Book_Natural_Gas_in_the_Netherlands.pdf.

D66 & VVD (2022, April). *Nederland als waterstofkoploper—actieplan VVD-D66.* Retrieved November 23, 2022 from https://d66.nl/wp-content/uploads/2022/04/Nederland-als-waterstof koploper-actieplan-VVD-D66.pdf.

Eppinga, A. (2021, August 28). *Germany is fully committed to hydrogen, and the Netherlands should follow suit.* Retrieved November 22, 2022 from https://innovationorigins.com/en/germany-is-fully-committed-to-hydrogen-and-the-netherlands-should-follow-suit/.

European Commission (2022, July 15). *State Aid: Commission approves up to 5.4 billion euro of public support by fifteen Member States for an Important Project of Common European Interest in the hydrogen technology value chain.* Retrieved November 23, 2022 from https://ec.europa.eu/commission/presscorner/detail/en/ip_22_4544.

European Hydrogen Backbone. *Country Narratives.* Retrieved November 22, 2022 from https://ehb.eu/page/country-specific-developments.

European Parliament (2023, April). *EU Rules for Renewable Hydrogen.* Retrieved November 6, 2023 from https://www.europarl.europa.eu/RegData/etudes/BRIE/2023/747085/EPRS_BRI(2023)747085_EN.pdf.

Government of the Netherlands. *Afbouw gaswinning Groningen.* Retrieved November 22, 2022 from https://www.rijksoverheid.nl/onderwerpen/gaswinning-in-groningen/afbouw-gaswinning-groningen#:~:text=Afbouw%20gaswinning%20naar%20nul&text=Vanaf%20oktober%202022%20staat%20het,winter%20of%20bij%20grote%20leveringsproblemen.

Government of the Netherlands (2015, December 15). *Policy Document on the North Sea 2016–2021.* Retrieved November 22, 2022 from https://www.government.nl/binaries/government/documenten/policy-notes/2015/12/15/policy-document-on-the-north-sea-2016-2021/nz-eng-beeldscherm.pdf.

Government of the Netherlands (2019, June 28). *Klimaatakkoord.* Retrieved November 22, 2022 from https://www.klimaatakkoord.nl/binaries/klimaatakkoord/documenten/publicaties/2019/06/28/klimaatakkoord/klimaatakkoord.pdf.

Government of the Netherlands (2022a, July 15), *Kabinet trekt extra 1,3 miljard uit voor waterstofprojecten.* Retrieved November 22, 2022 from https://www.rijksoverheid.nl/actueel/nieuws/2022/07/15/kabinet-trekt-extra-13-miljard-uit-voor-waterstofprojecten.

Government of the Netherlands (2022b, September 26). *Gaskraan Groningen op waakvlam.* Retrieved November 22, 2022 from https://www.rijksoverheid.nl/actueel/nieuws/2022/09/26/gaskraan-groningen-op-waakvlam.

Government of the Netherlands (2023, November 14). *Duitsland en Nederland versterken samenwerking op gebied waterstof.* Retrieved November 16, 2023 from https://www.rijksoverheid.nl/actueel/nieuws/2023/11/14/duitsland-en-nederland-versterken-samenwerking-op-gebied-waterstof#:~:text=Nederland%20en%20Duitsland%20leggen%20elk,van%20hernieuwbare%20waterstof%20vanaf%202027.

Hydrogen Europe. *EU projects—Hydrogen Europe.* Retrieved November 22, 2022 from https://hydrogeneurope.eu/eu-projects/.

HyXchange. *About us.* Retrieved November 22, 2022 from https://hyxchange.nl/about/.

HY3. *Decarbonizing the Dutch & German Industry through hydrogen.* Retrieved November 22, 2022 from https://hy3.eu/.

H-vision. *Using hydrogen to make large and fast cuts in carbon emissions.* Retrieved November 22, 2022 from https://www.h-vision.nl/en.

IEA & CIEP (2021, April). *Hydrogen in North-Western Europe, A vision towards 2030.* Retrieved November 14, 2023 from https://www.clingendaelenergy.com/inc/upload/files/NW-Europe-Hydrogen-Final.pdf.

IRENA (2022, January 15). *Geopolitics of the Energy Transformation: The Hydrogen Factor.* Retrieved November 22, 2022 from https://www.irena.org/publications/2022/Jan/Geopolitics-of-the-Energy-Transformation-Hydrogen.

Meijer, B. (2022, November 2). Dutch court carbon capture project ruling alarms building sector. *Reuter.* https://www.reuters.com/world/europe/dutch-court-says-permits-major-carbon-storage-project-are-uncertain-2022-11-02/.

MIEK (2021, November 22). *MIEK overview 2021 Multiyear programme Infrastructure Energy and Climate.* Retrieved December 8, 2022 from https://www.rijksoverheid.nl/documenten/rapporten/2021/11/26/meerjarenprogramma-infrastructuur-energie-en-klimaat---overzicht-2021.

Ministry of Economic Affairs and Climate (2020, April 6). *Government Strategy on Hydrogen.* Retrieved November 22, 2022 from https://www.government.nl/documents/publications/2020/04/06/government-strategy-on-hydrogen.

Ministry of Economic Affairs and Climate (2022a, June 28). *Nederlandse reactie op de publieke internetconsulatie door de Europese Commissie.* Retrieved November 22, 2022 from https://open.overheid.nl/repository/ronl-1c27dc1e0ab3ace66554a763d09eadb2c70d0362/1/pdf/22229490bijlage-3-nederlandse-reactie-op-europese-consultatie-gedelegeerde-handelingen-hernieuwbare-waterstof.pdf.

Ministry of Economic Affairs and Climate (2022b, June 29). *Kamerbrief over ontwikkeling transportnet voor waterstof.* Retrieved November 22, 2022, from https://open.overheid.nl/repository/ronl-5c57a9ba35fa907dcc805ca0da463dc33b036bb8/1/pdf/ontwikkeling-transportnet-voor-waterstof.pdf.

Ministry of Economic Affairs and Climate (2022c, June 29). *Kamerbrief voortgang ordening en ontwikkeling waterstofmarkt.* Retrieved November 22, 2022, from https://open.overheid.nl/repository/ronl-c369632dae73cb4ed9dd9e8e4af10e64e74ae1f9/1/pdf/voortgang-ordening-en-ontwikkeling-waterstofmarkt.pdf.

Ministry of Economic Affairs and Climate (2022d, December 9). *Kamerbrief over garantieregeling Porthos.* Retrieved November 22, 2022 from https://www.rijksoverheid.nl/onderwerpen/klimaatverandering/documenten/kamerstukken/2022/12/09/garantieregeling-porthos.

Ministry of Economic Affairs and Climate (2023a, July 3). *Concept-Nationaal Plan Energiesysteem: Hoofddocument.* Retrieved November 22, 2022 from https://www.rijksoverheid.nl/documenten/rapporten/2023/07/03/bijlage-1-hoofddocument-concept-npe.

Ministry of Economic Affairs and Climate (2023b, October 5). *Voortgang Delta Rhine Corridor.* Retrieved November 22, 2022 from https://www.tweedekamer.nl/kamerstukken/brieven_regering/detail?id=2023Z16805&did=2023D40820.

Mission Innovation. *Catalysing clean energy solutions for all.* Retrieved November 22, 2022 from http://mission-innovation.net/.

National Hydrogen Programme (2022, November). *Routekaart.* Retrieved November 22, 2022 from https://nationaalwaterstofprogramma.nl/documenten/handlerdownloadfiles.ashx?idnv=2339011.

National Hydrogen Programme (a*). Import en export.* Retrieved November 22, 2022 from https://nationaalwaterstofprogramma.nl/themas/thema+import+en+export/default.aspx.

National Hydrogen Programme (b). *Internationale Samenwerking.* Retrieved November 22, 2022 from https://www.nationaalwaterstofprogramma.nl/kennisbank/2543118.aspx?t=Internationale-samenwerking.

National Hydrogen Programme (c). *Subsidiemogelijkheden waterstof.* Retrieved November 22, 2022 from https://nationaalwaterstofprogramma.nl/kennisbank/2244223.aspx?t=Subsidiemogelijkheden-waterstof.

National Hydrogen Programme (d). *Thema's.* Retrieved November 22, 2022 from https://nationaalwaterstofprogramma.nl/themas/default.aspx.

National Hydrogen Programme (e). *Waterstofprojecten.* Retrieved November 22, 2022 from https://nationaalwaterstofprogramma.nl/kennisbank/2215926.aspx?t=Waterstofprojecten.

Netbeheer Nederland (2023). *Integrale Infrastructuur-Verkenning 2030–2050.* Retrieved November 22, 2022 from https://www.netbeheernederland.nl/_upload/RadFiles/New/Documents/A%20II3050%20Eindrapport%203.pdf.

Netherlands Enterprise Agency (2021a, March). *Excelling in Hydrogen.* Retrieved November 22, 2022 from https://www.rvo.nl/sites/default/files/2021/03/Dutchsolutionsforahydrogeneconomy.pdf.

Netherlands Enterprise Agency (2021b, April 26). *Programma Energyhoofdstructuur.* Retrieved November 22, 2022 from https://www.rvo.nl/sites/default/files/2021/04/Concept-Notitie-Reikwijdte-en-Detailniveau-Programma-Energiehoofdstructuur.pdf.

Netherlands Enterprise Agency (2021c, November 18). *Aramis.* Retrieved November 6, 2023 from https://www.rvo.nl/onderwerpen/bureau-energieprojecten/lopende-projecten/aramis#pro jectfase.

Netherlands Enterprise Agency (2023a, January 27). *Stimulering Duurzame Energieproductie en Klimaattransitie (SDE++).* Retrieved November 13, 2023 from https://www.rvo.nl/subsidies-financiering/sde?gclid=Cj0KCQjwxveXBhDDARIsAI0Q0x2yDvlyPafxqRxTifQs8-BsbzIC fJ3zU-cydvQk8IGC5Yxu00Pr45MaAsEAEALw_wcB.

Netherlands Enterprise Agency (2023b, February 3). *Delta Rhine Corridor.* Retrieved November 22, 2023 from https://www.rvo.nl/onderwerpen/bureau-energieprojecten/lopende-projecten/drc.

NortH2. *Kickstarting the green hydrogen economy.* Retrieved November 22, 2022 from https://www.north2.eu/en/.

Offshore Energy Magazine (2023, November 20). *Eneco plans to build 800 MW green hydrogen plant in Europoort.* Retrieved November 20, 2023 from https://www.offshore-energy.biz/eneco-plans-to-build-800-mw-green-hydrogen-plant-in-europoort/.

Offshore Energy magazine (2023, November 15). *The Netherlands and Germany bolster hydrogen ties.* Retrieved November 20, 2023 from https://www.offshore-energy.biz/the-netherlands-and-germany-bolster-hydrogen-ties/#:~:text=The%20aim%20is%20to%20secure,of%20green%20hydrogen%20from%202027.

Pentalateral Energy Forum (2020, May 11). *Joint political declaration of the pentalateral energy forum on the role of hydrogen to decarbonise the energy system in Europe.* Retrieved November 22, 2022 from https://zoek.officielebekendmakingen.nl/blg-939618.pdf.

Petroleum Intelligence Weekly (19 October 2023).

Port of Rotterdam (2022, July 18). *Renewable liquid hydrogen supply chain between Portugal and Netherlands on the horizon.* Retrieved December 13, 2022 from https://www.portofrotterdam.com/en/news-and-press-releases/renewable-liquid-hydrogen-supply-chain-between-portugal-and-netherlands-on.

Port of Rotterdam (a). *Import of hydrogen.* Retrieved November 22, 2022 from https://www.portofrotterdam.com/en/port-future/energy-transition/ongoing-projects/hydrogen-rotterdam/import-of-hydrogen.

Port of Rotterdam (b). *Refining and Chemicals.* Retrieved November 22, 2022 from https://www.portofrotterdam.com/en/setting/industry-port/refining-and-chemicals.

Programme sustainability industry. *Industrieclusters.* Retrieved November 22, 2022 from https://www.verduurzamingindustrie.nl/industrieclustersre/industrieclusters/default.aspx.

Raad van State (2022, November 2). *Bouw vrijstelling stikstof van tafel, maar geen algehele bouw stop.* Retrieved November 22, 2022 from https://www.raadvanstate.nl/actueel/nieuws/133608/bouwvrijstelling-stikstof-van-tafel/-highlight=porthos.

Raad van State (2023, Augustus 16). *Porthos-project Mag Doorgaan.* Retrieved November 6, 2022 from https://www.raadvanstate.nl/actueel/nieuws/augustus/porthos-project-uitspraak/.

Raza, R. (2022, October 13). Port of Rotterdam and Cespa Sign MoU for a Green Hydrogen Corridor. *Fleetmon.* https://www.fleetmon.com/maritime-news/2022/39829/port-rotterdam-and-cespa-sign-mou-green-hydrogen-c/.

RWE. *H2opZee.* Retrieved November 23, 2022 from https://www.rwe.com/en/research-and-development/hydrogen-projects/h2opzee.

Shell (2022, July 6). *Shell start bouw van Europa's grootste groene waterstoffabriek in Rotterdam.* Retrieved October 16, 2022 from https://www.shell.nl/media/nieuwsberichten/2022/holland-hydrogen-1.html.

Tata Steel (2021, October 30). *Klimaatambities.* Retrieved October 20, 2022 from https://www.tatasteeljobs.nl/nieuwsberichten/2021/klimaatambities.html.

TNO (2020, June). *The Dutch hydrogen balance, and the current and future representation of hydrogen in the energy statistics.* Retrieved October 20, 2022 from https://repository.tno.nl/island/object/uuid%3A77b361fb-0598-40aa-8be2-97e1f6e73ce5.

Topsector Energy (2020, May 1). *Overview of Hydrogen Projects in the Netherlands.* Retrieved October 20, 2022 from https://www.topsectorenergie.nl/sites/default/files/uploads/TKIGas/pub licaties/OverviewHydrogenprojectsintheNetherlandsversie1mei2020.pdf.

Topsector Energy (2022). G*eld uit het Nationaal Groeifonds voor innovatie waterstof en duurzame materialen.* Retrieved October 23, 2022 from https://www.topsectorenergie.nl/nieuws/geld-uit-het-nationaal-groeifonds-voor-innovatie-waterstof-en-duurzame-materialen.

van der Linde, C., & Stapersma, P. (2018). *The Dutch energy economy: the energy gateway to Northwest Europe.* Retrieved October 20, 2022 from https://spectator.clingendael.org/pub/2018/2/the-dutch-energy-economy/.

Hydrogen Strategy of Sweden: Unpacking the Multiple Drivers and Potential Barriers to Hydrogen Development

Stefan Ćetković and Janek Stockburger

Abstract This chapter investigates the main elements, drivers, and challenges of the hydrogen sector in Sweden. A particular focus is placed on the approach of the Swedish government to hydrogen development and its internal and external dimensions. The domestic interest in hydrogen in Sweden has in the past been primarily focused on the decarbonization of hard-to-abate industrial sectors, in particular the steel industry. Given the current surplus of low-carbon electricity supply, which relies on hydropower, nuclear and increasingly wind power, the attention was solely directed towards domestic production and use of low-carbon hydrogen for the industry. With the growing importance of hydrogen at the EU level, accompanied by the introduction of an EU hydrogen strategy, investment funds and common standards, there has been a rapid increase in interest by business actors in various hydrogen sectors (e-fuels, green hydrogen, ammonia) in Sweden. Individual regions in Sweden have also taken the initiative and made use of EU funds to try position themselves in and benefit economically from the emerging hydrogen sectors. As most private investors aim to use green hydrogen produced by renewable energy sources, the demand for green electricity, particularly in onshore and offshore wind, is expected to skyrocket. The government, however, has so far failed to enact credible plans and policies detailing where and how new wind power projects will be built and which sectors may gain priority access to renewable electricity. There has also been a lack of effort in facilitating the realization of infrastructure for the potential transport of hydrogen through pipelines or Swedish ports. In light of the growing interest in low-carbon hydrogen, the considerable industry know-how and the vast renewable energy potential in Sweden, there is a pressing need for a more comprehensive approach by the government and a stronger alignment with the efforts of the EU and other Member States.

S. Ćetković (✉)
Institute of Political Science, Leiden University, Turfmarkt 99, 2511 DP The Hague, Netherlands
e-mail: s.cetkovic@fsw.leidenuniv.nl

J. Stockburger
Munich School of Politics and Public Policy, Technical University of Munich, Georgenstraße 99, 80798 Munich, Germany

© Helmholtz-Zentrum Potsdam, Deutsches GeoForschungsZentrum GFZ 2024
R. Quitzow and Y. Zabanova (eds.), *The Geopolitics of Hydrogen*, Studies in Energy, Resource and Environmental Economics,
https://doi.org/10.1007/978-3-031-59515-8_10

1   Introduction

Sweden has been a global leader when it comes to embracing low-carbon technologies and decarbonizing its economy. Over the past years, low-carbon hydrogen has
seen a remarkable boost in interest by government and business actors in Sweden
as a much-needed and promising technology on the path towards climate neutrality.
Sweden is widely perceived as one of the most competitive countries for the development and utilization of low-carbon hydrogen, although fulfilling this potential is
not without its challenges. There have been multiple domestic and external factors
that have shaped the recent dynamic yet fluid development of hydrogen in Sweden.
This paper provides an overview of the hydrogen sector in Sweden, focusing on the
main domestic and external drivers and challenges. A particular focus is placed on
the government approach to hydrogen development in Sweden and its relation to EU
hydrogen policy.

This paper starts by presenting the domestic development of the hydrogen sector
in Sweden and identifying the driving forces and possible barriers to the successful
growth and system integration of hydrogen. This is done by looking at the interplay
between public and private sector activities in a broader political-economic and
material context. In the second part, the external dimension of the Swedish hydrogen
strategy is discussed in more detail. The main focus is on how hydrogen development
in Sweden, including both public and private sector, is related to the European policy
and market context.

2   Domestic Hydrogen Development in Sweden

2.1   The Swedish Energy Mix and Sources of Emissions

Sweden has successfully established itself as a global leader in decarbonization
efforts (IEA, 2019), while the country maintains one of the highest energy consumption rates per capita in the EU and internationally (TheGlobalEconomy.com, 2022).
The electricity and heating sectors in Sweden have largely been decarbonized; the
primary challenge remaining has been to curb emissions from industrial processes
and transportation. In 2022, 41% of the electricity supply came from hydropower,
29% from nuclear, 19% from wind and 1% from solar power, while the remaining
9% was generated in combined heat and power plants and industrial processes (see
Fig. 1). For heating, the residential and service sector rely dominantly on electricity
and district heating, the latter being largely based on biomass (Swedish Energy
Agency, 2022). The industry and transportation sectors are the key emitters, and
together account for 64% of Swedish emissions. Road traffic is by far the dominant source of emissions in the transportation sector, with a share of 92%. In the
industrial sector, the iron and steel industry is responsible for the highest amount
of emissions (34%), followed by the minerals industry (19%) and refineries (18%)

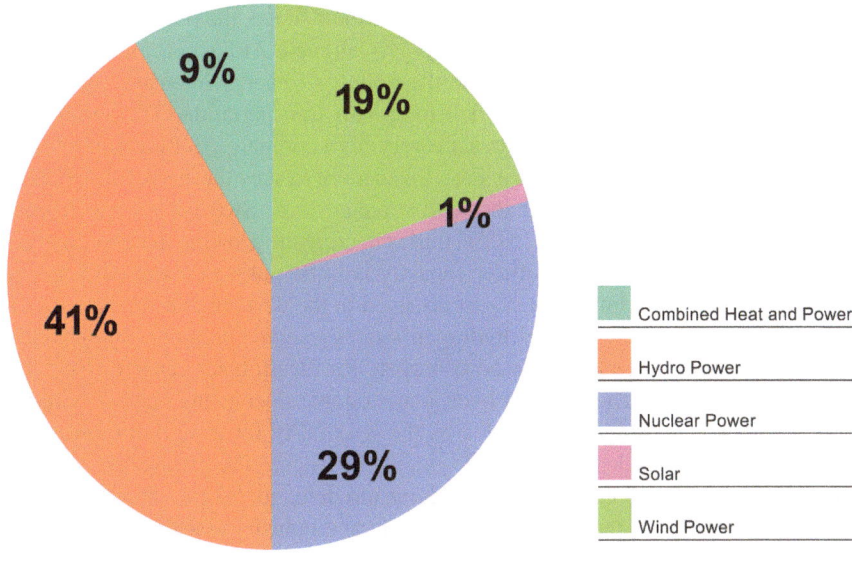

Electricity Mix of Sweden in 2022

Fig. 1 Electricity production in Sweden in 2022. *Source* Based on the data from Statistics Sweden (2023), Electricity supply and use 2001–2022 (GWh). Retrieved November 13, 2023, from https://www.scb.se/en/finding-statistics/statistics-by-subject-area/energy/energy-supply-and-use/annual-energy-statistics-electricity-gas-and-district-heating/pong/tables-and-graphs/electricity-supply-and-use-20012022-gwh/

(Ministry of the Environment, 2020). It is against this background that the government has intensified efforts and activities to promote the utilization of fossil-free hydrogen in industrial processes and transportation in Sweden over recent years. This has been strongly influenced by the growing political interest and the accelerated market and technology developments in fossil-free hydrogen in the EU and around the world.

2.2 Developing a Swedish Hydrogen Strategy

In developing the country's hydrogen strategy, the government first tasked Fossil Free Sweden, an industry-based government initiative, with the formulation of a strategic plan for hydrogen development. In January 2021, Fossil Free Sweden published its hydrogen strategy (Fossil Free Sweden, 2021). The government then commissioned the Swedish Energy Agency to develop a comprehensive proposal for a national hydrogen strategy. The Swedish Energy Agency presented its draft of the hydrogen strategy in November 2021 (Swedish Energy Agency, 2021). Although the two published strategies differ somewhat in their scope, the outlined vision and priorities for hydrogen development are largely aligned. At the core of the interest in hydrogen is the decarbonization of the country's energy-intensive industries, particularly steel production, alongside long-haul transportation. The formulated targets for new electrolyzer capacities are very high compared to European targets. The document published by Fossil Free Sweden envisages the need for 3 GW in new electrolyzers by 2030, while the draft strategy of the Swedish Energy Agency proposed an even higher target of 5 GW, thus reflecting the growing industry interest. An additional 10 GW in electrolyzer capacity is to be put into operation between 2030 and 2045. According to an expert involved in the development of the national hydrogen strategy, while the EU hydrogen strategy served as an important background document, the specific Swedish targets for electrolyzer capacity were not based on EU targets. The national hydrogen strategic documents operate with the term ´fossil-free´ hydrogen, which leaves the possibility for using both renewable energy sources and nuclear power for producing hydrogen. The planned electrolyzer capacities are intended entirely to meet domestic demand. So far, there have been no articulated plans by the Swedish government and industry associations to import or export hydrogen. The main vision has instead been to export goods produced with fossil-free hydrogen, such as ´green steel´, and to enter the global supply chain for fossil-free hydrogen production technologies. Nevertheless, the draft of the national hydrogen strategy suggests that Sweden should assess whether or not to become a hydrogen exporter (Swedish Energy Agency, 2021, p. 9). Although an external hydrogen strategy has been lacking, the Swedish government and particularly Swedish industry stakeholders have increased their international engagement in this area.

As of October 2023, the official adoption of the national hydrogen strategy by the government is still pending. Sweden held general elections on 11 September 2022, which resulted in the formation of the first-ever minority center-right government supported by the far-right populist party Sweden Democrats. While it remains to be seen whether the new government will introduce substantial changes to the national hydrogen strategy, some preliminary assessments about the changing policy environment based on the first steps taken by the new government are included in this contribution.

2.3 The Swedish Government as a Driver of Hydrogen Development

The increasing interest in fossil-free hydrogen in Sweden is driven primarily by the government's goals of achieving climate neutrality by 2045 and reducing transport emissions (excluding aviation) by 70% until 2030, relative to 2010. These goals are formulated in the Climate Policy Framework, which was adopted by an overwhelming majority in the Swedish parliament in 2017 (Ministry of the Environment,

2021). Fossil-free hydrogen is seen as a key solution in the decarbonization of energy-intensive industries and in securing their competitiveness in the future global low-carbon economy. Hydrogen is also expected to contribute to the decarbonization of the transportation sector, especially for heavy vehicles and long-haul transportation including trucks, trains, ships, and airplanes. Furthermore, hydrogen is recognized as a potential means to help balance the electricity grid given the growing share of intermittent renewable electricity, particularly wind power.

Several government support measures have been in place that indirectly or directly encourage the production and utilization of fossil-free hydrogen. A key climate policy instrument in Sweden has been the carbon tax, introduced in 1991. Since then, carbon tax rates per emitted ton of CO_2 have incrementally increased from €25 to €118 in 2022 (Ministry of Finance, 2022). Given that energy-intensive industries are already covered by the European Union Emission Trading Scheme, they are excluded from domestic carbon taxation. The carbon tax now primarily targets the transportation sector, with 95% of carbon tax revenues coming from motor fuels. Alongside the carbon tax, the government introduced the bonus-malus scheme in 2018, which provides favorable tax rates for low-emission vehicles, including both battery and fuel-cell vehicles. Thanks to government support, the purchase of low-emission vehicles in Sweden has surged in recent years, although this increase mainly pertains to rechargeable battery and hybrid vehicles. The deployment of fuel-cell vehicles, by contrast, has been negligible. By the end of 2021, 320,000 electrically powered vehicles were registered on Swedish roads, of which nearly 300,000 were passenger cars. In the same period, only 42 hydrogen-powered passenger vehicles and two lorries were registered (Trafikanalys, 2022).

The targeted government support for hydrogen production and infrastructure in Sweden has taken the form of two main support programs: Climate Leap and Industrial Leap. Climate Leap was introduced in 2015 and is administered by the Swedish Environmental Protection Agency. Its purpose is to provide financial support for low-carbon investments in the sectors not included in the EU Emission Trading Scheme. The funding is provided as a grant that covers a substantial share of the investment. The Swedish Environmental Protection Agency has noted a significant recent increase in interest in hydrogen projects under Climate Leap, which is attributed to the publication of the EU hydrogen strategy (Swedish Environmental Protection Agency, 2022, p. 40). In the period from 2015 to 2022 (March), 74 project applications relating to hydrogen were received, while 41 projects were awarded funding. In 2021 alone, 49 hydrogen-related project applications were submitted, with 29 receiving funding (Swedish Environmental Protection Agency, 2022). The typical hydrogen-related projects that have secured funding are refueling stations for hydrogen-powered vehicles, although production facilities for green hydrogen have also been supported. For example, in 2011 the company Strandmöllen AB received funding for its investment in 3 MW of electrolyzer capacity for the production of green hydrogen as transportation fuel (Strandmöllen, 2022). Industrial Leap, the second major support program, was launched in 2018 and is designed to support research, pilot, and demonstration projects for the decarbonization of industry. The program is administered by the Swedish Energy Agency and is expected to run through 2040 as

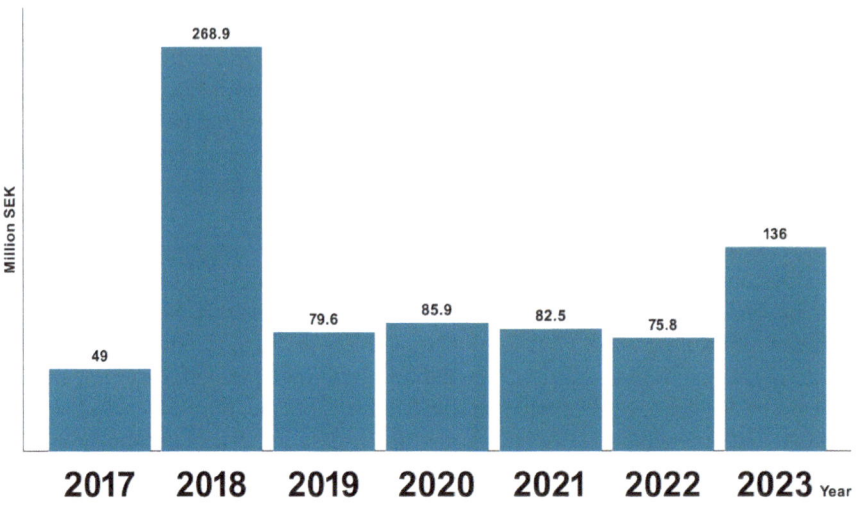

Approved Funding for Hydrogen Projects by Industrial Leap

Fig. 2 Approved funding for hydrogen projects under industrial leap. *Source* Based on the data from the Swedish Energy Agency (2023). Beviljade projekt inom Industriklivet. Retrieved November 13, 2023, from https://www.energimyndigheten.se/forskning-och-innovation/forskning/industri/industriklivet/

part of the effort to facilitate the achievement of the climate-neutrality goal by 2045. In the period up to 2020, 73% of the entire program budget was allocated to decarbonization investments in the iron and steel industry, while the mining sector was the second largest beneficiary, accounting for 7.8% of the budget (Government Offices of Sweden, 2021). Both Climate Leap and Industrial Leap have been integrated into the Swedish National Recovery and Resilience Plan (NRRP) and will thus be funded by the Next Generation EU recovery instrument. Climate Leap has been granted the largest share of the Swedish NRPP budget (24.6%) with €810 million while €287 million (8.7%) has been allocated for Industrial Leap (Binder, 2022). Figures 2 and 3 display the total amount of government funding allocated to hydrogen projects through Industrial Leap and Climate Leap. The funding data for Climate Leap is inclusive of information up to July 2023, and for Industrial Leap up to October 2023.

2.4   Industrial Interests in Hydrogen Development

Enhanced industrial competitiveness is an integral part and the desired outcome of Sweden's ambitious decarbonization strategy through 2045. The export of fossil-free products and services that are produced in a fossil-free value chain is seen as a major opportunity for Swedish companies. Swedish industry has emphasized fossil-free

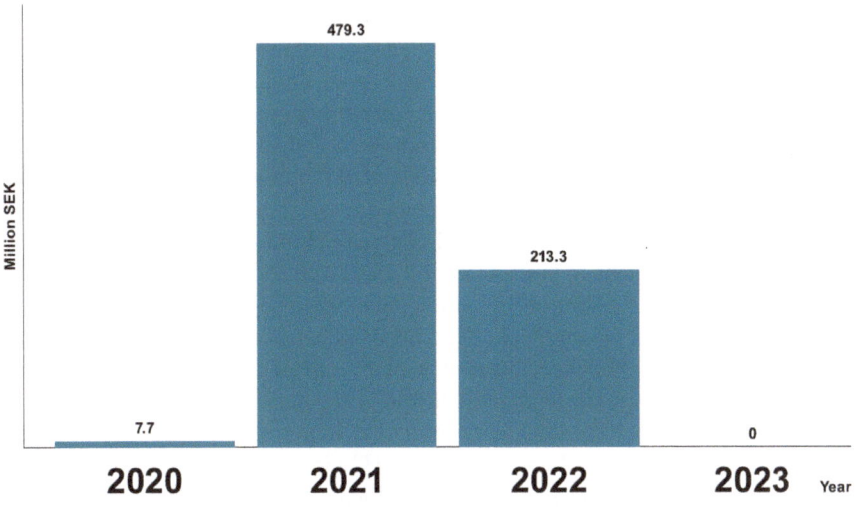

Approved Funding for Hydrogen Projects by Climate Leap

Fig. 3 Approved funding for hydrogen projects under climate leap. *Source* Based on the data from the Swedish Environmental Protection Agency (2023). Klimatklivets samlade resultat. Retrieved November 13, 2023, https://www.naturvardsverket.se/amnesomraden/klimatomstallningen/klimat klivet/resultat-for-klimatklivet/

hydrogen, with little attention paid to natural gas-based blue hydrogen. The strategy published by Fossil Free Sweden recommends that blue or gray hydrogen should not be banned, but instead that government regulation and support should focus primarily on fossil-free hydrogen (Fossil Free Sweden, 2021, p. 49). Several major industrial sectors, including steel production, the chemical industry and refineries, plan to switch their production, either completely or in large part, to fossil-free hydrogen or its derivates. In addition, the automotive industry in Sweden has been actively developing fuel-cell vehicles for both domestic and international markets. A growing number of Swedish companies are also considering entering the global hydrogen production supply chain as technology providers, subcontractors, or project developers. We will turn to these activities by external industries in later sections.

The iron and steel industry plays an important role in the Swedish economy. Sweden has rich and high-quality iron ore resources that are located mainly in the north of the country. Over 90% of the EU´s iron ore production originates from Sweden (SGU, 2020). Given the high emissions associated with the production of steel from iron ore, the transition to fossil-free hydrogen has been embraced as the main instrument for decarbonizing the Swedish steel industry. Hydrogen can be used as a substitute for fossil fuels at different stages of iron and steel production, including in the conversion of iron ore to iron sponge as well as in steelmaking from iron sponge or steel scrap. Several major projects for using fossil-free hydrogen in steel production have been initiated in Sweden. The most advanced project is

HYBRIT, a cooperation of the steelmaking company SSAB, the iron ore producer LKAB and the energy company Vattenfall. The goal of the project is to develop iron and steel production based on fossil-free hydrogen, which eventually is to become the main production method at SSAB's manufacturing units in Sweden and abroad. The HYBRIT project received support from the Swedish government through the Industrial Leap program and in 2022 was awarded funding from the EU Innovation Fund (Vattenfall, 2022a). A further large-scale fossil-free steel production project has been initiated by the newly founded company H2 Green Steel. The company, established in 2020, plans to build an entirely new iron and steel production facility in the city of Boden in northern Sweden, to be powered by an electrolyzer plant of 800 MW capacity (H2 Green Steel, 2022; S&P Global Commodity Insights, 2022). The announcement by H2 Green Steel of its plans for fossil-free steel production led the Swedish Energy Agency to increase the proposed national target for electrolyzer capacity to 5 GW by 2030, thus confirming the importance of industry plans for the government strategy. In September 2023, H2 Green Steel raised additional 1,5 billion euros in equity funds amounting to the total of 5,3 billion euros raised by the company but the final investment decision has not yet been taken (Parkes, 2023). Another Swedish steel manufacturer, Ovako, received the green light in November 2022 to construct an electrolyzer plant for producing fossil-free hydrogen which will be used to generate heat at its plant in Hofors. This project is supported by several companies including Volvo Group, Hitachi Energy, H_2 Green Steel and NEL Hydrogen (Steel Times International, 2022). The electrolyzer of 20 MW capacity was put into operation in September 2023. The excess hydrogen produced at this plant will be provided to Volvo for use in fossil-fuel trucks. According to the company officials, Ovako plans to roll-out the production and use of fossil-free hydrogen in all of its plants by 2030 under the condition that sufficient supply of fossil-free electricity is available (Martin, 2023).

The production of synthetic or electro fuels is another fossil-free hydrogen application that has attracted a growing interest in Sweden. Electro fuels are typically produced using fossil-free hydrogen and captured CO_2 of biogenic nature. Electro fuels can be used as feedstock in the industry, like eMethanol in the chemical industry, or in maritime transportation and aviation. Perstorp, one of the major Swedish chemical companies, has been part of Project Air, which aims to develop sustainable production of synthetic methanol using green hydrogen and carbon capture and utilization technology. Carbon will be captured from several sources, including the company´s operations as well as biomethane. The project, implemented together with the Finnish-owned energy companies Fortum and Uniper, received support from the Swedish Energy Agency and has most recently also been awarded funding from the EU Innovation Fund (Project Air, 2022). Sweden's largest refiner, Preem, located on the west coast of Sweden, has teamed up with Vattenfall to produce green hydrogen to decarbonize Preem´s operations and use it as an input for manufacturing synthetic fuels. The major offshore wind potential located on the Swedish west coast is to supply the majority of the renewable energy for producing green hydrogen (Vattenfall, 2022b). The Swedish company Liquid Wind, founded in 2017, has announced ambitious plans to scale up the manufacturing of eMethanol, which is to be used primarily as a transportation fuel in the maritime sector, for example (Liquid Wind, 2020).

The major automotive industry players in Sweden, Scania AB and Volvo AB, have been engaged in the development of hydrogen-based vehicles. While Scania AB has been particularly active in bringing fuel-cell trucks to market, Volvo AB has mainly focused on battery-based vehicles, although it has also been involved in testing the production of fuel-cell trucks for long-haul transport. Both companies are well-placed to become competitive manufacturers of fuel-cell vehicles in the global market. Hydrogen trucks, however, are still in the development stage, and many future challenges for mainstreaming light and heavy hydrogen vehicles remain. The representatives of Scania, for instance, have raised concerns regarding hydrogen-based heavy vehicles pertaining to issues of overall efficiency, maintenance costs and the availability of green hydrogen in Europe (Scania, 2021). In 2022, Scania officials announced a project to develop 20 fuel-cell trucks in cooperation with the US company Cummins Inc, but the company restated that fully battery powered trucks remain Scania's main strategy (Scania, 2022).

The hydrogen strategy of Fossil Free Sweden states that Sweden does not have competitive manufacturers of electrolyzers but has many companies that are increasingly taking part in the global hydrogen supply chain as technology providers and subcontractors (Fossil Free Sweden, 2021, p. 42).

2.5 Low-Carbon Electricity Mix and Different Bidding Zones

Sweden not only has an almost entirely carbon-free electricity supply system, it is also a net exporter of electricity. In 2021 it exported 25 TWh of electricity, mostly to neighboring countries including Finland, Denmark, Poland and Lithuania (Swedish Energy Agency, 2022). During the first half of 2022, Sweden became the largest electricity exporter in the EU (EnAppSys, 2022). The government, along with a growing number of industry players, see considerable potential in carbon-free electricity and a major comparative advantage for fossil-free hydrogen production. To this one should add a significant potential of untapped wind power Sweden, particularly with regard to offshore wind. The political and industry interest in offshore wind increased during the previous government, also driven by the growing need for renewable electricity in the production of green hydrogen in the future. As already noted, wind power has been identified by the Swedish industry as the preferable source for producing fossil-free hydrogen (RISE Research Institutes of Sweden, 2022). The previous government announced in 2022 the plans to ramping-up offshore wind deployment with the ultimate goal of achieving 120 TWh of electricity from offshore wind (Durakovic, 2022). To achieve this goal, the government proposed a more centralized approach to dedicate suitable areas for offshore wind and promote projects to interested investors. Previously, it was a sole responsibility of offshore wind investors to find a suitable location and apply for multiple permits. The transmission system operator was also tasked by the government with expanding the necessary grid capacity for connecting offshore wind parks. Offshore wind investors, based on the outlined policy, would be required to only bear the costs of connecting their projects to the grid connection points rather than investing in grid connection points themselves (Baltic Wind, 2022).

However, the new government in its coalition agreement pledged to revert that decision and prevent any government support for covering the connection costs for offshore wind projects (Moderata samlingspartiet et al., 2022). The new coalition agreement also does not define any concrete goals for the expansion of wind power. In its hydrogen strategy, Fossil Free Sweden called for the government to adopt an offshore wind strategy to provide long-term certainty for the planned hydrogen projects (Fossil Free Sweden, 2021). Despite the significant potential, Sweden has so far built only 192 MW of offshore wind, while as much as 15 GW of projects are in the pipeline and could become operational before 2030 (Baltic Wind, 2022). See Figs. 4 and 5 for an overview of the deployment of onshore and offshore wind power in Sweden. Torn between the demands to increase the supply of low-carbon electricity by tapping into the vast offshore wind power potential, on the one hand, and the local opposition and negative stance of the Sweden Democrats towards wind power, the current government commissioned a new study for assessing the potential for improving the development of offshore wind projects expected to be published in July 2024. This may risk further delays in the development of offshore wind projects (Baltic Wind, 2023). In the meantime, the government decided in May 2023 to authorize the construction of two offshore wind parks off the West coast despite the vocal opposition of the Sweden Democrats and local policymakers (Szumski, 2023).

In contrast to the somewhat ambiguous position on wind power, the new government has been unified in its support for maintaining and expanding nuclear power. The Framework Agreement on energy policy, adopted in 2016 by all parliamentary parties except for the far-right Sweden Democrats, set the target of achieving a 100% renewables-based electricity system by 2040. Importantly, however, the document does not envisage a ban on new nuclear power plants, nor does it allow for closing the existing nuclear power plants based on political decisions (Swedish Nuclear Society & Analysgruppen, 2016). The status of nuclear power has been a contested issue in Swedish politics. The more left-leaning parties have insisted on transitioning towards renewable energy sources, while the conservative and right-leaning parties have remained supportive of nuclear power. After the right-leaning parties won a slim majority following general elections in September 2022, they promised to facilitate a fast expansion of nuclear power capacities in Sweden through favorable administrative and financial conditions (Moderata samlingspartiet et al., 2022). In the coalition agreement, the newly elected government also replaced the goal of 100% renewable to 100% fossil-free electricity by 2040, clearly implying the important place of nuclear power in the future electricity mix. The implications of the war in Ukraine on energy supply chains and the necessity to make the domestic electricity supply more reliable have been used as arguments by the government for the accelerated deployment of nuclear power. In September 2023, the government tabled a new legislative

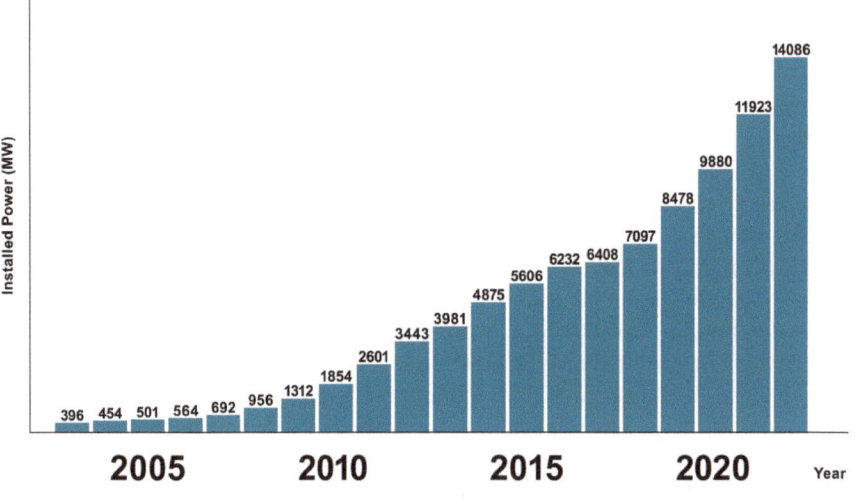

Installed Onshore Wind Power in Sweden

Fig. 4 Cumulative installed capacity of onshore wind in Sweden. *Source* Swedish Energy Agency (2023). Number of Wind Turbines, Installed Capacity and Wind Power Production by Installation Type, Whole Country, 2003-, Retrieved November 13, 2023. https://pxexternal.energimyndigheten. se/pxweb/en/Vindkraftsstatistik/Vindkraftsstatistik/EN0105_5.px/

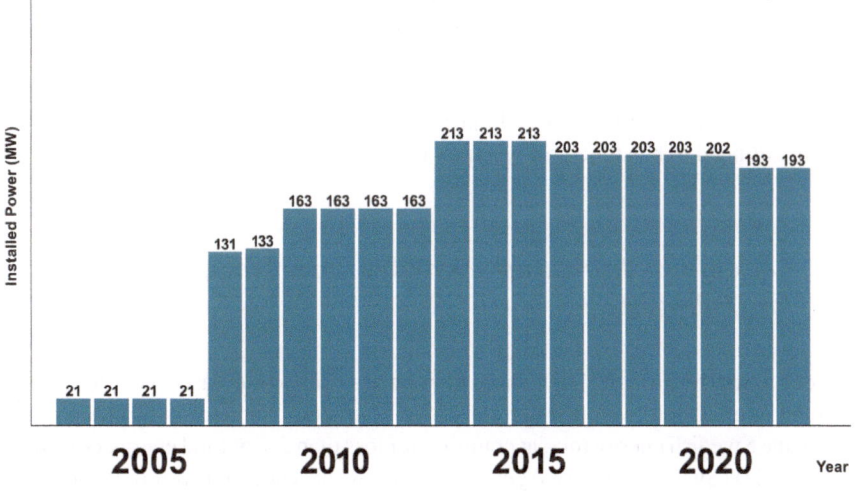

Installed Offshore Wind Power in Sweden

Fig. 5 Cumulative installed capacity of offshore wind in Sweden. *Source* Swedish Energy Agency (2023). Number of Wind Turbines, Installed Capacity and Wind Power Production by Installation Type, Whole Country, 2003-, Retrieved November 13, 2023. https://pxexternal.energimyndigheten. se/pxweb/en/Vindkraftsstatistik/Vindkraftsstatistik/EN0105_5.px/

proposal which lifts all restrictions to the construction of new power plants in Sweden (Government Offices of Sweden, 2023).

An important feature of the Swedish electricity market, which strongly affects fossil-free hydrogen investment decisions, is the separation into four bidding zones for electricity, as established in 2011. The reasoning behind the four different price zones is to encourage the construction of production capacities in areas affected by electricity shortages (Hansson et al., 2017). In the northern parts of the country (bidding zones SE1 and SE2), electricity has been historically cheaper due to surplus electricity production from hydropower and increasingly wind power, combined with lower levels of electricity consumption (Hansson et al., 2017). In the southern part of Sweden (bidding zone SE3), electricity production comes mostly from nuclear power, which in combination with higher levels of consumption leads to higher electricity prices. The existence of four different bidding zones with different electricity generation structures has implications for hydrogen investments and will lead to regionally specific hydrogen development trajectories. This will further be influenced by the EU additionality criteria and the necessity for hydrogen production facilities to meet spatial correlation requirements by purchasing renewable electricity from the same bidding zone. The major green steel projects are located in the northern part of the country, due to the appeal of cheaper and renewables-based electricity. H2 Green Steel, for instance, signed a long-term Power Purchasing Agreement with Statkraft for 2 TWh of annual hydropower-based electricity for its green steel project in Boden (Statkraft, 2022). Northern Sweden also offers large available areas that can be used for the generation of renewable energy from wind power (Fossil Free Sweden industry strategy, 2021). The company Svevind plans to build a large onshore wind power park with a total capacity of up to 4 GW in northern Sweden (Svevind, 2021). The production projects of renewable hydrogen and its derivatives in bidding zones SE3 and SE4 will need to rely primarily on onshore and offshore wind power which will need to be significantly expanded.

2.6   Lack of Gas Infrastructure and Natural Storage for Hydrogen

Sweden initiated an early transition to biomass and waste-based heating, which resulted in an underdeveloped natural gas infrastructure. Natural gas plays a minor role in the Swedish energy mix, accounting for less than 2% of total energy consumption in 2020 (Swedish Energy Agency, 2022). The absence of a well-developed gas infrastructure in Sweden limits the possibility to store and transport hydrogen. The Swedish Energy Agency, in its hydrogen strategy draft, indicates the possibility of transporting hydrogen via train, truck or ship in cases where no pipelines exist (Swedish Energy Agency, 2021). However, the Swedish industry estimates that the absence of gas infrastructure will reduce the possibilities of cost-effective transport of hydrogen over long distances (Fossil Free Sweden, 2021). Additionally, Sweden lacks hydrogen storage opportunities because geographically the country has only few salt caverns that can be used for future hydrogen storage. One interviewee said that Sweden could benefit from a broad connection to the EU gas infrastructure, as the country's storage possibilities would be expanded. A plan exists for Sweden to become a part of the so-called "European Hydrogen Backbone" over the coming decades (EHB, 2022). The Hydrogen Backbone is an initiative by 33 infrastructure operators with the goal, among other things, of accelerating decarbonization and connecting regions "with abundant supply potential with centers of demand" (EHB, 2022, p. 4). In 2022, gas infrastructure operators from Finland and Sweden, Gasgrid Finland and Nordion Energi, launched the Nordic Hydrogen Route initiative to build a cross-border infrastructure for transporting hydrogen through pipelines in the Bothnian Bay (Nordic Hydrogen Route, 2022). The Nordic Hydrogen Route is designed as part of the European Hydrogen Backbone Initiative, based on the idea to link the hydrogen sectors in Sweden and Finland with the rest of Europe. However, the Swedish industry assessment is that, because the country simply lacks the necessary infrastructure, it is "not a realistic scenario" for Sweden to become part of the European Hydrogen Backbone (Fossil Free Sweden, 2021, p. 8).

2.7   Summary of Opportunities and Trade-Offs

Ambitious climate goals, a highly competitive and future-oriented industry and available emission-free electricity constitute the main opportunities for Sweden to build a successful hydrogen sector. Sweden possesses both a high emerging domestic demand and attractive domestic supply conditions for fossil-free hydrogen. So far, the Swedish government has not taken a particularly proactive role when it comes to making long-term strategic decisions for spurring the domestic hydrogen market. While a bottom-up, industry-driven approach proved fruitful in the early experimentation phase, the mainstreaming of fossil-free hydrogen may require more active government planning. This concerns in particular the question of securing a sufficient supply of low-carbon electricity and the related grid capacities in the future. Given the plans for massive electrification of the Swedish economy, electricity demand is set to rise significantly. The scenarios calculated by the Swedish electricity grid operator project that electricity demand will almost double by 2050 from the current 140–266 TWh, under the scenario in which the electricity mix is based on a combination of nuclear and renewable energy (Svenska Kraftnät, 2021). The study notes that the future electricity demand may turn out to be even higher. The coalition agreement of the new government foresees that the electricity demand will already increase to at least 300 TWh by 2045 (Moderata samlingspartiet et al., 2022). Both the electricity price and renewable energy share will be central to building a domestic green hydrogen industry that is capable of entering the global supply chains and foreign markets. The question of whether Swedish companies will be in a position to export hydrogen and its derivatives also remains to be addressed. In the absence of a government vision for the future electricity mix and support for renewable energy infrastructure, important conflicts, and bottlenecks in the supply of low-carbon electricity may emerge. Given the lack of a suitable gas infrastructure, the issues of whether and how hydrogen will be transported within and outside Sweden, and under what conditions hydrogen gas infrastructure can be built, have also remained largely unresolved (RISE Research Institutes of Sweden, 2022).

3 External Dimension of Hydrogen Development in Sweden

So far, Sweden has lacked a coherent external hydrogen strategy, due to the fact that hydrogen development has largely evolved in a bottom-up fashion, driven by private sector decisions and market impulses. This is not to say that the hydrogen sector in Sweden will not gradually become more embedded into regional and international supply chains, on the contrary. Sweden as an open market economy and an EU Member State is deeply integrated into the Nordic and European markets and policy frameworks. In that sense, the future of the hydrogen sector in Sweden is closely intertwined with the regulatory market and infrastructural developments in the Nordic region and the EU. Many companies active in the growing hydrogen market in Sweden have already established partnerships and launched joint projects with business partners from abroad in a bid to strengthen their business case in a promising but uncertain investment environment.

3.1 Strategic Objectives of the External Hydrogen Dimension in Sweden

Given the ambitious plans for industrial advances through the utilization of fossil-free hydrogen in Sweden, the priority for Swedish industry and government has increasingly been to achieve favorable regulatory conditions for the emerging hydrogen industry. The industry and the government have turned their attention in particular to the evolving EU hydrogen standards to ensure that the regulatory framework is not an obstacle to ongoing and planned investments in fossil-free hydrogen. On the industry side, the objective has also been to explore new partnerships for projects in hydrogen production, infrastructure, and transport. This is mainly occurring within the Nordic regional cooperation, while new avenues for international cooperation and trade in hydrogen and its derivatives are also being explored. The Swedish government has been less active in steering and supporting the internationalization of the domestic hydrogen sector.

3.2 Lobbying for Favorable EU Hydrogen Standards

The draft of the national hydrogen strategy states that Sweden intends to take an active role in the dialogue in European and international fora regarding technical, economic and regulatory standards on hydrogen (Swedish Energy Agency, 2021, pp. 8–9). Public and private sector efforts have been particularly focused on negotiations over the new EU hydrogen rules. The most contested provision has been the ´additionality criteria´, which is meant to ensure that electricity deployed to produce green hydrogen is generated through additional renewable energy installations and not from existing capacities. The Swedish industry has been vocal in making the case that such a rule should not apply to countries like Sweden that already have a high share of renewable energy and an oversupply of electricity. The stringent additionality criteria seem particularly challenging for the decarbonization of the steel industry, which requires a substantial electricity supply for generating fossil-free hydrogen. Swedish industry actors have thus proposed that the additionality criteria should not apply to countries that have a higher than 50% share of renewable energy in their electricity mix (Swedenergy, 2021). Such a position has also been officially backed by the Swedish government, which has opposed any requirements that fossil-free hydrogen production facilities be directly connected to newly constructed renewable electricity plants (Ministry of Infrastructure, 2021). Swedish industry associations have further opposed any mandatory renewable energy targets for the industry in the new Renewable Energy Directive, as such targets would discriminate against the use of fossil-free electricity that may be generated by nuclear power plants (Swedenergy, 2021). The final adopted rules in the hydrogen delegated act which relaxes the additionality requirement for bidding zones where renewable electricity share exceeds 90% or is below the defined threshold of CO_2 intensity largely address the concerns of the Swedish industry.

3.3 External Actions of the Swedish Industry

The Swedish industrial players have increasingly been exploring opportunities to transfer their acquired expertise in green hydrogen production to external markets and develop commercial applications for hydrogen and its derivatives. There is, for instance, an emerging cooperation between Swedish and Spanish companies, as both countries share highly ambitious green hydrogen strategies. In 2021, H2 Green Steel signed an agreement with the major Spanish energy company Iberdrola to build an electrolyzer plant in Spain for green steel production (Iberdrola, 2021). In the opposite direction, the Spanish fertilizer producer Grupo Fertiberia, reached an agreement in 2021 with the authorities of the northern Swedish region of Norrbotten to construct a green ammonia and fertilizer plant featuring a 600 MW electrolyzer. The plant could become operational by 2026 and is also expected to export green ammonia abroad (Ammonia Energy Association, 2021).

In collaboration between industrial actors, regional government authorities and scientific institutes, hydrogen development in Sweden has increasingly taken the form of clusters centered around regional ports. Each regional cluster seeks to attract hydrogen investments and benefit economically from the emerging fossil-free hydrogen sector. Some of the clusters have been actively supported by the EU Hydrogen Valleys program, which is part of the EU Smart Specialization Platform (European Commission, 2022). The previously mentioned northern region of Norrbotten, which is host to the two major green steel projects, is one of the main hydrogen clusters. In the port city of Lulea, Uniper and ABB have launched the Bothnian Bay H2 initiative, which will enhance the development of an additional hydrogen hub (Uniper, 2021). The plans provide for construction of green hydrogen production capacities based on wind power. Green hydrogen is meant for industrial processes or as an input for producing maritime fuels. The export of hydrogen derivatives is also planned. The central-eastern region of Gävleborg has also invested in efforts to establish another hydrogen cluster, the so-called Mid Sweden Hydrogen Valley, supported by the EU Hydrogen Valleys program. The remaining two regional clusters are located in the south, around the port city of Trelleborg, and in the central-western region with a focal point in Göteborg (Region Gävleborg, 2022).

The import of hydrogen has not been high on the agenda, given the good domestic conditions for fossil-free hydrogen production and limited gas infrastructure for transporting hydrogen. However, under the previously mentioned Nordic Hydrogen Route, plans for constructing a hydrogen pipeline infrastructure have been announced. Such an infrastructure may create entirely new possibilities in the internationalization of the Swedish hydrogen sector, although its realization remains highly uncertain.

4   Conclusion

Sweden has witnessed a very dynamic and diverse development of the fossil-free hydrogen sector in recent years. With its ambitious decarbonization targets and various support programs for low-carbon technologies, the Swedish government has put in place an overall support framework while leaving it to the private sector to make technology and investment choices on the path to decarbonization. With the current surplus supply of carbon-free electricity, the utilization of fossil-free hydrogen has advanced domestically as a logical choice for cutting emissions in strategic energy-intensive industries such as iron and steel. The actions of Swedish stakeholders towards the utilization of fossil-free hydrogen in steel production have preceded the EU hydrogen strategy and policy framework. In many other sectors, including hydrogen-based transportation and synthetic fuels, the interest of industry and regional governments has largely been fueled by the growing support for fossil-free hydrogen at EU level. The development of a charging infrastructure for hydrogen vehicles, industrial hydrogen pilot projects as well as the establishment of regional hydrogen clusters in Sweden have all been decisively supported through EU funds.

Both of the main government support programs, Industrial Leap and Climate Leap, have been incorporated into the EU Next Generation Fund. The emerging EU hydrogen rules and standards will also have vast implications for the Swedish fossil-free hydrogen sector. At the same time, the draft of Sweden's national hydrogen strategy is only loosely linked to the EU hydrogen strategy.

The Swedish government has so far avoided embracing a stronger steering role in hydrogen development. On the one hand this has to do with the traditional preference by the Swedish government for a market-based approach, while on the other hand it is due to the inherited liberalized and regionally integrated electricity market, which makes it more difficult for the government to assume a steering role. The previous government coalition launched efforts to develop a national hydrogen strategy and accelerate the implementation of offshore wind projects. The newly elected government, however, has signaled less enthusiasm for wind power, instead placing the main emphasis on maintaining and expanding nuclear power. In the absence of clear and decisive governmental steering, the development paths of the fossil-free hydrogen sector in Sweden will remain fluid and largely driven by industry initiatives. Most importantly, without an organized and timely utilization of the Swedish renewable energy potential, mainly in wind power, the investments in fossil-free hydrogen will likely be at risk. This in turn indicates that the EU´s influence will continue to be of exceptional importance and that the EU may require a more holistic approach to fossil-free hydrogen which would include a more coordinated policy strategy along the entire hydrogen production and use cycle.

Acknowledgements Research for this chapter was financially supported by the German Federal Foreign Office within the framework of the project "Geopolitics of the Energy Transformation—Implications of an International Hydrogen Economy" (GET Hydrogen), funding reference number AA4521G125.

References

Ammonia Energy Association. (2021). Renewable ammonia in Sweden. Retrieved December 13, 2022 from https://www.ammoniaenergy.org/articles/renewable-ammonia-in-sweden/.

Baltic Wind. (2023). Government study evaluating offshore wind projects may delay their development in Sweden. Retrieved November 13, 2023 from https://balticwind.eu/government-study-evaluating-offshore-wind-projects-may-delay-their-development-in-sweden/.

Baltic Wind. (2022). Sweden's acceleration in offshore wind energy. Retrieved December 13, 2022 from https://balticwind.eu/swedens-acceleration-in-offshore-wind-energy/.

Binder, E. (2022). Sweden's National Recovery and Resilience Plan: Latest state of play. Retrieved December 13, 2022 from https://www.europarl.europa.eu/RegData/etudes/BRIE/2022/733581/EPRS_BRI(2022)733581_EN.pdf.

Durakovic, A. (2022). Sweden Launches Major Offshore Wind Push, Targets 120 TWh Annually. Retrieved December 13, 2022 from https://www.offshorewind.biz/2022/02/15/sweden-launches-major-offshore-wind-push-targets-120-twh-annually/.

EHB. (2022). European Hydrogen Backbone: A EUROPEAN HYDROGEN INFRASTRUCTURE VISION COVERING 28 COUNTRIES. Retrieved December 13, 2022 from https://ehb.eu/files/downloads/ehb-report-220428-17h00-interactive-1.pdf.

EnAppSys. (2022). Sweden overtakes France as Europe's biggest net power exporter. Retrieved December 13, 2022 from https://www.enappsys.com/sweden-overtakes-france-as-europes-biggest-net-power-exporter/.

European Commission. (2020). communication from the commission to the European parliament, the council, the European economic and social committee and the committee of the regions—A hydrogen strategy for a climate-neutral Europe. Retrieved December 13, 2022 from https://eur-lex.europa.eu/legal-content/EN/TXT/PDF/?uri=CELEX:52020DC0301&from=EN.

European Commission. (2022). Hydrogen valleys. Retrieved December 13, 2022 from https://s3platform.jrc.ec.europa.eu/hydrogen-valleys#fragment-89005-rvar.

Fossil Free Sweden. (2021). Strategy for fossil free competitiveness: Hydrogen. Retrieved December 13, 2022 from https://fossilfrittsverige.se/wp-content/uploads/2021/01/Hydrogen_strategy_for-_fossil_free_competitiveness_ENG.pdf.

Government Offices of Sweden. (2021). Sveriges återhämtningsplan. Retrieved December 13, 2022 from https://www.regeringen.se/49bfc1/contentassets/dad10f1743b64c78a1c5b2d71f81a6eb/sveriges-aterhamtningsplan.pdf.

Government Offices of Sweden. (2023). Ett första steg mot ny kärnkraft i Sverige – för första gången på 40 år. Retrieved November 13, 2023 from https://www.regeringen.se/pressmeddelanden/2023/09/ett-forsta-steg-mot-ny-karnkraft-i-sverige--for-forsta-gangen-pa-40-ar/.

H2 Green Steel. (2022). A pioneering venture – starting in Boden. Retrieved December 13, 2022 from https://www.h2greensteel.com/boden.

Hansson, J., Hackl, R., Taljegard, M., Brynolf, S., & Grahn, M. (2017). The potential for electrofuels production in Sweden utilizing fossil and biogenic CO_2 Point Sources. *Frontiers in Energy Research, 5,* 4. https://doi.org/10.3389/fenrg.2017.00004

Iberdrola. (2021). Iberdrola and H2 Green Steel sign 2.3 billion euros green hydrogen deal. Retrieved December 13, 2022 from https://www.iberdrola.com/press-room/news/detail/deal-green-hydrogen-iberdrola-h2-green-steel.

IEA. (2019). Sweden is a leader in the energy transition, according to latest IEA country review. Retrieved December 13, 2022 from https://www.iea.org/news/sweden-is-a-leader-in-the-energy-transition-according-to-latest-iea-country-review.

Liquid Wind. (2020). Flagships. Retrieved December 13, 2022 from https://www.liquidwind.se/flagships.

Martin, P. (2023). Sweden's largest electrolyser project inaugurated to produce hydrogen for green steelmaking. Retrieved November 13, 2023 from https://www.hydrogeninsight.com/industrial/swedens-largest-electrolyser-project-inaugurated-to-produce-hydrogen-for-green-steelmaking/2-1-1512389.

Ministry of Finance. (2022). Carbon Taxation in Sweden. Retrieved December 13, 2022 from https://www.government.se/498550/globalassets/government/bilder/finansdepartementet/carbon-taxes/220422-carbon-tax-sweden---general-info.pdf.

Ministry of Infrastructure. (2021). Rådets möte (energiministrarna) den 11 juni 2021. Retrieved December 13, 2022 from https://www.regeringen.se/49cb21/contentassets/612beacd7cad4f07a8ae9a879122da6c/kommenterad-dagordning-tte-energi-11-juni-2021.

Ministry of the Environment. (2020). Sweden's long-term strategy for reducing greenhouse gas emissions. Retrieved December 13, 2022 from https://unfccc.int/documents/267243.

Ministry of the Environment. (2021). Sweden's climate policy framework. Retrieved December 13, 2022 from https://www.government.se/articles/2021/03/swedens-climate-policy-framework/#:~:text=By%202045%2C%20Sweden%20is%20to,greenhouse%20gases%20into%20the%20atmosphere.&text=The%20framework%20contains%20ambitious%20climate,and%20stability%20in%20climate%20policy.

Moderata samlingspartiet, Kristdemokraterna, Liberalerna, & Sverigedemokraterna. (2022). Tidöavtalet: Överenskommelse för Sverige. Retrieved December 13, 2022 from https://kristdemokraterna.se/download/18.715f6f45183890627fcf689/1665737287060/Tid%C3%B6avtalet%20-%20%C3%96verenskommelse%20f%C3%B6r%20Sverige.pdf.

Nordic Hydrogen Route. (2022). ABOUT THE PROJECT. Retrieved December 13, 2022 from https://nordichydrogenroute.com/project/.

Parkes, R. (2023). H2 Green Steel secures 1,5 bn in equity to fund hydrogen based iron and steel project. Retrieved November 13, 2023 from https://www.hydrogeninsight.com/industrial/h2-green-steel-secures-1-5bn-in-equity-to-fund-hydrogen-based-iron-and-steel-project/2-1-151 4193.

Project Air. (2022). Project Air selected by EU Innovation Fund as one of 17 projects to share funding of EUR 1.8 billion. Retrieved December 13, 2022 from https://projectair.se/.

Region Gävleborg. (2022). Vätgas en drivkraft för framtiden. Retrieved December 13, 2022 from https://www.regiongavleborg.se/regional-utveckling/naringsliv-och-innovation/mid-swe den-hydrogen-valley/.

RISE Research Institutes of Sweden. (2022). Prestudy H2ESIN: Hydrogen, energy system and infrastructure in Northern Scandinavia and Finland. Retrieved December 13, 2022 from https://www.ri.se/sites/default/files/2022-11/Project%20report%20-%20Prestudy%20H2ESIN_1.pdf.

S&P Global Commodity Insights. (2022). Sweden's H2 Green Steel signs 14-TWh PPA to power planned electrolyzer. Retrieved December 13, 2022 from https://www.spglobal.com/commodityinsights/en/market-insights/latest-news/electric-power/060822-swedens-h2-green-steel-signs-14-twh-ppa-to-power-planned-electrolyzer.

Scania. (2021). Remissvar från Scania CV AB gällande EU-kommissionens förslag till ändring av direktiv (2014/94/EU) om utbyggnad av infrastrukturen för alternativa bränslen: DNR I2021/02043. Retrieved December 13, 2022 from https://www.regeringen.se/4a5665/contentassets/fdf 25f423ebc494db94b95497b566e2f/scania-ab.pdf.

SGU. (2020). Ore production and trends. Retrieved December 13, 2022 from https://www.sgu.se/en/mineral-resources/mineral-statistics/ore-production-and-trends/.

Statkraft. (2022). H2 Green Steel partners with Statkraft for 14 TWh of renewable electricity. Retrieved December 13, 2022 from https://www.statkraft.com/newsroom/news-and-stories/arc hive/2022/statkraft-signs-ppa-with-h2-green-steel/.

Statistics Sweden. (2023). Electricity supply and use 2001–2022 (GWh). Retrieved November 13, 2023 from https://www.scb.se/en/finding-statistics/statistics-by-subject-area/energy/energy-supply-and-use/annual-energy-statistics-electricity-gas-and-district-heating/pong/tables-and-graphs/electricity-supply-and-use-20012022-gwh/.

Steel Times International. (2022). Sweden's largest electrolyzer approved for construction. Retrieved December 13, 2022 from https://www.steeltimesint.com/news/swedens-largest-ele ctrolyzer-approved-for-construction.

Strandmöllen. (2022). Strandmöllen AB has placed a purchase order for a 3MW electrolyser from FEST GmbH. Retrieved December 13, 2022 from https://www.strandmollen.se/las-mer/strand mollen-electrolyser.

Svenska Kraftnät. (2021). Långsiktig marknadsanalys 2021: Scenarier för elsystemets utveckling fram till 2050. Sundbyberg. Retrieved December 13, 2022 from https://www.svk.se/siteassets/om-oss/rapporter/2021/langsiktig-marknadsanalys-2021.pdf.

Svevind. (2021). This wind project can still hit 4GW. Retrieved December 13, 2022 from https://svevind.se/en/2021/07/02/this-wind-project-can-still-hit-4gw/.

Swedenergy. (2021). Swedenergy, Swedish Windenergy and Swedish Gas Association's common position on hydrogen in the EU RED directive. Retrieved December 13, 2022 from https://www.energiforetagen.se/globalassets/energiforetagen/sa-tycker-vi/internationellt-arbeteu/red-iii-position-paper-hydrogen-swedenergy-feb-2022-final.pdf.

Swedish Energy Agency. (2021). Förslag till Sveriges nationella strategi för vätgas, elektrobränslen och ammoniak. Retrieved December 13, 2022 from https://www.energimyndigheten.se/remiss var-och-uppdrag/?query=v%C3%A4tgas&cat=1&year=&recipient=.

Swedish Energy Agency. (2022). Energy in Sweden 2022: An Overview. Retrieved December 13, 2022 from https://energimyndigheten.a-w2m.se/FolderContents.mvc/Download?ResourceId= 208766#:~:text=In%20Sweden%20we%20use%20domestic,divided%20into%20supply% 20and%20consumption.

Swedish Environmental Protection Agency. (2022). Lägesbeskrivning för Klimatklivet. Retrieved December 13, 2022 from https://www.naturvardsverket.se/globalassets/amnen/klimat/klimatkli vet/lagesbeskrivning-for-klimatklivet-2022-04-13.pdf.

Swedish Nuclear Society, & Analysgruppen. (2016). The Swedish energy policy agreement of 10 June 2016 – unofficial english translation. Retrieved December 13, 2022 from https://analys.se/ wp-content/uploads/2016/06/swedish-political-energy-agreement-2016.pdf.

Szumski, C. (2023). Swedish far-right outraged over governments wind power expansion plans. Retrieved November 13, 2023 from https://www.euractiv.com/section/politics/news/swedish-far-right-outraged-over-governments-wind-power-expansion-plans/.

TheGlobalEconomy.com. (2022). Energy use per capita, 2015—Country rankings. Retrieved December 13, 2022 from https://www.theglobaleconomy.com/rankings/energy_use_per_cap ita/.

Trafikanalys. (2022). Eldrivna vägfordon—ägande, regional analys och möjlig utveckling till 2030: Rapport 2022:12. Stockholm. Retrieved December 13, 2022 from https://www.trafa.se/global assets/rapporter/2022/rapport-2022_12-eldrivna-vagfordon---agande-regional-analys-och-en-mojlig-utveckling-till-2030.pdf.

Uniper. (2021). BotnialänkenH2. Retrieved December 13, 2022 from https://www.uniper.energy/ sverige/nyheter/botnialanken/.

Vattenfall. (2022a). HYBRIT receives support from the EU Innovation Fund. Retrieved December 13, 2022 from https://group.vattenfall.com/press-and-media/pressreleases/2022/hyb rit-receives-support-from-the-eu-innovation-fund.

Vattenfall. (2022b). Vattenfall and Preem to investigate large scale decarbonization using offshore wind and hydrogen. Retrieved December 13, 2022 from https://group.vattenfall.com/press-and-media/pressreleases/2022/vattenfall-and-preem-to-investigate-large-scale-decarbonizat ion-using-offshore-wind-and-hydrogen.

Stefan Ćetković is an assistant professor of energy and environmental politics and policy at Leiden University in the Netherlands. Previously, he was a senior researcher and lecturer at the Munich School of Politics and Public Policy at the Technical University of Munich, Germany. He has a long experience as a researcher, lecturer and policy advisor on energy and climate issues at the intersection between politics and technology.

Janek Stockburger is a graduate student at the TUM School of Governance in Munich. He is currently enrolled in the master's program "Politics and Technology" (M.Sc.). Previously, Janek completed his Bachelor of Science in electrical engineering and information technology at the Karlsruhe Institute of Technology (KIT). In his master's degree, Janek specialized in environmental and climate policy. Janek's focus includes energy policy, industrial policy, and the geopolitics of energy transition.

Norway's Hydrogen Strategy: Unveiling Green Opportunities and Blue Export Ambitions

Jon Birger Skjærseth, Per Ove Eikeland, Tor Håkon Jackson Inderberg, and Mari Lie Larsen

Abstract This chapter examines the challenges and prospects for Norway's internal and external hydrogen strategy from around 2019, when Norway's low-carbon hydrogen policies and activities began to gain traction. Norway has taken a technology-neutral approach to 'green' and 'blue' hydrogen technologies linked to reducing emissions. Two end-use sectors have been prioritized: maritime transport and energy-intensive industries. This strategy is based on Norway's energy mix, industry structure/interest and research competence. Climate concerns appear as the predominant motivation underlying the Norwegian government's low-carbon hydrogen strategy, with industrial value creation as an additional key goal. Political priorities roughly align with actual funding priorities, as there has been a massive increase in direct state aid to low-carbon hydrogen projects. Externally, Norway's hydrogen strategy has potential significance for Europe, particularly for countries with maritime interests and high hydrogen import needs. However, Norway's technology-neutral approach differs from those of most other European countries. What Norway's hydrogen strategy will mean for Europe remains to be seen—but its main interests centre on the export of 'blue' hydrogen, with 'green' hydrogen reserved primarily for meeting domestic needs.

1 Background and Introduction

In 1927, the Norwegian company Norsk Hydro started production of 'green' hydrogen as input to ammonia and fertiliser production from electrolysis, based on Norway's rich hydropower resources. Norsk Hydro's production of hydrogen from electrolysis was gradually replaced by the cheaper method of producing hydrogen from re-formation of natural gas, as sizable gas resources had been discovered in the North Sea. However, Norsk Hydro continued to develop and manufacture electrolysis technologies for the world market through the company Nel, which has since

J. B. Skjærseth (✉) · P. O. Eikeland · T. H. J. Inderberg · M. L. Larsen
The Fridtjof Nansen Institute, Lysaker, Norway
e-mail: jbskjaerseth@fni.no

© Helmholtz-Zentrum Potsdam, Deutsches GeoForschungsZentrum GFZ 2024
R. Quitzow and Y. Zabanova (eds.), *The Geopolitics of Hydrogen*, Studies in Energy, Resource and Environmental Economics,
https://doi.org/10.1007/978-3-031-59515-8_11

become a leading global hydrogen technology company based on renewable sources (Hydro, 2021). In 1996, the Norwegian Hydrogen Association (NHA) was established—currently it has some 60 members from industry, universities, and research institutes. These actors and institutions have been involved in international hydrogen research programmes under the International Energy Agency (IEA) and the EU's framework programmes since their inception (SINTEF, 2020).

Since the turn of the twenty-first century, Norway's hydrogen strategy and activities have undergone at least three phases. In the first phase (2000–2010), several hydrogen initiatives were introduced, with support from the Norwegian Research Council and state-aid programmes (NFR, 2020). In 2004, Norsk Hydro and Enercon established the world's first 'hydrogen society' on the island of Utsira off the southwestern coast, with the aim of demonstrating how isolated communities could become energy self-sufficient. This demonstration project, based on wind power and fuel cells, attracted worldwide attention, but was terminated in 2010 due to technical inefficiency and poor commercial prospects (DN, 2008; SINTEF, 2020). Perhaps the most notable project in this phase was HyNor, a joint Statoil/Equinor and Norsk Hydro undertaking aimed at demonstrating various production technologies in the use of hydrogen cars and fuelling stations connecting Oslo and the city of Stavanger. However, with the financial crisis and the emerging market for electric vehicles in Norway, among other factors, HyNor was terminated in 2009 (SINTEF, 2020).

Hydrogen technologies continued in niche companies, but the second phase (2010–2019) was marked by low levels of political attention and industrial activity. For example, a major 2016 White Paper on Norwegian energy policy towards 2030 paid scant attention to hydrogen (Norwegian Government, 2016). The government's strategy focused mainly on maintaining hydrogen research and on following the international development of hydrogen technologies and markets. For most uses, direct electrification based on Norway's surplus of renewable energy generated mainly by hydropower was seen as preferable to hydrogen; in the road transport sector, electric vehicles and biofuels were emerging as the main climate solution.

By 2019, Norway was producing some 225,000 tons of 'grey' hydrogen from gas reforming in industry, mainly for its own production of ammonia and methanol (DNV-GL, 2019). Two companies (Nippon Gases and Ineos/Rafnes) produced and distributed 'green' hydrogen to the small-transport market, with only five hydrogen-fuelled buses and 140 registered hydrogen cars, and a mere five filling stations nationwide. In June 2019, a hydrogen filling station near Oslo exploded and injured two people, literally 'fuelling' concerns over safety issues (TU, 2019). Low-carbon solutions for ships were underway, but hydrogen-based maritime coastal transport was yet to start operation.

As yet, there is no regular market for low-carbon hydrogen—neither in Norway nor abroad. Norway's major oil and gas industry has the potential to supply 'blue' hydrogen based on reforming natural gas with carbon capture and storage (CCS), but such production has not yet been established. Norway's natural gas export through pipelines amounted to around 117 billion m^3 (2017) of gas intended for terminals in Belgium, France, Germany, and the UK, enough to cover about a quarter of Europe's needs, and potentially enough to produce approximately 25 million tons of hydrogen

(DNV-GL, 2019). Norway has been among the few European countries to advance its CCS development, extracting CO_2 from natural gas and storing it on the continental shelf. The world's largest test centre for CCS technologies was established in 2012 at Mongstad, with Equinor and Shell as key partners. After some political vacillation, the government co-funded the establishment of the Northern Lights project on infrastructure for CO_2 storage on the Norwegian continental shelf, designed to develop the world's first open-source CO_2 transport and infrastructure for delivering carbon storage as a service (Gassnova, 2023). This was expanded into the 'Longship' wider policy initiative, which in addition to Northern Lights included carbon capture (Ministry of Petroleum & Energy, 2020).

From around 2019, Norway's low-carbon hydrogen policies and activities began to gain traction, in line with increasingly ambitious climate policy agendas in the EU and Norway. Here we focus on the challenges and prospects for Norway's internal and external hydrogen strategy in this third phase. The key internal issues to be examined are the policies, political priorities of technologies and end-use sectors, and their alignment with actual state-aid priorities. Further, we examine Norway's international approach, its domestic basis and what this means for Europe.

Data for this study stem mainly from national expert reports, governmental white papers, research papers, media articles and interviews with the Norwegian Ministry of Energy.

2 Norway's Hydrogen Strategy: Internal Dimension

2.1 Strategy and Policies

In June 2020, the Norwegian government published a national hydrogen strategy, immediately prior to the launch of the EU's hydrogen strategy (Norwegian Government, 2020). The Norwegian strategy is based on several perceived competitive advantages for production and distribution of hydrogen. These include industrial experience along the entire hydrogen value chain; large gas resources and the potential for increasing the production of renewable energy (hydropower in particular); the Norwegian petroleum industry's expertise in handling large-scale industry projects; the CO_2 storage potential at the Norwegian continental shelf, and the extensive experience in maritime industries along the value chain.

However, the government's 2020 strategy was criticised for being merely a description of the status quo rather than a full-fledged action plan with specific goals; it was also noted that no new policy measures were proposed (TU, 2020a). The government responded by issuing a hydrogen roadmap in 2021, backed up by increased state-funding for hydrogen research and industry projects (Norwegian Government, 2021a). This hydrogen strategy and roadmap laid out the development of low-carbon hydrogen (emission-free or close to emission-free) from electrolysis of water from renewable energy ('green'), and natural gas with carbon capture and storage (CCS) ('blue'). In contrast to the EU's 'green' hydrogen approach, Norway has taken a technology-neutral approach linked to reducing emissions. Two end-use sectors have been prioritised: maritime transport and energy-intensive process industries. In the maritime sector, the government and industry have pursued various technologies for different types of vessels, including full electrification and biogas. Hydrogen and ammonia are deemed most suitable for large, long-distance vessels (Norwegian Government, 2019a).

Climate concerns appear as the predominant motivation underlying the Norwegian hydrogen strategy and roadmap, with industrial value creation as an additional key goal. The hydrogen strategy was initiated by the Conservative Party-led government—more specifically, the Ministry of Climate and the Environment, and the Ministry of Energy. The strategy was based on Norway's energy mix, industry structure and research competence, and was placed in the context of the hydrogen strategies of European and other countries, the European Green Deal and the Next Generation EU recovery plan in response to Covid-19. Industrial interests also shaped Norway's hydrogen initiatives. Equinor, with strong interests in natural gas and CCS, is by far the largest company in Norway, with 67% state ownership. In the first 2019 public consultation on the hydrogen strategy, Equinor argued that that the strategy should prioritise the maritime sector and contribute to large-scale 'blue' hydrogen based on natural gas and CCS (Equinor, 2019).[1] The company followed up in its 2020 position paper to the roadmap, stressing that a necessary condition for 'blue' hydrogen is a value chain for CCS—under preparation within the Longship and Northern Lights projects (Equinor, 2020).[2] Moreover, large stakeholders such as Equinor and Statkraft have ownership interests in both 'blue' and 'green' hydrogen technologies, and represent key drivers alongside technology providers.

In 2021, as noted, the government published the hydrogen roadmap, intended to address the shortcomings of the criticised 'thin' strategy. For the short term, by 2025, this roadmap aims at the establishment of: (1) five hydrogen hubs for maritime transport; (2) one or two industrial projects for hydrogen, to demonstrate value chains with global technology diffusion potential; and (3) five to ten pilot projects for demonstration of new, more cost-effective hydrogen solutions and technologies. The Norwegian Parliament also requested the government to focus more on facilitating large-scale 'green' hydrogen production and to explore how ammonia production could be electrified (Norwegian Government, 2022a). We return to Norway's long-term export-oriented strategy in the next section.

When a Labour Party-led government took office in October 2021, replacing the Conservative Party-led government, its main contribution to the hydrogen strategy and roadmap was to link national hydrogen production and consumption more

---

[1] Both the strategy and the roadmap were subject to public consultations—receiving 49 and 53 responses, respectively, from a range of hydrogen stakeholders.

[2] If, according to the company, 10% of Norway's natural gas export were used for production of 'blue' hydrogen, 20 million tons of CO_2 would need storage annually (four times the Northern Lights CO_2 handling capacity).

directly to ambitious new national climate targets (Norwegian Government, 2022a).[3] Process industries represent a significant source of national emissions with a major potential for reduction through low-carbon hydrogen; by contrast, the maritime emissions from ferry boats and high-speed water vessels for civilian use are relatively minor, which reflects technology development ambitions particularly in the maritime sector (Norwegian Government, 2019b, and Norwegian Government, 2022e).[4] In 2022, the government also strengthened its industrial value-creation ambitions by releasing a roadmap and future vision for 'green' industry development. Here, hydrogen figured on the list of seven priority areas, which included offshore wind, batteries, maritime industries, CCS, bioeconomy and process industries (Norwegian Government b; c, 2022b, 2022c).

The main policy instruments envisaged for attaining the targets and visions consist of a combination of energy, climate, industry and research, and innovation policies (Norwegian Government, 2022a): increase public funding for the whole innovation chain based on existing institutions[5]; increase the CO_2 tax from approx. € 60/ton– € 200/ton by 2030,[6] in line with the national climate plan; develop zero-emissions public procurement standards, particularly for long-range ferries and other ships not suited for electrification; retain current tax benefits for hydrogen cars (as for electric cars) and exemption from consumer tax for electricity used to produce hydrogen through electrolysis; and increase funding to research, innovation and market introduction.

The government has also announced that it is giving consideration to establishing a system of Contracts for Difference to stimulate hydrogen value-chain development. Under such a system, the state would guarantee steady income to hydrogen frontrunners. Hydrogen stakeholders have increasingly called for more measures and programmes for scaling up and commercializing technologies—in particular, Contracts for Difference to help to promote a hydrogen market (Energi og Klima, 2020). Contracts for Difference have so far not been applied in Norway.

2.2 Funding Activities and Challenges

Political priorities roughly align with actual funding priorities. All the new political initiatives have propelled state-aided hydrogen activities and projects across the

---

[3] New climate policy targets included domestic reduction of GHGs by 55% by 2030—covering both ETS and non-ETS sectors.

[4] In 2020, process industry represented ca. 23% of total CO_2 emissions. Ferry boats and high-speed water vessels for civilian use account for only 3% of total CO_2 emissions from the transport sector (2016).

[5] Norwegian Research Council, Enova (energy transition state fund), Gassnova (gas pipeline operator) and Innovation Norway.

[6] The CO_2 tax covers both the EU ETS and non ETS sectors. For the petroleum industry, which is part of the EU ETS, the CO_2 tax is added to the ETS allowance price to equal the carbon price in non ETS sectors.

country (E-24, 2021). With regard to research, further upscaling came with the 2022 State Budget, when €31 million in funding was granted for two major new hydrogen R&D centres. Since 2020, there has been a massive increase in direct state aid to hydrogen projects—mostly to industry and maritime transport, with a declining trend in funding for road-based transport (see Fig. 1).

In addition to industrial 'green' and 'blue' hydrogen demonstration projects, maritime transport is prioritised. In June 2022 for instance, Enova—the state fund for the green transition—granted €120 million in support to hydrogen investments in the maritime industries. This funding comprised the establishment of five production/infrastructure facilities along the coast, intended to facilitate further technological advancement, in addition to end-use by seven coastal vessels powered by hydrogen or ammonia (Enova, 2022a).

In January 2022, the first contract for the procurement of two long-range zero-emissions ferry boats was signed. These vessels were mandated to operate with a minimum hydrogen content of 85%. Additionally, a milestone was achieved in 2023, when a 15-year delivery agreement was established between a shipping company and a green hydrogen producer (Fauna, 2023).

Enova has granted support for 16 larger vessels with hydrogen or ammonia as their primary fuel sources, along with the establishment of five hydrogen hubs (Enova, 2023). It is noteworthy, however, that despite these concerted efforts, the industry remains challenged by the perception of elevated risks associated with hydrogen investments. The support scheme has yet (October 2023) to stimulate a substantial influx of investment decisions.

Enova now seeks to realize *profitable* value chains through two new support schemes 'Hydrogen in vessels' and 'Ammonia in vessels' (Enova, 2023). The

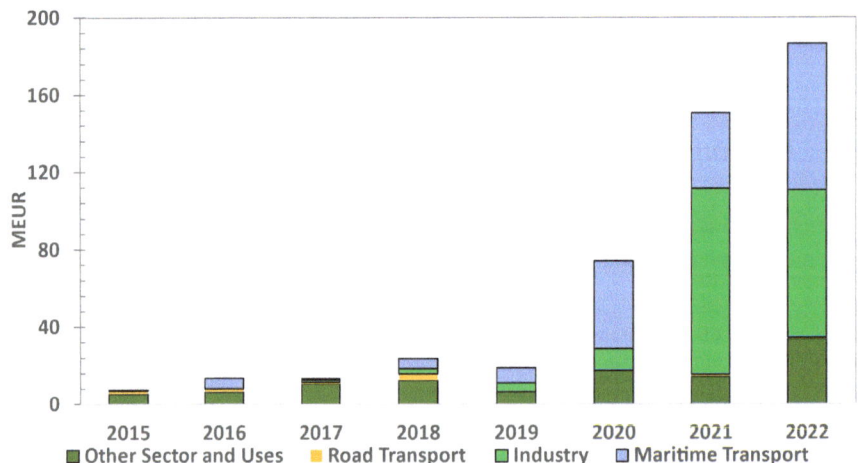

Fig. 1 State aid to hydrogen projects in Norway 2015–2022. *Source* Enova SF/Heilo, https://www.enova.no/heilo/hydrogen/stotte-til-hydrogenprosjekter/

schemes are planned to continue until 2026, with two calls per scheme each year, emphasising cost efficiency through 'competitive bidding'.

In the revised National Budget (June 2023) there were no signs of any firm hydrogen commitments; no funding has been set aside for a plan for the introduction of contracts for difference, nor for a nationwide network of hydrogen filling stations and hubs (Norwegian Hydrogen Forum, 2023).

Uncertainties about low-carbon hydrogen competitiveness, technologies and costs, infrastructure, markets, transport, and safety permeate the hydrogen strategy. Due to Norway's already high degree of electrification based on renewable energy, the demand for low-carbon hydrogen to reduce emissions may be more limited than the case in many EU member states. Furthermore, electrolysis is significantly more energy-demanding than direct electrification (Home & Hole, 2019).[7] Thus, 'green' hydrogen production will increase the demand for renewable power and grids, and raise electricity prices (Larsen and Dupuy, 2023). Earlier forecasts and expectations of Norway's renewable power surplus have changed rapidly with more ambitious climate targets and net-zero industry ambitions, along with political pressure to develop more renewable electricity. However, there has been strong public opposition to building more renewable energy in Norway, due to trade-offs with land use and nature protection. This has been clearly illustrated by the massive local opposition to land-based wind power (Skjærseth & Rosendal, 2022).

3   Norway's Hydrogen Strategy: External Dimension

Norway's international hydrogen strategy mirrors the domestic strategy in that it is directed towards export of 'blue' hydrogen in the long term and development of technology related mainly to the maritime and industrial sectors. Norway's 'blue' hydrogen focus is on European markets, with initiatives linked to the North Sea area. In 2022, there was virtually no production of 'blue' hydrogen in Norway.

By 2050, Norway's hydrogen roadmap envisions a well-established market for hydrogen serving the sectors of maritime transport, process industry and heavy vehicles. By that timepoint, Norwegian industry is expected to be a world-leading exporter of hydrogen, technologies and solutions (Norwegian Government, 2021a). Establishing international markets and export are seen as crucial priorities, particularly for the development of 'blue' hydrogen value chains requiring large investments in major facilities and infrastructure for both gas and CCS. As noted above, the state-sponsored CCS projects for capture, transport, and storage of CO_2 (Longship and Northern Lights) might ultimately facilitate production of 'blue' hydrogen for a European market. This initiative has received some €2.5 billion in support, which is likely to increase. The goal here is to establish an operational transport and storage network by 2025, with national CO_2 storage customers as well as European users. By contrast,

---

[7] Given 60–65% efficiency, 50–55 kWh is needed to produce one kg of hydrogen gas with 33 kWh energy content.

'green' hydrogen production is regarded as involving quite different economies of scale, with smaller production units serving national demand. A basic premise for this long-term strategy is that future demand will come mostly from international markets outside of Norway (Norwegian Government, 2020). Main priorities here are engagement in the emerging EU regulatory framework, the involvement of industry in projects abroad, and international cooperation on research, standard-setting, transport and markets.

3.1 The EU and Norway

European markets are highly important to Norway. The EU's hydrogen strategy and emerging policies both align with and diverge from Norway's approach. Although Norway is not an EU member, it is closely tied to EU climate and energy policies through the European Economic Area agreement. Indeed, Norway has a long-standing energy dialogue with the EU. Both the EU and Norway emphasise their mutual interests in achieving climate neutrality and promoting 'green' industrial growth based on the EEA agreement. However, Norway was not included as a priority partner in the initial EU hydrogen strategy that prioritised renewable 'green' hydrogen, which provoked some domestic political reactions (TU, 2020c). Recently, however, the EU and Norway have intensified their energy dialogue through a Norway-EU Green Alliance which includes 'blue' hydrogen as a transitional mode of production (Green Alliance, 2023). In February 2022, Norwegian Prime Minister Jonas Gahr Støre met with the President of the European Commission Ursula von der Leyen and EU Executive Vice President Frans Timmermans to discuss enhanced cooperation on climate, energy and industrial transformation to leverage green industry, renewable energy, hydrogen and CCS (EnergyLive, 2022).[8] This Green Alliance initiative was later included in the EU's external strategy for international energy cooperation linked to the REPowerEU plan in response to the war in Ukraine (Norwegian Commission, 2022; Parliament, 2022).

However, Norway has argued that the European Commission should adopt a more flexible approach to define when hydrogen production should be considered renewable with references to the 'additionality' principle proposed for a delegated act under the 2019 Renewable Energy Directive (whereby renewable hydrogen should be produced by additional renewable electricity). The government held that this principle could threaten Norway's hydrogen projects as Norway remains a net exporter of land-based renewable electricity (Montel, 2022). In June 2023, the European Commission adopted the act with the principle retained, but with the opportunity to count electricity as fully renewable if the hydrogen plant is located in a bidding zone where the average proportion of renewable electricity exceeds 90% (European

---

[8] The EU and Norway agreed on the Green Alliance in April 2023.

Union, 2023a).[9] Norway is one of few countries that follows the 90%-rule in all its bidding zones and may thus attract attention as location of hydrogen production for Europe (Collins, 2023).

Also, other recent EU policies apparently align well with Norway's priorities. A series of new EU policies aims at decarbonizing the shipping industry and kick-starting the uptake of renewable and low-carbon fuels. The revised EU ETS Directive requires shipping companies to buy allowances for parts of their emissions (40%) in 2025, increasing to 100% from 2027.[10] Recent agreement on the FuelEU Maritime Regulation establishes a framework for increased supply and demand of low-carbon maritime fuels based on targets set for reduction in the carbon-intensity of energy used onboard ships.[11] To promote the use of renewable fuels of non-biological origin (RFNBOs), the Council and Parliament agreed to allow a 'multiplier' of two to be used when calculating the GHG intensity of the energy used onboard ships.[12] The revised Alternative Fuels Infrastructure Regulation set targets for distance between electric charging and hydrogen refuelling infrastructure in road transport. However, no mandatory requirements are included for hydrogen infrastructure for the maritime sector in ports, leaving it up to the Member States' discretion (European Union, 2023b). These EU regulatory initiatives may add some drivers, supportive of Norway's vision of using its competitive advantage in development of alternative maritime fuels, including hydrogen and ammonia (Norwegian Government, 2021b).

Cooperation on research and innovation is seen as vital for cutting costs and making low-carbon hydrogen commercially attractive. Norwegian actors participated in several projects under the Horizon 2020 programme for hydrogen called the Fuel Cells and Hydrogen Joint Undertaking (FCH JU). In 2021, the EU launched the Clean Hydrogen Joint Undertaking (usually referred to as the Clean Hydrogen Partnership), a public–private partnership that is the successor of FCH JU and is funded by Horizon Europe. Norwegian research and technology actors participate in this partnership, which is closely aligned with the goals of the European Hydrogen Strategy and supports research and innovation activities in the production, distribution and use of hydrogen in transport, industry and buildings. Of these, hydrogen use in transport and industry are especially relevant to Norway.

Norwegian hydrogen developers are eligible to apply for funding from various EU programmes—among them, Horizon Europe (including specific funds reserved

---

[9] In the previous calendar year and hydrogen production does not exceed a maximum number of hours set in relation to the proportion of renewable electricity in the bidding zone.

[10] All emissions from travels within the EU and half of emissions for travels from and to the EU are covered.

[11] Reflecting expected lack of hydrogen production in the short-term, targets agreed for 2025 (2%) and 2030 (6%) are relatively low, while 80% was politically agreed for 2050.

[12] Subject to certain conditions, such as sufficient production capacity and availability to the maritime sector, a 2% RFNBO usage target would take effect as of 2034 if the Commission reports that in 2031 RFNBO amount to less than 1% in the fuel mix (Council of the European Union, 2023a, 2023b).

for regional Hydrogen Valleys and the Clean Hydrogen Partnership), and the Innovation Fund on low-emission technologies which has been opened for projects to decarbonize the maritime sector. Emerging from the RePowerEU Plan to accelerate the uptake of hydrogen in the EU, the Commission in 2023 established a new funding instrument based on the Innovation Fund, the European Hydrogen Bank. It aims at unlocking private investments in green hydrogen value chains (not blue hydrogen projects), both domestically and in third countries. In autumn 2023, a pilot auction (competitive bidding) will be launched to support the production of renewable hydrogen, with indicative budget of EUR 800 million (European Commission, 2023b).

EU funding programmes may be important but most of the total funding is likely to be provided by the member states. The EU has adopted specific state-aid guidelines as part of its competition policy. Under these guidelines, specific rules have been adopted for the category defined as Important Projects of Common Interest (IPCEI), which provide better opportunities for national state aid to facilitate European technological leadership, also in the field of hydrogen. State aid to IPCEI projects may cover up to 100% of costs. EU state-aid guidelines adopted in 2021 include support to both 'green' and 'blue' hydrogen; the EU taxonomy also includes sustainability criteria for investments in natural gas and the production, distribution, and storage of low-emission hydrogen. Norway has committed funds to two hydrogen projects under the IPCEI platform (see below).

Other EU policy initiatives that may affect Norway's hydrogen strategy include the Net-Zero Industry Act, the ocean energy strategy on offshore wind power production and infrastructure in the North Sea, and the revision of the EU gas market, including markets for renewable gases, natural gas and hydrogen.

3.2 Private Sector Involvement

The private sector is increasingly engaged in the implementation of international hydrogen projects. Several major state-sponsored industrial demonstration projects indicate how the domestic hydrogen strategy includes the positioning of Norway in the emerging international hydrogen value chain, including 'green' and 'blue' hydrogen production, process industries and maritime end-users:

TiZir's hydrogen project at the smelter in Tyssedal has received €26.1 million from Enova and has been granted status as an IPCEI project. This project, which aims to replace coal with 'green' hydrogen in the production of titanium dioxide, may have an international potential (Enova, 2021a).

The Barents Blue project is led by Horisont Energi in cooperation with Equinor and Vår Energi. The project has received €48.2 million from Enova and has also been granted status as an IPCEI project. Barents Blue aims to develop the world's first ammonia plant with zero CO_2 emissions and a daily production of 600 tons of hydrogen to be transformed into 3000 tons ammonia. The CO_2 will be stored at the shelf outside Finnmark in northern Norway (Enova, 2021b).

Yara has received €28.3 million for constructing a 'green' hydrogen demonstration plant to show that ammonia produced using renewable energy can reduce emissions of CO_2 in fertiliser production (Yara, 2022).[13]

The 2019 the Aurora project aimed to develop a complete liquid hydrogen supply chain for commercial shipping, in cooperation with BKK, Air Liquide and Equinor, involving hydrogen produced by electrolysers. However, in March 2022 the parties decided to cancel this project. They announced that they will need to evaluate how the market for liquid hydrogen develops, and noted the need for further support schemes for hydrogen production in the form of contracts for difference to make hydrogen available at an affordable price (Eviny, 2022).

To date, the largest energy company Equinor is most active internationally—participating in at least six international hydrogen projects at various points in the innovation chain (Equinor, 2022). In 2019, Equinor partnered with the French energy company Engie to promote the production of 'blue' hydrogen, focusing on consumers and the public authorities in Belgium, the Netherlands and France (Equinor, 2021). In 2020, it launched a 'blue' hydrogen project outside Hull in the UK (H2H Saltend), also to demonstrate the potential of 'blue' hydrogen to the EU and Germany (BT, 2020). In 2021, Equinor started a feasibility study for producing 'blue' hydrogen in Norway for export through new pipelines to Europe (E24, 2021a), and in 2023, they announced plans to build hydrogen pipeline between Norway and Germany by 2030. The plan is part of a new strategic partnership with the German utility RWE for the development of large-scale value chains for clean hydrogen (Collins & Radowitz, 2023). Thus, the first part of the plan involves construction of new gas power plants in Germany. Equinor also launched a major €35 billion euro vison towards 2035 for ocean wind, 'blue' hydrogen and CCS, with the company pledging to cover approximately one-third of the costs (E24, 2021b). However, these 'blue' hydrogen plans and projects are surrounded by uncertainty and the company is not yet producing any 'blue' hydrogen.

Also other Norwegian companies are active internationally in hydrogen sector, such as Hydro and Yara (with roots to Norsk Hydro). Yara is engaged in several international projects: Firstly, Yara with others launched the Green H2 Catapult– a global private initiative supported by the UN Climate Change High-Level Champions initiative. They also teamed up with clean energy companies, including ENGIE, Idemitsu Kosan, Jera, Kyushu Electric Power, Trafigura and Ørsted, to produce clean ammonia (Yara, 2023). Hydro has 'produced the world's first successful batch of aluminium using green hydrogen fuelled production in Navarra, Spain (Hydro, 2023). Furthermore, the Norwegian Maritime Authority granted temporary approval for the use of ammonia as a fuel, for the first-ever ammonia-fuelled bulk carrier—a Norwegian ship which will be set afloat in 2025 (Norwegian Maritime Authority, 2023).

Several Norwegian companies also export hydrogen technology (e.g., NEL, HydrogenPro, HYON, Zeg Power). With roots back to Norsk Hydro's production of 'green' hydrogen, NEL has developed into a global hydrogen company with leading

---

[13] Two major companies have withdrawn from the project—Aker Clean Carbon and Statkraft.

technologies in the production, storage and distribution of hydrogen based on renewable sources. In 2017, NEL became one of the world's largest electrolyser companies, with the largest manufacturing plant for hydrogen fuelling stations and plans for the world's largest electrolyser manufacturing plant (NEL, 2022). In 2022, the company acquired its largest contract ever for electrolyser equipment to a US customer (DN, 2022), and in 2023, NEL signed a contract with Statkraft for the delivery of 40 MW electrolysers for production of green hydrogen in Norway (Statkraft, 2023). The current estimate for the cost of green hydrogen production in Norway is €5.20 per kg hydrogen, of which the cost for power and grid is about 60% (Enova, 2023b).

3.3   *International Policy Dimensions*

Norway does not have a distinct, unified foreign policy on hydrogen. The Ministry of Energy plays a key role in coordinating Norway's international engagement. However, several sector ministries are involved in various aspects of the hydrogen chain: The Ministry of Climate and Environment 'owns' the state fund for transition (Enova), with main responsibility for the IPCEI projects; the Ministry of Trade, Industry and Fisheries has responsibility for fuels, industrial value creation and 'green' shipping; the Ministry of Transport is in charge of fuelling stations and ferry boats; and the Ministry of Education and Research is responsible for research and innovation. This administrative fragmentation may pose coordination challenges.

In addition, the Ministry of Energy, with Enova and Innovation Norway, is part of the European Clean Hydrogen Alliance for developing hydrogen value chains in the EU. This alliance keeps track of potential hydrogen investment projects.

With some 40 years of energy cooperation, Germany is a main bilateral partner on hydrogen, and indeed the only country that works with Norway on specific hydrogen projects. According to a joint Norwegian–German statement from January 2022 on enhancing the dialogue on energy and industrial transformation, Norway wants to 'contribute actively to the rapid development of the hydrogen market in Germany and the EU', and Germany would like to see 'Norway become a future partner for the production and supply of hydrogen' (Norwegian Government, 2022f p.1). This includes a joint plan for the use of 'blue' hydrogen for a transition period, to 'ensure environmental and climate integrity by establishing for example the highest possible standards for Carbon Capture and Storage'—indicating that CCS-sceptical Germany accepts 'blue' hydrogen.

In June 2022, Norway signed an agreement with several industrial actors in Northern Germany which is set to run until 2025. This industrial partnership includes offshore wind, hydrogen and CCS, and aims at developing new projects and funding in these areas (Norwegian Government, 2022d). Germany also represents a potential market for exporting Norwegian maritime hydrogen technology. Through Innovation Norway's High Potential Opportunities programme for major export initiatives, Norway focuses on exporting hydrogen technology to northern Germany for the development of hydrogen value chains in the maritime sector (Innovasjon Norge,

2022). In January 2023, cooperation between Norway and Germany took a new step forward. During a meeting between German Vice-Chancellor Robert Habeck and Norwegian Prime Minster Jonas Gahr Støre, Equinor and RWE announced joint plans that included the production of hydrogen and infrastructure for export to Germany. Nel also announced agreements with Statkraft and the German HH2E on production of electrolysers and hydrogen production facilities.

Also important are France, Belgium, the UK and the Nordic countries. Norway has close contact with France on information sharing and innovation: The Global Growth Hydrogen programme with France focuses on private actors and 'green' hydrogen, which is also part of Innovation Norway's new export strategy (Innovasjon Norge, 2023). Norway's MoU on energy with Belgium includes knowledge sharing on hydrogen. There is also regular bilateral contact with the UK and established cooperation with Sweden and Denmark, particularly on 'green' hydrogen and infrastructure (Hydrogen24, 2022). This is underpinned by the Nordic Council of Ministers' Nordic Energy Research, which has facilitated several hydrogen-related initiatives. Moreover, the Nordic climate and environment ministers have agreed that their countries will work together to establish green corridors for emissions-free shipping between ports in the region (Nordic Co-operation, 2022).

Beyond Europe, Norway participates in several ministerial meeting platforms in North America and Japan, including the Japan-led Energy Ministerial Meetings, where hydrogen frontrunners discuss hydrogen solutions to facilitate a commercial breakthrough, share information and develop standards and regulations. These are followed up in other arenas where Norway is involved, such as Mission Innovation (MI) linked to the Paris Agreement and the International Partnership for the Hydrogen Economy, which is involved in setting standards for hydrogen. Norway also participates in the IEA technology network and has joined the Clean Hydrogen Mission led by Australia, Chile, the EU, the UK and the USA, and the Zero-Emission Shipping Mission led by Norway, Denmark and the USA. The aim of the Clean Hydrogen Mission is for 5% of the global deep-sea fleet to be emissions-free by 2030.

3.4   International Markets, Financing, Capacity Development and Challenges

The European market is, as noted, central to Norway's 'blue' hydrogen export ambitions. Spurred by the war in Ukraine, the REPowerEU plan aims at rapidly reducing the EU's dependence on Russian fossil fuels (mainly gas) and accelerating the green transition, thereby significantly raising the EU's hydrogen ambitions (European Commission, 2022; Skjærseth, 2023). The previous goal of 10 million tons of annual hydrogen production within the EU is to be complemented by 10 million tons of annual imports by 2030, in order to meet the EU's climate target. According to European hydrogen industry estimates, there will be a need for ca. 120 GW electrolysis capacity in the EU to meet the 10 million ton domestic hydrogen production target. This is a significantly higher amount than the 40 GW originally envisaged in the 2020 European Hydrogen Strategy (Ansari et al., 2022).

These EU import ambitions generally align with Norway's 'blue' hydrogen export ambitions, but whether Norway will be able to produce and export significant amounts of low-carbon hydrogen to Europe by 2030 remains uncertain. Norway and Germany have initiated a feasibility study on large-scale transportation of hydrogen, mainly via pipelines from Norway to Germany (Norwegian Government, 2022a). Gassco—the Norwegian system operator for the gas pipelines between Norway and Europe— has assessed alternative use of the pipelines, and technical barriers to using them for hydrogen transport. However, capacity will probably not be available before 2030. Mixed transport of natural gas and hydrogen through pipelines is also being considered. Other options include LNG and natural gas export in the form of 'blue' ammonia.

Investments and capacity building beyond Europe have lower priority. However, in August 2022, the Norwegian Agency for Development Cooperation (Norad) for the first time allocated some support to hydrogen projects in Africa. Through the company Scatec, €8.5 million was allocated to production of 'green' ammonia in Egypt, Morocco, South Africa and Tunisia. The funding was partly motivated by climate concerns in Africa, and by the need to support hydrogen production to replace Russian gas export to Europe—in line with the REPowerEU plan (Norad, 2022). Moreover, the Norwegian state-owned renewable fund 'Nysnø', together with Equinor and Yara, has invested in the hydrogen fund AP ventures (DN, 2021).

The main external challenge to Norway's export ambitions is developing a commercial market for low-carbon hydrogen in Europe. Despite more ambitious EU targets, major challenges that have been identified include demand for electricity and lack of production capacity for electrolysers (Ansari et al., 2022). Regarding competitiveness of 'blue' hydrogen, Norway as a gas producer may not enjoy competitive advantages as gas market prices is expected to develop similarly in Norway and other parts of the European gas market (Norwegian Government, 2021a:116).

A commercial hydrogen marked may prove controversial also from a sustainability perspective. Ongoing research examines its potential impact on the atmosphere including the ozone layer (Forskning.no, 2021). The sustainability of 'blue' hydrogen specifically is also heavily debated (see NRK, 2021). From a life-cycle perspective, it has been held that 'blue' hydrogen will not be low-carbon unless the release of fugitive methane in the production process is dealt with (Howarth & Jacobsen, 2021). In the EU, carbon capture and storage has a long and politically contentious history, also with regard to the issue of whether captured CO_2 can be stored indefinitely. CO_2 storage also requires large amounts of electricity (Espegren et al., 2021).

4   Conclusions

Norway's hydrogen strategy has potential significance for Europe, given its focus on low-carbon hydrogen technologies for industry and maritime end-users, development of electrolysers, and production of 'blue' hydrogen for export to the continent. These points concern most EU member states with low-carbon hydrogen ambitions, particularly those with maritime interests, high hydrogen import needs and gas pipelines to Norway, such as Germany. The external dimension of Norway's hydrogen strategy largely mirrors the internal dimensions—international industry initiatives and policy agreements focus on establishing infrastructure and a European market for exports of blue hydrogen produced in Norway. Norway's technology-neutral approach to 'green' and 'blue' hydrogen linked to emissions deviates from the EU and most other European countries. However, the Norway-EU Green Alliance and recent changes in Germany's hydrogen strategy indicate higher acceptance for imports of 'blue' hydrogen as a transitional technology (Reuters, 2023).

However, there remain several challenges for realizing the Norwegian hydrogen strategy. Regular domestic or international commercial markets for low-carbon hydrogen have not yet emerged. Uncertainties regarding the competitiveness of low-carbon hydrogen, technologies and costs, infra-structure, markets, transport and safety permeate Norway's hydrogen strategy. Key internal challenges include growing demands for high-priced, increasingly scarce renewable electricity, and trade-offs with land use, such as nature protection. Other challenges include technical barriers, sustainability issues, uncertain commercial prospects and the short-to-medium term perspective of 'blue' hydrogen as a transitional technology versus Norway's long-term export strategy.

Regulatory frameworks, markets and technology for blue and green hydrogen in Europe are still under development. Thus, it is too early to conclude exactly what Norway's hydrogen strategy will mean for Europe, and whether European policies and markets will be supportive of Norway's main priorities: export of 'blue' hydrogen with 'green' hydrogen primarily covering domestic needs, and technology development to enable hydrogen use in maritime and industrial sectors.

Acknowledgements Research for this chapter was financially supported by the German Federal Foreign Office within the framework of the project "Geopolitics of the Energy Transformation—Implications of an International Hydrogen Economy" (GET Hydrogen), funding reference number AA4521G125.

References

Ansari, D., J. Grinschgl, J., and Pepe, J. M. (2022). Electrolysers for the hydrogen revolution: Challenges, dependencies and solutions. Stiftung Wissenschaft und Politik, German Institute for International and Security Affairs. SWP Comment 2022/C 57. doi:https://doi.org/10.18449/2022C57

BT (Bergens Tidende). (2020). Equinor skal produsere hydrogen fra naturgass:—Håper dette kan overbevise Europa. 2 July. https://www.bt.no/nyheter/okonomi/i/zGvwB1/equinor-skal-produs erehydrogen-fra-naturgass-haaper-dette-kan-overbevise-europa

DN (Dagens Næringsliv). (2008). Kraftlaboratoriet, February 3, 2008. https://ekstern.filer.uib.no/ matnat/gfi/media/PDF/2008/Utsira_DN.pdf

DN (Dagens Næringsliv). (2021). Statens fornybarfond Nysnø satser på hydrogen. April 4, 2021. https://www.dn.no/marked/statens-fornybarfond-nysno-satser-pa-hydrogenfond/2-1-1003215

DN (Dagens Næringsliv). (2022). Nel fikk sin største ordre noensinne. https://www.dn.no/bors/ nel/oslo-bors/nel-fikk-sin-storste-ordre-noensinne-aksjen-stiger-kraftig-fra-start-pa-borsen/2-1-1262004

DVV-GL. (2019). Produksjon og bruk av hydrogen i Norge. Høvik: Rapport 2019–0039. https:// www.regjeringen.no/contentassets/0762c0682ad04e6abd66a9555e7468df/hydrogen-i-norge---synteserapport.pdf

E24. (2021a). Equinor vurderer hydrogenrør til Europa:—Hvis vi gjør dette, så blir det gigantisk. https://e24.no/energi-og-klima/i/M3jaX5/equinor-vurderer-hydrogenroer-til-europa-hvis-vi-gjoer-dette-saa-blir-det-gigantisk

E24. (2021b). Equinor-sjefens gigantplan: Satser 350 milliarder på grønn industri. https://e24.no/ olje-og-energi/i/G3jzVV/equinor-sjefens-gigantplan-satser-350-milliarder-paa-groenn-industri

Energi og Klima. (2020). Dette er bransjens innspill til hydrogen-veikartet. November 11, 2020. https://energiogklima.no/nyhet/dette-er-bransjens-innspill-til-hydrogen-veikartet/

EnergyLive. (2022). EU and Norway announce plans to establish Green Alliance. https://www.ene rgylivenews.com/2022/02/24/eu-and-norway-announce-plans-to-establish-green-alliance/

Enova. (2021a). TIZIRs hydrogensatsing ved smelteverket i Tyssedal får inntil 261 millioner kroner fra Enova og blir IPCEI-prosjekt. December 17, 2021. Press Release. https://presse.enova.no/ pressreleases/tizirs-hydrogensatsing-ved-smelteverket-i-tyssedal-faar-inntil-261-millioner-kro ner-fra-enova-og-blir-ipcei-prosjekt-3151660

Enova. (2021b). Barents blue. https://www.enova.no/om-enova/om-organisasjonen/teknologipor tefoljen/barents-blue/?gclid=CjwKCAjwi8iXBhBeEiwAKbUofSG96pSPMAnk9JCqW41Js IokaIdF68N0xkHz-FGo_lpSB_jC6oqv4BoC-nwQAvD_BwE

Enova. (2022a). Støtte til hydrogen i maritim sektor. June 13, 2022. Press Release https:// presse.enova.no/pressreleases/enova-stoetter-hydrogenprosjekter-i-maritim-sektor-med-112-milliarder-kroner-3190840?_ga=2.172091107.1509209179.1661506617-299855189.164190 1971&_gac=1.89728745.1661506619.CjwKCAjw3qGYBhBSEiwAcnTRLvRtY--oMGLzC ZIzVjdyqKwjta0vTWykjNo-5Wu0ZbuhdR0lydNNgxoCQpEQAvD_BwE

Enova. (2022b). Støtte til hydrogenprosjekter. https://www.enova.no/heilo/hydrogen/stotte-til-hyd rogenprosjekter/. Accessed October 25, 2023

Enova. (2023a). Enova planlegger nye støtteprogrammer for 'hydrogen i fartøy' og 'Amoniakk i fartøy"n—gi dine innspill nå. https://info.enova.no/nb/haf. Accessed August 30, 2023.

Enova. (2023b). Kostnader for hydrogenproduksjon fra kraft i Norge. 22.06.2023. https://info.enova. no/kostnader-for-hydrogen-rapport

Equinor. (2019). Innspill til regjeringens helhetlige hydrogenstrategi. https://www.regjeringen.no/ contentassets/0762c0682ad04e6abd66a9555e7468df/equinor---innspill-til-helhetlig-hydrog enstrategi.pdf

Equinor. (2020). Innpill til 'Veikart for hydrogen'. https://www.regjeringen.no/contentassets/66d e7ddcf7a6494694202b760fa3f50f/equinor-asa.pdf

Equinor. (2021). ENGIE og Equinor inngår samarbeid om utvikling av blått hydrogen. https://www. equinor.com/no/news/archive/20210218-join-forces-engie-hydrogen

Equinor. (2022). Hydrogen. https://www.equinor.com/no/energi/hydrogen?utm_source=google& utm_medium=cpc&utm_campaign=hydrogen&gclid=CjwKCAjw3qGYBhBSEiwAcnTRLj3 Fmd_fw4uUCUPbyWe9Ob3pz_B3HZgejqPzxY3swx9AvICmiNZqPRoCLwQQAvD_BwE

Espegren, K., Damman, S., Pisciella, I. G., & Tomasgard, A. (2021). The role of hydrogen in the transition from a petroleum economy to a low-carbon society. *International Journal of Hydrogen Energy, 46*, 2315–23138.

Eviny. (2022). Hydrogenprosjektet Aurora lagt på is. https://kommunikasjon.ntb.no/pressemelding/hydrogenprosjektet-aurora-lagt-pa-is?publisherId=17847929&releaseId=17927880

European Commission. (2022). European Commission, RePowerEU. https://ec.europa.eu/commission/presscorner/detail/en/IP_22_3131

European Commission. (2023a). European green deal: Ambitious new law agreed to deploy sufficient alternative fuels infrastructure, Press release, March 28, 2023, Brussels https://ec.europa.eu/commission/presscorner/detail/en/ip_23_1867

European Commission. (2023b). Upcoming EU hydrogen bank pilot auction: European commission publishes terms & conditions, Brussels, August 30, 2023. https://climate.ec.europa.eu/news-your-voice/news/upcoming-eu-hydrogen-bank-pilot-auction-european-commission-publishes-terms-conditions-2023-08-30_en

European Commission. (2023c). Net-Zero Industry Act: Making the EU the home of clean technologies manufacturing and green jobs, Press Release March 16, 2023, Brussels. https://ec.europa.eu/commission/presscorner/detail/en/IP_23_1665

European Commission, undated. Reducing emissions from the shipping sector. https://climate.ec.europa.eu/eu-action/transport/reducing-emissions-shipping-sector_en

European Union. (2023a). REGULATION (EU) 2023/1184 of 10 February 2023 supplementing Directive (EU) 2018/2001 of the European Parliament and of the Council by establishing a Union methodology setting out detailed rules for the production of renewable liquid and gaseous transport fuels of non-biological origin, Official Journal, L 157/11. https://eur-lex.europa.eu/legal-content/EN/TXT/PDF/?uri=CELEX:32023R1184

European Union. (2023b). REGULATION (EU) 2023/1804 of 13 September 2023 on the deployment of alternative fuels infrastructure, and repealing Directive 2014/94/EU, Official Journal, L 234/1. https://eur-lex.europa.eu/legal-content/EN/TXT/PDF/?uri=CELEX:32023R1804

Fauna. (2023). Inngår grønn avtale for fergene i Lofoten. August 23, 2023. https://fauna.no/gronn-avtale-for-lofotfergene/

Forskning.no. (2021). Norge skal satse på hydrogen. Men hva skjer når gassen lekker ut? https://forskning.no/hydrogen-klima-teknologi/norge-skal-satse-pa-hydrogen-men-hva-skjer-nar-gassen-lekker-ut/1810484

Gassnova. (2023). Historien. https://gassnova.no/historie

Green Alliance, Norway-EU 2023. https://www.regjeringen.no/contentassets/debc1b0c0a2f47d1b77beb96d896cf45/20230420eu-norway-green-alliance-final.pdf

Home, H. and J. Hole. (2019). Hydrogen i det moderne energisystemet i Norge. *NVE-faktaark*, 12/2–19. https://publikasjoner.nve.no/faktaark/2019/faktaark2019_12.pdf

Howarth, R. W., & Jacobsen, M. Z. (2021). How green is blue hydrogen? *Energy Science and Engineering*. https://doi.org/10.1002/ese3.956

Hydro. (2021). Hydrogen siden 1926. https://www.hydro.com/no-NO/energy/about-hydro-energy/kraften-bak-hydro-energis-historie/hydrogen-siden-1926/

Hydro. (2023). World's first batch of recycled aluminium using hydrogen fueled production. June 15, 2023. https://www.hydro.com/en-NO/media/news/2023/worlds-first-batch-of-recycled-aluminium-using-hydrogen-fueled-production/

Hydrogen24. (2022). Skal styrke næringslivs-samarbeidet med Sverige. https://hydrogen24.no/2022/05/02/norge-skal-styrke-naeringslivs-samarbeidet-med-sverige/

Collins, L. (2023, February 16). European Commission finally publishes definition for green hydrogen, providing the certainty required for investment to begin. *Hydrogen Insight*. https://www.hydrogeninsight.com/policy/european-commission-finally-publishes-definition-for-green-hydrogen-providing-the-certainty-required-for-investment-to-begin/2-1-1403136

Collins, L., & Radowitz, B. (2023, January 5). Norway and Germany announce plan to build hydrogen pipeline between the two countries by 2030. *Hydrogen Insight*. https://www.hydrogeninsight.com/policy/norway-and-germany-announce-plan-to-build-hydrogen-pipeline-between-the-two-countries-by-2030/2-1-1383658

Innovasjon Norge. (2022). Hydrogen and maritime solutions Germany. https://www.innovasjonno
    rge.no/no/tjenester/internasjonal-satsing/kurs-og-kompetanse/High-Potential-Opportunities-
    HPO/High-Potential-Opportunities-HPO-eng/hydrogen-and-maritime-solutions-germany/
Innovasjon Norge. (2023). Global growth hydrogen and CCUS France. https://www.innovasjonno
    rge.no/no/tjenester/internasjonal-satsing/kurs-og-kompetanse/global-growth/global-growth-
    hydrogen-frankrike/global-growth-hydrogen-france/
Larsen, M. L., & Dupuy, K. (2023). Greening industry: Opportunities and challenges in electricity
    access for Norwegian industry firms. *Journal of Cleaner Production.*
Ministry of Petroleum and Energy. (2020). Langskip—fangst og lagring av CO_2. Meld. St. 33
    (2019–2020) Melding til Stortinget. https://www.regjeringen.no/contentassets/943cb244091d
    4b2fb3782f395d69b05b/nn-no/pdfs/stm201920200033000dddpdfs.pdf
Montel. (2022). Regjeringen: EU-krav til grønt hydrogen truer Norges satsing. https://www.montel
    news.com/no/news/1317323/regjeringen-eu-krav-til-grnt-hydrogen-truer-norges-satsing
NEL. (2022). Electrolysers & fueling stations. https://nelhydrogen.com/about/
Nordic Co-operation. (2022). Clear to proceed—green shipping corridors in the Nordic region.
    https://www.norden.org/en/news/clear-proceed-green-shipping-corridors-nordic-region
NFR. (2020). Status for hydrogensforskning i Norge. https://www.forskningsradet.no/utlysninger/
    hydrogensatsing-2021/status-for-hydrogenforskning-i-norge/NOU, Norwegian Official Publi-
    cation, 2012. Energiutredningen—verdiskaping, forsyningssikkerhet og miljø. NOU 2012:9,
    Oslo. Ministry of Petroleum and Energy, 2012. https://www.regjeingen.no/contetsets/eb90bf
    50e63b4df7ae472b75a1d4a71c/no/pdfs/nou201220120009000dddpdfs.pdf
Norad. (2022). Norad støtter grønn hydrogen i Afrika. https://www.norad.no/aktuelt/nyheter/2022/
    norad-stotter-gronn-hydrogen-i-afrika/
Norwegian Government. (2016). Kraft til endring. Energipolitikken mot 2030. *Meld.St.25*
    (2015–2016). Oslo, April 15, 2016. https://www.regjeringen.no/no/dokumenter/meld.-st.-25-
    20152016/id2482952/
Norwegian Government. (2019a). Regjeringens handlingsplan for grønn skipsfart. https://www.
    regjeringen.no/contentassets/2ccd2f4e14d44bc88c93ac4effe78b2f/handlingsplan-for-gronn-
    skipsfart.pdf
Norwegian Government. (2019b). Plan for fossilfri kollektivtrafikk i 2025. https://www.regjer
    ingen.no/contentassets/383ec46d92b54c02af488558e2dbe0c1/handlingsplan-for-fossilfri-kol
    lektivtransport.pdf
Norwegian Government. (2020). The Government's hydrogen strategy. https://www.regjeringen.
    no/contentassets/40026db2148e41eda8e3792d259efb6b/y-0127b.pdf
Norwegian Government. (2021a). Energi til arbeid—langsiktig verdiskapning fra norske energires-
    surser. *Meld. St.36* (2020–2021). Oslo: June 11, 2021. https://www.regjeringen.no/no/dokume
    nter/meld.-st.-36-20202021/id2860081/
Norwegian Government. (2021b). FuelEU Maritime. https://www.regjeringen.no/no/sub/eos-not
    atbasen/notatene/2021/jan/fueleu-maritime/id2828701/
Norwegian Government. (2022a). Meld. St.11 (2021–2022) Tilleggsmelding til *Meld.St.*36. https://
    www.regjeringen.no/contentassets/e38e9f5393fc4f109b6394f61bd750f8/no/pdfs/stm202120
    220011000dddpdfs.pdf
Norwegian Government. (2022b). Slik skal Norge bli en grønn energikjempe. Press Release, June
    23, 2022. https://www.regjeringen.no/no/aktuelt/slik-skal-norge-bli-en-gronn-industrikjempe/
    id2920522/
Norwegian Government. (2022c). Grønn industrikjempe—Tiltak for veikart for grønt indus-
    triløft. https://www.regjeringen.no/contentassets/1c3d3319e6a946f2b57633c0c5fcc25b/tiltak-
    i-veikart-for-gront-industriloft.pdf
Norwegian Government. (2022d). Norwegian/German energy partnership. Press Release,
    June 13, 2022. https://www.regjeringen.no/no/aktuelt/styrket-hydrogensamarbeid-med-nord-
    tyskland/id2918423/
Norwegian Government. (2022e). Prosessindustrien. https://www.regjeringen.no/no/tema/naring
    sliv/gront-industriloft/prosessindustrien/id2920300/

Norwegian Government. (2022f). Joint statement Germany–Norway. https://www.regjeringen.no/contentassets/e895093224b641199d68e4cb5c73ae79/joint-statement-germany-norway.pdf
Norwegian Hydrogen Forum. (2023). Små hydrogendrypp i revidert nasjonalbudsjett. https://www.hydrogen.no/aktuelt/nyheter/sma-hydrogendrypp-i-revidert-nasjonalbudsjett. Accessed 05.09.2023.
Norwegian Maritime Authority. (2023). Mileåæl mot amoniakk-godkjenning. 07.06.2023. https://www.sdir.no/aktuelt/nyheter/forelopig-godkjent-for-ammoniakk-som-drivstoff/
Norwegian Parliament. (2022). Norge og EUs nye strategi for energisamarbeid. https://www.stortinget.no/no/Hva-skjer-pa-Stortinget/EU-EOS-informasjon/EU-EOS-nytt/2022/eueos-nytt---25.-mai-2022/norge-og-eus-nye-strategi-for-energisamarbeid/
NRK. (2021). Ny studie: Blått hydrogen er ikke noen klimaredning. September 17, 2022. https://www.nrk.no/tromsogfinnmark/ny-studie-slar-fast-at-blatt-hydrogen-er-ikke-noen-klimaredning-_-norge-mener-det-er-miljovennlig-1.15645150NTB, 2023. Hydrogen blir for dyrt. https://kommunikasjon.ntb.no/pressemelding/hydrogen-blir-for-dyrt?publisherId=17824355&releaseId=17971515 Accessed 05.09. 2023.
Reuters. (2023). Germany's updated hydrogen strategy sees heavy reliance on imported fuel in future, 26 July. www.reuters.com/business/energy/german-cabinet-approves-updated-national-hydrogen-strategy-2023-07-26/
Sand, M. (2021). Hva er hydrogen? Klima/CICERO, January 18, 2021. https://cicero.oslo.no/no/posts/klima/hva-er-hydrogen
SINTEF. (2020). Largescale hydrogen production in Norway-possible transition pathways towards 2050 (S. Damman, E. Sandberg, E. Rosenberg, P. Pisciella, U. Johansen). https://ntnuopen.ntnu.no/ntnu-xmlui/bitstream/handle/11250/2649737/Final%2Breport%2B2020-00179.pdf
Skjærseth, J. B. (2023). The war in Ukraine and EU climate leadership. *Czech Journal of International Relations, 58*(2), 93–107.
Skjærseth, J. B.,& Rosendal, K. (2022). Implementing the EU renewable energy directive in Norway: From tailwind to headwind. *Environmental Politics.* doi:https://doi.org/10.1080/09644016.2022.2075153 (with K. Rosendal).
Statkraft. (2023). Nel and Statkraft pave the way for a green hydrogen value chain in Norway. 06.01.2023. https://www.statkraft.com/newsroom/news-and-stories/2023/nel-and-statkraft-pave-the-way-for-a-green-hydrogen-value-chain-in-norway/
TU (Teknisk Ukeblad). (2019). Årsaken bak Sandvika-eksplosjonen. https://www.tu.no/artikler/slik-startet-lekkasjen-som-forte-til-hydrogen-eksplosjonen-i-sandvika/468765
TU. (2020a). Svært uferdig strategi: Regjeringen vet ikke hva den vil med hydrogen. https://www.tu.no/artikler/svaert-uferdig-strategi-regjeringen-vet-ikke-hva-den-vil-med-hydrogen/493332
TU. (2020b). Yaras planer: Fra gass til (mye) strøm. https://www.tu.no/artikler/yaras-planer-fra-gass-til-mye-strom/503987
TU. (2020c). EU varsler storsatsning på hydrogen. https://www.tu.no/artikler/eu-varsler-storsatsing-pa-hydrogen-glemmer-norge/493327
TU. (2021). Nettmangel i bergensområdet: Ti industriprosjekter har fått nei. https://www.tu.no/artikler/kraftmangel-i-bergensomradet-ti-industriprosjekter-har-fatt-nei/507748?key=KHt575F4
TU. (2022). Statkraft og Aker trekker seg fra Yaras prestisjeprosjekt. https://www.tu.no/artikler/statkraft-og-aker-trekker-seg-fra-yaras-prestisjeprosjekt/518447
Ursin Reed, Eilif. (2021). Historien om hydrogen. *Magasinet KLIMA,* January 12, 2021. https://cicero.oslo.no/no/artikler/historien-om-hydrogen
Yara. (2022). Corporate release, January 28, 2022. https://www.yara.com/corporate-releases/yara-and-linde-engineering-agree-to-build-a-24-mw-green-hydrogen-demonstration-plant-in-norway.-both-companies-aim-to-achieve-a-significant-carbon-dioxide-reduction-in-the-production-of-fertilizers-in-norway/
Yara. (2023). Enabling the hydrogen economy. https://www.yara.com/yara-clean-ammonia/

Jon Birger Skjærseth is Research Professor at the Fridtjof Nansen Institute. His research interests include international environmental cooperation, European climate and energy policies and corporate strategies.

Per Ove Eikeland is Senior Researcher at the Fridtjof Nansen Institute. His research interests include European climate, energy and technology policies and energy sector corporate transition strategies.

Tor Håkon Jackson Inderberg is Research Professor at the Fridtjof Nansen Institute. He works with energy and climate policies in wide jurisdictional settings and has research interests in energy transition, political feasibility, and anchoring policies.

Mari Lie Larsen is a Researcher at the Fridtjof Nansen Institute. Her research interests are within the fields of global environmental governance, sustainable resource management and energy transition strategies.

The Geopolitics of Hydrogen in Europe: The Interplay between EU and Member State Policies

Rainer Quitzow and Yana Zabanova

Abstract Drawing on the findings of the case studies presented in this edited volume, this final chapter summarizes and discusses the geopolitical challenges of hydrogen development in the European Union. The chapter provides a review of how the interplay of national and EU-level politics and policies is shaping the EU's domestic and international hydrogen policy. It presents key insights from the evolution of hydrogen policy in the EU, as well as at the national level in Germany, France, Poland, Hungary, Spain, Italy, the Netherlands, Sweden, and Norway (as a member of the European Economic Area). After reviewing important commonalities and differences across these cases, the chapter examines their interplay with policies at the EU level as well as potential synergies and sources of tension between the selected countries. It discusses how domestic politics and energy policy legacies shape differing policy approaches and priorities, including chosen technology pathways for hydrogen production, priority hydrogen uses and positions towards the development of cross-border infrastructure and trade. The chapter concludes with a reflection on how Europe's strengths and vulnerabilities shape its role in the global geopolitics of hydrogen and inform its international engagement on the transition to net zero more broadly.

1 Geopolitical Challenges of Hydrogen in the EU

Hydrogen has been in the global spotlight, and nations and blocs around the world are rushing to stake their claim in the future hydrogen economy. As a region, Europe has been at the forefront of these developments. The EU and a number of its Member States are investing large amounts of public resources in hydrogen R&D and early

R. Quitzow · Y. Zabanova (✉)
Research Institute for Sustainability (RIFS), Helmholtz Centre Potsdam, Berliner Str. 130, 14467 Potsdam, Germany
e-mail: yana.zabanova@rifs-potsdam.de

R. Quitzow
Technische Universität Berlin, Strasse des 17. Juni 135, 10623 Berlin, Germany

© Helmholtz-Zentrum Potsdam, Deutsches GeoForschungsZentrum GFZ 2024
R. Quitzow and Y. Zabanova (eds.), *The Geopolitics of Hydrogen*, Studies in Energy, Resource and Environmental Economics,
https://doi.org/10.1007/978-3-031-59515-8_12

industrial deployment. They are developing ambitious policies and are collectively shaping international hydrogen governance. The EU and selected Member States, most notably Germany, are also making hydrogen a prominent component of their external energy and climate policy, eager to develop hydrogen partnerships with a wide range of partners.

The reason for this is clear. While the EU has been a global frontrunner in energy and climate policy, it also exhibits important vulnerabilities within a future net-zero economy. Most significantly, Europe as a continent is scarce in renewable energy resources (Eicke & De Blasio, 2022). The EU—without the relatively abundant wind, hydropower and biomass resources in the UK, Norway and Ukraine, respectively—is even more constrained. Hence, developing reliable access to renewable hydrogen resources for the future development of the EU's net-zero economy is a crucial pillar of its future energy and economic security (Ansari & Pepe, 2023; Quitzow, Mewes et al., 2023). In this, the EU differs from the US and China who have the potential to produce sufficient domestic renewable energy to decarbonize their economies.

While only a few years ago, the EU relied on the efficiency of open global markets to secure the needed resources to ensure its standing in a global economy, this has fundamentally changed following, first, the supply chain crises of the Covid-19 pandemic and then the energy crisis in the wake of the invasion of Ukraine by Russia (Quitzow et al., 2022). This not only raises the stakes for developing a diversified and reliable supply of renewable hydrogen. It also further highlights the need to develop a leadership position in hydrogen and other net-zero technologies. Given the EU's scarcity in renewable resources, this is the only pathway towards securing a strong economic stake in a climate-friendly energy future (Quitzow, Triki et al., 2023). But also here, the EU faces vulnerabilities. After Europe led the world in the build-up of renewable energy markets in the 2000s, China has emerged as the undisputed leader in solar energy technologies, while Europe is struggling to preserve its competitive edge in the wind industry (Quitzow & Hughes, 2018; Quitzow et al., 2017). This further underlines the urgency to retain Europe's leadership in hydrogen-related technologies.

Moreover, these developments are raising awareness within the EU and its Member States that the EU's past reliance on open markets—and its ability to influence these markets via its unrivalled regulatory power—has reached its limits. Concepts like open strategic autonomy have emerged in the effort to reconcile the EU's prized single market as the driver of prosperity within a liberal economic order, on the one hand, with the need to secure its economic, technological and energy security in an increasingly hostile and insecure geopolitical context, on the other (Helwig & Sinkkonen, 2022; Miró, 2023; Prontera & Quitzow, 2022). The EU faces the difficult task of maintaining the benefits of European economic integration, while responding to the challenges coming from the state-dominated Chinese economic model as well as the emerging subsidy-driven climate and energy policy in the US (Kleimann et al., 2023).

This delicate balance is also reflected in EU hydrogen policy. The EU's nature as a supranational entity and a union of 27 Member States is both an asset and a liability in this context. The EU's large and attractive single market is essential for generating the needed economies of scale for its innovative electrolyzer manufacturers, and

EU regulatory development retains an important role in setting global trends. There is a developed cross-border gas infrastructure, which can be repurposed to carry hydrogen. Its well-developed R&D funding landscape and public–private cooperation on technology and standards are also key assets. All these factors make the EU more than the sum of its parts and help place the bloc on a strong footing vis-a-vis other large players such as the US or China.

However, it is increasingly clear that this will not suffice to sustain its competitive advantage in an emerging hydrogen economy. The EU's limited ability to generate its own resources and the need to find a compromise among 27 Member States on key policy directions have undercut the ambition of some of its plans, resulting in internal disagreements and delays (Quitzow, Triki et al., 2023). This is particularly pronounced in all energy-related areas, where the EU lacks the competence to interfere with national decision-making. In this vein, the interests and positions among the EU Member States are crucial to the development of the EU hydrogen policy, emerging at the interface between national and EU-level decision-making.

Building on the case studies presented in this book, this final chapter provides a review of how this interplay of national and EU-level politics and policies are shaping the EU's domestic and international hydrogen policy. It highlights how the positions and interests of key Member States are both enabling and constraining the EU's ambitions to develop its position as a major player in a global hydrogen economy. It draws on in-depth insights on the evolution of EU-level policy development and national developments in Germany, France, Poland, Hungary, Spain, Italy, the Netherlands, Sweden, as well as Norway. Though Norway is not an EU member state, it is a member of the European Economic Area and is closely aligned with EU energy policy. Its abundant gas and renewable energy resources give Norway a central role in European hydrogen developments. After reviewing important commonalities and differences across these cases, the chapter will examine their interplay with policies at the EU level. It then identifies potential synergies as well as sources of tension between the selected countries. Finally, it discusses implications for Europe's role in the emerging global hydrogen economy.

2 Domestic Politics and Energy Policy Legacies

While the various Member States discussed in this volume are developing—at different paces—their own national hydrogen policy visions and strategies, these policy trajectories are embedded in pre-existing energy policy legacies, domestic climate and energy politics as well as broader political developments shaping their relationship to the EU and other Member States. These play a decisive role in shaping emerging policy approaches.

France, where nuclear power has a high political and symbolic importance, has been lobbying hard for including nuclear-based technologies and hydrogen as strategically important technologies at the EU level. France's desire to provide state support to its nuclear power plants has led to a row with Germany over the final shape of the electricity market reform in the EU, a key element of the EU's Fit-for-55 policy package (Simon & Kurmayer, 2023). Its insistence on energy sovereignty has created an aversion to imports and trade, including its intra-European dimension.

In Poland and Hungary, right-wing populist governments—before the change in government in Poland after its 2023 elections—have frequently confronted the European Commission and disrupted EU decision-making on climate-related issues (Huber et al., 2021). Both countries voted against the updated Renewable Energy Directive (RED III), citing concerns that compliance with the more ambitious targets would place too heavy a burden on their economies (Council of the European Union, 2023, p. 4). In addition, challenges to democratic governance and the rule of law have constrained their access to funds from the EU's Recovery and Resilience Facility (RRF). As the RRF is a major source of funding for green transition measures, including hydrogen-related ones, this has a direct impact on the development of a hydrogen economy in these countries.[1] Yet, despite important commonalities between Hungary and Poland, their positions on the relationship to Russia and related energy security issues are at opposite ends of the spectrum (Kopper et al., 2023; Szabo & Fabok, 2020). Poland has been very critical of European dependence on Russian natural gas and is currently not willing to entertain the prospect of new hydrogen import dependencies. Hungary in turn has gone out of its way to retain its energy ties to Russia and is considering blue hydrogen production based on Russian natural gas.

In Italy, the economic crises of recent years have led the government to roll back support for renewable projects, resulting in lagging deployment rates. This is further exacerbated by cumbersome local permitting procedures and emerging public resistance to wind energy deployment. Public opposition related to land use issues or other concerns is common in other countries, too. Norway, in particular, has witnessed a public backlash against wind power deployment, as citizens are concerned that large wind farms would negatively affect the country's prized natural landscapes, endanger the local fauna, and disrupt the indigenous lifestyles of Sami reindeer herders. Coupled with its abundant natural gas reserves, this has made Norway a strong proponent of blue hydrogen as it builds hydrogen partnerships with its Northern European neighbors, in particular Germany. In the Netherlands, the environmental legislation aimed at tackling the country's high nitrogen emissions from agriculture and the construction industry has repeatedly led to protests by farmers and construction workers. It also emerged as a serious legal obstacle to the construction of Porthos, the country's largest and most important CO_2 transport and storage project at the Port of Rotterdam with implications for blue hydrogen production. In the end, the project did receive the green light in August 2023 following a ruling by the Dutch Council of State.

Spain, on the other hand, sees its large renewable energy potential as an important asset within a future net-zero economy, both for the development of exports and for the build-up of climate-friendly industrial production. Despite its limited renewable energy resources, Germany has a similarly positive stance, reflecting its

---

[1] The change of government in Poland following the autumn 2023 elections will bring important changes in this regard.

strong renewable energy policy legacy and public opinion opposing both nuclear and carbon capture and storage technologies (Brunnengräber & Di Nucci, 2014; Sturm, 2020). This, however, is counterbalanced by a vocal industry that is concerned about the availability of hydrogen, especially in the short term. This has created openings for more flexible German positions. Sweden, due to a combination of ample renewable potential, the continued use of nuclear power and a robust industrial base, is interested in developing renewable and nuclear-based hydrogen production but lacks strong domestic drivers in support of hydrogen trade.

3 Competing Hydrogen Technology Pathways in Europe

As already alluded to, these distinct policy legacies and domestic political constellations have translated not only into different hydrogen strategies but also at times diverging positions at the EU-level. The European Commission has invested significant political capital in prioritizing renewable hydrogen both for domestic production and import. As part of this effort, it has developed a stringent set of criteria for renewable hydrogen and derivatives within its Renewable Energy Directives (RED) and related delegated acts (Directorate-General for Energy, 2023). By contrast, nuclear-based hydrogen or blue hydrogen produced with natural gas with carbon capture and storage still lack a detailed definition in the EU regulatory framework. The only benchmark is included in the Gas Markets and Hydrogen Directive adopted in May 2024, which states that low-carbon hydrogen should ensure GHG emissions savings of at least 70% as compared to the use of unabated fossil fuel alternatives. The methodology for calculating emissions savings is still under development and will be adopted in a separate delegated act at a later stage (Martin, 2024).

This EU policy approach, however, is met by a diverse set of technology preferences among individual member states. Spain, with its large renewable energy endowment and a developed renewable energy industry, has consistently advocated an emphasis on renewable hydrogen—complete with strict sustainability requirements. Driven by its ambitious climate agenda—and the imperative to decarbonize its hard-to-abate industrial sectors—Germany has also supported renewable hydrogen. It has also been apprehensive about the potential role of blue hydrogen. This stems both from its major miscalculation regarding its reliance on natural gas from Russia as well as its longstanding scepticism about carbon capture and storage technologies. Nevertheless, the prospects of sourcing such hydrogen from Norway, considered a safe partner, have led to an increasing willingness to not only accept but even support the use of blue hydrogen in Germany.

Many states with a strong role of fossil fuels—including Hungary, Italy and Norway—envision a much more prominent role for blue hydrogen. Hungary, which is an EU outlier in continuing to maintain close energy relations with Russia, aims at blue hydrogen production at the scale that even somewhat exceeds the projected levels of renewable and nuclear hydrogen production. Italy is also interested both in renewable and blue hydrogen. Norway, in turn, has a two-pronged strategy. It aims to export blue hydrogen—or even natural gas as a feedstock for blue hydrogen—to the EU, while producing green hydrogen for domestic decarbonization. Germany's gradually growing openness to considering blue hydrogen imports serves as a reinforcing factor in these plans. In addition, the demand targets adopted by the EU as part of the updated Renewable Energy Directive (RED III) in October 2023 are likely to increase demand for hydrogen-based maritime fuels, which matches Norway's interest in providing green solutions for the maritime sector.

Finally, there is France, which has been leading efforts in the EU to endorse a special status for nuclear-produced hydrogen. When the EU was still debating the definition of renewable hydrogen, France mobilized a coalition of allies—including Finland and selected Central and Eastern European countries, where nuclear power contributes a significant share of electricity production—to support nuclear-based hydrogen. In response, Austria led the formation of an informal rival "renewable" alliance. The European Commission has not gone so far as to recognize nuclear-based hydrogen as renewable. However, France succeeded in obtaining significant concessions. For instance, the RED III's 42% renewable hydrogen use target in industry by 2030 can be discounted by 20% if the share of hydrogen from fossil fuels in the member state does not exceed 23% in 2030 and 20% in 2035 (Martin, 2023b). In other words, if a Member State uses a sufficiently large share of non-fossil (e.g., nuclear-based) hydrogen, it will only have to meet a reduced target of renewable hydrogen use in industry. Hungary with its strong reliance on nuclear power generation and plans for expanding its nuclear fleet is also interested in having nuclear-based hydrogen play an important role.

These diverging positions are part of a larger debate on the role of nuclear energy in the decarbonization of Europe's energy sector. France in particular has been leading efforts in gaining recognition for nuclear power as a strategically important net-zero technology eligible for support at the European level. These efforts have been opposed by Germany, Austria and several other Member States. It was largely through French efforts that the European Parliament and the Council of Ministers reinstated nuclear power on the list of "strategically important technologies" in the Net-Zero Industry Act in December 2023, in contrast to the Commission's original proposal dated March 2023 (Messad, 2023).

4 Prioritizing Different Types of Hydrogen Use in the EU

At the EU-level, the use of hydrogen to decarbonize the industrial sector is clearly the priority. The revised Renewable Energy Directive adopted in October 2023 (RED III) includes an ambitious target of 42% of renewable hydrogen use (as a fuel and feedstock) in industry by 2030. The transport target, by contrast, is much more modest, aiming for only 1% of renewable fuels of non-biological origin (RFNBO) by 2030. Over the past years, the range of priority uses identified has been reduced, as there is a growing concern that the scaling-up of renewable hydrogen will take considerably longer than initially expected and than suggested by current targets for the year 2030 (Martin, 2023a, 2023b). As a result, a number of hydrogen uses considered in the past—such as in residential heating, power storage or personal mobility—are now increasingly viewed as inefficient, given expected supply constraints.

Yet at the Member State level, the clear prioritization that is evident in EU policies is often missing. Some countries do largely follow the EU's priorities. These include Germany and the Netherlands with their focus on the steel and chemical industry as well as Sweden, which is intent on decarbonizing steel production. Spain, in turn, combines a strong emphasis on decarbonizing existing uses—such as hydrogen use in refineries and the fertilizer sector—but it is also supporting the steel industry. Italy and France also prioritize decarbonizing existing uses, such as in refineries and ammonia production.

Mobility has been a more contentious arena. While support for hydrogen-based mobility was a common element of national support schemes in Europe in the past, there is growing evidence that battery electric mobility may well come to dominate even sectors previously viewed as hard-to-electrify, such as heavy-duty trucks. Still, both France and Italy maintain a strong focus on hydrogen for mobility. This includes trains in Italy's case and light duty vehicles in France, which are meant to complement battery electric mobility. Furthermore, a number of countries, especially those where hydrogen ambitions are relatively recent, either list a very wide range of potential applications without clear prioritization or are facing significant domestic divisions regarding the role of hydrogen in the future. Some examples in the first group include Poland and Hungary. Finally, Spain faces a slightly different question. It has to consider the trade-offs of concentrating on decarbonizing its domestic industry, which was the original motivation behind its hydrogen strategy, and developing exports of renewable hydrogen to the rest of Europe.

5 Renewable Energy Deployment and Hydrogen: Up to Speed?

Producing renewable hydrogen at scale requires large amounts of renewable energy capacity installed and accelerated deployment rates. Yet the degree to which hydrogen plans and ambitions are integrated into energy and climate policy planning at the national level differs significantly from country to country. In a number of cases, hydrogen ambitions are essentially declarative and are not reflected in the plans for adequate expansion of the renewable energy capacity. Poland, which lists a wide range of potential hydrogen uses, is a case in point. Introduced in 2016, the highly restrictive distancing rules for onshore wind ("10 h") have largely stalled wind power development in the country. The rules have since been revised only slightly, falling short of the hopes of energy transition supporters. In addition, Poland's right-wing Law and Justice (PiS) government, in power between 2015 and 2023, was generally sluggish on energy transition. It placed itself in opposition to the EU's ambitious goals in the Fit-for-55 package, going as far as challenging some of its aspects in the European Court of Justice (Weise & Posaner, 2023). Italy has also suffered from a slow pace in developing new renewable energy capacity for years. In the wake of the economic crisis in the mid-2010s, the government dismantled several support measures for renewable energy deployment (Prontera, 2021). Sweden also faces the challenge of integrating its hydrogen plans with the rest of the energy sector. Sweden's hydrogen plans have been driven by business actors, including very prominently the steel industry, with a lack of coordination and steering efforts by the government. While Sweden has abundant renewable energy resources, there is still a question of where new renewable energy capacity (such as wind power) will be built and which actors will get preferential access to hydrogen production sites. Finally, there is the issue of technology competition. Even in Sweden and Norway—two countries with the most decarbonized grids in Europe and ambitious national climate policies–demand for green electricity is expected to rise steeply in coming years. This raises the question of how much renewable energy will be available for green hydrogen production, given the growing electrification demands (Kilpeläinen et al., 2023).

6   Funding Hydrogen Policy in the EU: Up to the Task?

A key source of contention in the EU has been about the funding for green industrial policy measures, especially against the backdrop of the intensifying subsidy race internationally. The Recovery and Resilience Facility (RRF), the central part of the NextGenerationEU pandemic recovery package, has emerged as an important source of funding for the green transition in Member States. RRF funding includes approximately 10 billion EUR for hydrogen development in Member States (European Commission, 2023). In some countries, such as Poland or Hungary, the opportunity to receive EU funding was a strong motivator behind these nations' interest in clean hydrogen development. Even though RRF funds come from joint EU borrowing, they are allocated and spent at the Member State level with little coordination taking place.

Despite the intensifying cleantech competition following the United States' Inflation Reduction Act (2022), the EU has not taken significant steps towards scaling-up EU-level funding. While some Member States have called for more external debt to better equip the EU to face the key challenges, leading countries like Germany and several Nordic states have opposed new EU-level borrowing at a large scale. Instead, they have called for a more efficient use of remaining funds in the NextGenerationEU package (Martinez & Strupczewski, 2023). As a result, the Commission's attempt to create a "European Sovereignty Fund" to fund investments in net-zero technologies did not materialize and has been replaced by the strongly reduced concept of a Strategic Technologies for Europe Platform with hardly any new funding (Bourgery-Gonse, 2023).

Instead, the EU's main response to the global subsidy race has been to relax its traditionally strict rules on granting national subsidies (called state aids) (Quitzow, Triki et al., 2023). This has been supported by countries like Germany and France, which see this as an opportunity to support domestic industry. Available data on spending corroborate the highly unequal distribution of subsidies. Under the Temporary Crisis Framework 2022, which allowed Member States to grant state aids to help overcome the economic impact of the war in Ukraine, Germany alone accounted for 53% of the total volume of state aid approved, France for 24% and Italy for 7% (Allenbach-Ammann, 2023). The resulting risk is that new subsidies will undermine the European single market and that less well-to-do countries will not be able to compete, skewing the playing field within the EU. For this reason, a number of Member States—including Central and Eastern European states but also Sweden, the Netherlands and Belgium—have repeatedly called on the Commission to proceed with utter caution. In their position paper dated February 2023, they warned of likely negative effects, such as "the fragmentation of internal market, harmful subsidy races and weakening of regional development" (Reuters, 2023). As the RRF funding is scheduled to run out in 2026, the issue of financing European industrial policy is likely to remain on the agenda for years to come.

7 The Politics of Connectivity: Hydrogen Infrastructure in the EU

To bring green hydrogen from renewables-rich sites—whether in the EU or from abroad—to demand centres, the EU will need to develop some form of cross-border hydrogen infrastructure. Constructing such infrastructure will require large private and public investments on the scale of many tens of billions of euros, with estimates routinely revised upwards. Among Member States, the momentum behind the push for a European hydrogen infrastructure comes mainly from the Member States with an interest in hydrogen trade and their gas transmission system operators (TSO). This includes potential importers (Germany), future hydrogen hubs (Netherlands, Italy) and hydrogen export hopefuls (Spain, Norway). These countries are already now mapping out plans for their national hydrogen backbones and their interconnections to neighbouring states.

This stands in contrast to French interests. Similar to the government's stance on gas interconnections with Spain (Crisan-Revol, 2017), France initially did not support plans to develop hydrogen infrastructure between the two countries. Rather, its aim is to establish its own hydrogen production based on domestic nuclear and renewable energy generation and to avoid exposure to imports of hydrogen from outside the EU altogether. In this vein, it opposed Spanish efforts to promote the H2Med pipeline. The planned project would connect the renewables-rich Iberian peninsula with central and northern Europe and is the linchpin of Spain's efforts to become a key hydrogen supplier for Europe. It is also strongly supported by Germany who sees it as the key to importing hydrogen, not only from Spain but also Morocco and potentially other North African countries, at a later stage. France finally agreed to the project, though only after plans were presented that would have rerouted the pipeline project via Italy, thus avoiding French territory altogether.

An important pioneer in pipeline development in Northern Europe is the Netherlands, which is targeting a role as a European hydrogen hub. Its state-owned company Gasunie has been given the task to develop a national hydrogen grid, consistent with the Dutch ambitions to position the country as a clean energy hub for Northwest Europe. The Netherlands is thus investing both in adapting port infrastructure and repurposing gas pipelines to import decarbonized fuels and to export hydrogen to neighbouring states such as Germany or Belgium. In October 2023, Gasunie launched the construction of the national backbone. Germany is following suit with its vision of a core hydrogen grid (*Wasserstoff-Kernnetz*), which would be some 9700 km long and consist of some 60% repurposed pipelines and 40% new pipelines. Importantly, Germany is emphasizing the future network's European dimension. It is required that projects within the *Wasserstoff-Kernnetz* have the status of a European Project of Common Interest (PCI) or Important Project of Common European Interest (IPCEI). Moreover, Germany and Norway have already launched concrete plans for the construction of pipelines, not only for the export of hydrogen to Germany but also the export of CO_2 to Norway for subsurface storage in the North Sea (Kilpeläinen et al., 2023).

Like the Netherlands, Italy hopes to become a hub for clean energy imports from the African continent. It has, therefore, shown interest in developing stronger infrastructure links with non-EU countries. However, given Italy's dependence on gas imports, there is a risk that these pipelines, while nominally required to be hydrogen-ready, are mainly used for gas imports for years to come. Overall, Italy's level of engagement does not match that of the Netherlands or Germany. Similarly, Sweden, which has significant export potential, has for now prioritized efforts to develop hydrogen for domestic decarbonization and localized use for the development of new, climate-friendly industries. The lack of a pre-existing gas distribution grid also favour this stance by the Scandinavian country (Kilpeläinen et al., 2023).

8   The Politics of International Hydrogen Trade

These positions on infrastructure development are mirrored in how Member States have approached the development of international hydrogen trade more broadly. In the REPowerEU (2022) package, the EU unveiled its plans to import large amounts of clean hydrogen and derivatives, in addition to increasing domestic production (European Commission, 2022). Unsurprisingly, this is most relevant to Member States positioning themselves as future importers or import hubs. Germany is expected to be the largest future hydrogen importer in the EU. In its updated national hydrogen strategy (BMWK, 2023), Germany estimates that it will need to import some 50–70% of its future hydrogen demand. It was in anticipation of these imports that Germany has initiated its national H2Global scheme to procure clean hydrogen and derivatives from abroad. The Netherlands, which has a significant energy-intensive industry, has also acknowledged that it will likely need hydrogen imports and has been investing into the relevant infrastructure.

Both countries have spearheaded European efforts to develop an international hydrogen market by investing actively in bilateral and multilateral hydrogen diplomacy. Germany has signed a wide range of agreements with nations all around the globe, including both likely early suppliers and hydrogen hopefuls that are only making their first steps. It has also developed a prominent import scheme for hydrogen and derivatives called H_2 Global. From the outset, the Netherlands became a partner in the H_2 Global scheme, and the Netherlands has been an important ally of Germany in the sphere of international hydrogen diplomacy. It has also signed similar agreements with a wide range of partners, both within Europe and beyond. German and Dutch hydrogen diplomacy and bilateral outreach have also preceded the EU's own fledgling efforts in this area. It was only in late 2022 that the EU signed cooperation agreements with several countries like Egypt, Namibia, Kazakhstan, and Japan. Since then, however, the pace has accelerated, as the EU began actively integrating hydrogen into its bilateral cooperation with a number of African and Latin American countries, as well as with Ukraine and Norway.

Next to Germany and the Netherlands, there are a number of hydrogen export hopefuls, most notably Spain and Norway, who support hydrogen trade, albeit with a more European approach. Spain, with its ambitious plans to supply hydrogen to other EU member states, has put a strong emphasis on cooperation with fellow Member States, including Portugal and Germany. It has also been critical of what it perceives to be Germany's choice to prioritize non-EU countries as partners in the emerging hydrogen market. Indeed, despite its geographic proximity to Morocco, Spain has not yet developed a strong engagement with its neighbor, preferring to concentrate on EU partners. As alluded to above, Norway has pursued an active policy to promote exports of blue hydrogen to Germany, accompanied by a corresponding pipeline project. It is not strongly supporting the development of renewable hydrogen exports, signalling the potential for future tensions with Germany's ambition to replace blue with renewable hydrogen.

On the other end of the spectrum is France. With its emphasis on energy sovereignty and nuclear-based hydrogen, it is opposed to large-scale hydrogen imports and has instead put an emphasis on the development of regional hydrogen valleys or clusters that pool demand locally. Similarly, its hydrogen diplomacy in non-EU countries focuses on the development of hydrogen for domestic use rather than exports. While this may contradict the German import-oriented diplomacy in principle, it can also be seen as complementary, filling a gap left by Germany's primary focus on supporting larger, export-oriented investments.

Poland has not positioned itself openly against hydrogen imports but has largely side-stepped the discussion in its domestic hydrogen strategy, as this could spark fears of new energy dependencies. This does not mean that such imports may not be necessary in the future to pursue the decarbonization of hard-to-abate sectors. However, the domestic politics is not conducive to a pro-active development of import-oriented policies. Other countries, like Sweden, lack gas infrastructure links to neighbouring countries. Its plans, therefore, prioritize on-site, domestic uses, coupled with visions to export green industrial goods produced with domestic clean hydrogen.

Finally, in selected countries, it is large corporate actors from the oil and gas sector that are engaged in international outreach, seeing this as a future substitute for their business in international oil and gas markets. In France, this compensates for the scepticism of their national governments with its emphasis on energy sovereignty. Leading energy companies like Engie and Total are actively involved in exploring opportunities for hydrogen development in Africa and other potential export regions. In Italy, the idea of imports is generally viewed positively. Similar to the past idea of becoming a natural gas hub—which did not materialize—Italy is now seeking to position itself as a hub for energy imports from the Southern Mediterranean region. However, instead of strong government-led diplomacy, Italian energy corporations, like Snam, Enel and Eni, are pursuing their own corporate hydrogen diplomacy in Latin America and Northern Africa.

9   European Hydrogen Politics: The Art of the Possible

The wide diversity of positions, approaches and priorities attached to hydrogen by Member States has been a challenge for hydrogen policy making in the EU, often resulting in delayed and complex regulatory outcomes and in a relatively slow-moving international hydrogen policy. However, the EU's unique nature as a supra-national body and its large single market can also be viewed as a source of synergies that need to be identified and proactively harnessed. For instance, some of the lower-income Southern Member States, such as Greece, Bulgaria and Romania, have excellent renewable energy potentials that remain underutilized. With the right amounts of investment, they could help produce competitively priced renewable hydrogen for the European market. Specialization along the hydrogen value chain could also generate benefits for the entire EU. For instance, Norway's emphasis on decarbonizing marine transport and development of clean marine fuels can spur cooperation in the Nordic region and help Member States achieve their climate goals in the area of maritime shipping.

More broadly, experimentation and policy learning at the national level can be an important asset for EU hydrogen policy. Several proactive Member States are already experimenting with support schemes that are yet to be adopted at the EU level. The Netherlands and Germany have introduced their national versions of carbon contracts for difference to decarbonize their industry. This is something the EU is planning to do as part of the ongoing reform of its emissions trading system (EU ETS). The Netherlands is laying the groundwork for meeting the RFNBO quotas for industry adopted as part of RED III by introducing so-called purchase obligations. Norway, as a member of the European Economic Area, is implementing national zero-emission procurement standards, which have the potential to stimulate demand for green, climate-friendly products. Similar public procurement schemes for low-carbon products have been under discussion in the EU but have not been introduced yet. Given the large volumes of public procurement within the Union, they could have a significant impact on creating lead markets for green products in Europe. Finally,

Germany's innovative hydrogen import scheme H2Global has an important role in testing and improving mechanisms for concluding offtake contracts. It is now being extended to other Member States, with plans to fully Europeanize the mechanism in the future.

Internationally, the EU can and does already benefit from the activities of selected Member States. Germany's successful track record of active international engagement on energy transition in the Global South (Quitzow & Thielges, 2022) and in emerging economies can prepare the ground for follow-up engagement on behalf of Team Europe (i.e., the EU, Member States and European financial institutions). Germany's national development bank, KfW, is a prominent financial institution and a partner in Team Europe initiatives. It invests into a hydrogen economy in several countries, including Chile. Similarly, France has launched efforts in a number of African countries, albeit with a primary focus on domestic hydrogen production and use. As alluded to above, this may offer an important complementarity to Germany's import-oriented strategy.

That said, harnessing synergies for clean hydrogen and the broader energy transition in the EU will require skilful governance. One of the most important challenges the EU is facing is the fact that its climate governance is not well-aligned with that of energy policy. While climate policy is a shared domain of the EU and Member States and has been strongly Europeanized since the 1990s, energy policies fall mainly within the remit of Member States (Pisani-Ferry et al., 2023, p. 3). According to Article 194 of the Treaty on the Functioning of the European Union, each Member State has the right to "determine the conditions for exploiting its energy resources, its choice between different energy sources and the general structure of its energy supply." In principle, hydrogen policy, whose main thrust has been to decarbonize hard-to-abate sectors, is rooted in European climate policy. At the same time, hydrogen development exhibits important overlaps with energy policy and thus leads to tensions between the two levels.

These governance challenges are a broader issue that affects not only the EU's climate and energy policy (Pisani-Ferry et al., 2023). One of the biggest obstacles to developing a strong green industrial policy in the EU is the lack of coordination (Tagliapietra et al., 2023). At the moment, instead of a unified green industrial policy, there is a fragmented landscape of various industrial policy initiatives at EU, member state and regional level. They may come into conflict with each other or even undermine the single market. The EU's recent turn towards relaxing guidelines on granting state aids pushes it further along this trajectory, as Member States implement national measures to protect and support their industry, with little to no coordination taking place (Quitzow, Triki et al., 2023).

This raises more fundamental, political questions regarding the limits of the EU in its current form as a vehicle for harnessing the opportunities of the hydrogen sector for building a net-zero economy. Countries with strong renewable energy potential have an interest in attracting investments, not only in the production of renewable hydrogen but also in new climate-friendly industrial production. Conversely, existing centres of industrial production, notably in Germany, are keen on retaining industrial value creation in existing locations. These diverging interests add to the manifold uncertainties surrounding the hydrogen economy, from the development of transport infrastructure to the prioritization of hydrogen uses or production pathways. As outlined in this volume, many of these aspects are subject to intense political negotiations within the European Union. In the sphere of investment support, the EU has not been able to assume a prominent role, instead relaxing its restrictions on state-aid as a means of relegating this to the level of Member States.

The result has been the concentration of investment support in those countries with the largest fiscal space (Quitzow, Nunez et al., 2023). Where this coincides with a relatively abundant potential for renewable energy generation, this points to a virtuous combination of assets, opening up a clear pathway to net-zero industrial production. Within the EU, Sweden represents such a case. In the absence of such a fortuitous convergence of fiscal and natural resources, however, this points to the need for political negotiation across countries over the distribution of costs and benefits within a future hydrogen economy. Such a politicization of supply chain development is not only likely to slow down the pace of investment and reduce the efficiency of final outcomes. It may also exacerbate pre-existing economic imbalances within the EU, with risks for political stability within the Union. A large-scale EU-level investment support scheme could overcome such risks and inefficiencies, but would likely suffer from the typical woes of EU policy making, i.e. complex yet highly politicized rules riddled with exceptions and exemptions catering to individual Member States. Nevertheless, even incremental steps in the direction of a European investment agenda are essential for securing the EU's leadership in a global hydrogen economy, while safeguarding the single market.

10   The EU in the Global Geopolitics of Hydrogen

Finally, these questions facing the European Union's internal governance closely mirror similar challenges at the global level. The EU and its Member States confront a similar set of questions when negotiating with potential hydrogen exporting countries. As these partners consider the promise of an international hydrogen economy, they will closely weigh the opportunities and risks and the future costs and benefits of renewable hydrogen production. For the time being, Germany has taken the lead in developing a host of bilateral partnerships focused on research and development, capacity building and knowledge and information sharing, while regulatory developments at the EU-level are largely defining the terms of hydrogen trade. Indeed, here the bloc's constrained renewable energy potential has pushed the EU and its largest Member State, Germany, to the forefront of the international hydrogen economy. Even US renewable hydrogen standards have followed key elements of the EU regulatory regime, providing an important indication of a continued "Brussels effect" (Bradford, 2019).

Indeed, the European Union's role as a renewable energy scarce region may well position it as a champion of international trade and continued economic interdependence. While an important vulnerability to be carefully managed, this also provides a strong impetus for international engagement that offers important synergies with other aspects of its international agenda. European demand for hydrogen from the Global South is not only a liability, but also an entry-point for broader engagement with these countries. Hydrogen trade can function as a vehicle for engaging on broader questions of decarbonization as well as cooperation on technology or critical minerals. Not least, it promises to position European technology suppliers as potential leaders in supplying its trading partners with hydrogen-related technologies, both upstream and downstream. Coupled with a strong and coordinated European approach to climate finance, this could provide the EU with important leverage in pursuing not only its narrow economic interests but also the development of a robust framework for promoting global decarbonization efforts.

Acknowledgements Research for this chapter was financially supported by the German Federal Foreign Office within the framework of the project "Geopolitics of the Energy Transformation— Implications of an International Hydrogen Economy" (GET Hydrogen), funding reference number AA4521G125.

Literature

Allenbach-Ammann, J. (2023, January 13). EU commission's Vestager proposes change to state aid rules. *Euractiv*. https://www.euractiv.com/section/economy-jobs/news/eu-commissions-ves tager-proposes-change-to-state-aid-rules/
Ansari, D., & Pepe, J. M. (2023). *Toward a hydrogen import strategy for Germany and the EU: Priorities, countries, and multilateral frameworks* (SWP Working Paper).
BMWK. (2023). *National Hydrogen Strategy Update*. Federal ministry for economic affairs and climate action (BMWK).
Bourgery-Gonse, T. (2023, June 21). Commission 'annihilated symbolic value' of EU sovereignty fund, leading MEP says. *Euractiv*. https://www.euractiv.com/section/economy-jobs/news/com mission-annihilated-symbolic-value-of-eu-sovereignty-fund-leading-mep-says/
Bradford, A. (2019). *The Brussels Effect: How the European Union Rules the World*. Oxford University Press.
Brunnengräber, A., & Di Nucci, M. R. (Eds.). (2014). *Im Hürdenlauf zur Energiewende: Von Transformationen, Reformen und Innovationen*. Springer Fachmedien. https://doi.org/10.1007/978-3-658-06788-5
Council of the European Union. (2023, September 29). *Interinstitutional File: 2021/0218(COD), 'I/A' Item Note*. https://data.consilium.europa.eu/doc/document/ST-13188-2023-ADD-1-REV-2/en/pdf
Crisan-Revol, A. (2017). The SouthWest Europe Regional initiative to connect the Iberian Peninsula to the EU gas market. In C. Jones (Ed.), *EU Energy Law*, Vol. 11: *The role of gas in the EU's energy union*. Claeys & Casteels Law Publishing.
Directorate-General for Energy. (2023, June 20). *Renewable hydrogen production: New rules formally adopted*. European Commission. https://energy.ec.europa.eu/news/renewable-hyd rogen-production-new-rules-formally-adopted-2023-06-20_en
Eicke, L., & De Blasio, N. (2022). Green hydrogen value chains in the industrial sector—Geopolitical and market implications. *Energy Research & Social Science, 93*, 102847. https://doi.org/10.1016/j.erss.2022.102847
European Commission. (2022). *REPowerEU Plan* (COM/2022/230 final). https://eur-lex.europa.eu/legal-content/EN/TXT/?uri=COM%3A2022%3A230%3AFIN&qid=1653033742483

European Commission. (2023, May 11). *Speech of Timmermans at the World Hydrogen Summit 2023*. European Commission: Press Corner. https://ec.europa.eu/commission/presscorner/det ail/en/speech_23_2704

Helwig, N., & Sinkkonen, V. (2022). Strategic autonomy and the EU as a global actor: The evolution, debate and theory of a contested term. *European Foreign Affairs Review, 27* (Special). https://www.kluwerlawonline.com/api/Product/CitationPDFURL?file= Journals\EERR\EERR2022009.pdf

Huber, R. A., Maltby, T., Szulecki, K., & Ćetković, S. (2021). Is populism a challenge to European energy and climate policy? Empirical evidence across varieties of populism. *Journal of European Public Policy, 28*(7), 998–1017. https://doi.org/10.1080/13501763.2021.1918214

Kilpeläinen, S., Quitzow, R., & Tsoumpa, M. (2023). *Hydrogen in the Nordics—Drivers of European Cooperation?* Friedrich-Ebert-Stiftung.

Kleimann, D., Poitiers, N., Sapir, A., Véron, N., Veugelers, R., & Zettelmeyer, J. (2023). *How Europe should answer the US Inflation Reduction Act* (Issue 04/23; Policy Contribution). Bruegel.

Kopper, A., Szalai, A., & Góra, M. (2023). Populist foreign policy in central and eastern Europe: Poland, Hungary and the shock of the Ukraine crisis. In P. Giurlando & D. F. Wajner (Eds.), *Populist foreign policy: Regional perspectives of populism in the international scene* (pp. 89–116). Springer International Publishing. https://doi.org/10.1007/978-3-031-22773-8_4

Martin, P. (2023a, September 25). Cost of green hydrogen unlikely to fall 'dramatically' in coming years, admit developers. *Hydrogen Insight*. https://www.hydrogeninsight.com/production/cost-of-green-hydrogen-unlikely-to-fall-dramatically-in-coming-years-admit-developers/2-1-152 3281

Martin, P. (2023b, October 31). EU's 2030 targets for green hydrogen use in industry and transport become law with publication in official journal. *Hydrogen Insight*. https://www.hydrogeninsi ght.com/policy/eus-2030-targets-for-green-hydrogen-use-in-industry-and-transport-become-law-with-publication-in-official-journal/2-1-1545432

Martin, P. (2024, May 21). EU's hydrogen and low-carbon gas markets package to become law after sign-off from member states. *Hydrogen Insight*. https://www.hydrogeninsight.com/policy/eus-hydrogen-and-low-carbon-gas-markets-package-to-become-law-after-sign-off-from-mem ber-states/2-1-1646971

Martinez, M., & Strupczewski, J. (2023, February 16). Germany dashes hopes for new EU common borrowing. *Reuters*. https://www.reuters.com/markets/europe/germany-dashes-hopes-new-eu-common-borrowing-2023-02-16/

Messad, P. (2023, December 8). EU countries reinstate nuclear among 'strategic' net-zero technolo-gies. *Euractiv*. https://www.euractiv.com/section/energy-environment/news/eu-countries-reinst ate-nuclear-among-strategic-net-zero-technologies/

Miró, J. (2023). Responding to the global disorder: The EU's quest for open strategic autonomy. *Global Society, 37*(3), 315–335. https://doi.org/10.1080/13600826.2022.2110042

Pisani-Ferry, J., Tagliapietra, S., & Zachmann, G. (2023). *A new governance framework to safeguard the European green deal* [Bruegel Policy Brief].

Prontera, A. (2021). The dismantling of renewable energy policy in Italy. *Environmental Politics, 30*(7), 1196–1216. https://doi.org/10.1080/09644016.2020.1868837

Prontera, A., & Quitzow, R. (2022). The EU as catalytic state? Rethinking European climate and energy governance. *New Political Economy, 27*(3), 517–531. https://doi.org/10.1080/13563467. 2021.1994539

Quitzow, R., Huenteler, J., & Asmussen, H. (2017). Development trajectories in China's wind and solar energy industries: How technology-related differences shape the dynamics of industry localization and catching up. *Journal of Cleaner Production, 158*, 122–133. https://doi.org/10. 1016/j.jclepro.2017.04.130

Quitzow, R., & Hughes, L. (2018). Low-carbon technologies. *National Innovation Systems, and Global Production Networks: THe State of Play*. https://doi.org/10.4337/9781783475636.00030

Quitzow, R., Mewes, C., Thielges, S., Tsoumpa, M., & Zabanova, Y. (2023). *Building partnerships for an international hydrogen economy—Entry-points for European policy action* (FES Diskurs). Friedrich-Ebert-Stiftung.

Quitzow, R., Nunez, A., & Marian, A. (2023). Positioning Germany in an international hydrogen economy. *Energy Strategy Reviews, 53* (May 2024), 101361 https://doi.org/10.1016/j.esr.2024.101361

Quitzow, R., Renn, O., & Zabanova, Y. (2022). The crisis in Ukraine: Another missed opportunity for building a more sustainable economic paradigm. *GAIA–Ecological Perspectives for Science and Society, 31*(3), 135–138. https://doi.org/10.14512/gaia.31.3.2

Quitzow, R., & Thielges, S. (2022). The German energy transition as soft power. *Review of International Political Economy, 29*(2), 598–623. https://doi.org/10.1080/09692290.2020.1813190

Quitzow, R., Triki, A., Wachsmuth, J., Fragoso García, J., Kramer, N., Lux, B., & Nunez, A. (2023). *Mobilizing Europe's full hydrogen potential: Entry-points for action by the EU and its member states* (HYPAT Discussion Paper). Fraunhofer ISI. https://publica.fraunhofer.de/handle/publica/451548

Reuters. (2023, February 15). Eleven EU countries urge 'great caution' in loosening state aid rules. *Euractiv.* https://www.euractiv.com/section/economy-jobs/news/eleven-eu-countries-urge-great-caution-in-loosening-state-aid-rules/

Simon, F., & Kurmayer, N. J. (2023, October 19). Deal on EU electricity market reform: What did Paris and Berlin obtain? *Euractiv.* https://www.euractiv.com/section/electricity/news/deal-on-eu-electricity-market-reform-what-did-paris-and-berlin-obtain/

Sturm, C. (2020). *Inside the energiewende: Twists and Turns on Germany's soft energy path.* Springer Nature.

Szabo, J., & Fabok, M. (2020). Infrastructures and state-building: Comparing the energy politics of the European commission with the governments of Hungary and Poland. *Energy Policy, 138,* 111253. https://doi.org/10.1016/j.enpol.2020.111253

Tagliapietra, S., Veugelers, R., & Zettelmeyer, J. (2023). *Rebooting the European Union's Net Zero Industry Act.* Bruegel.

Weise, Z., & Posaner, J. (2023, June 12). Poland to challenge EU climate laws before top court. *POLITICO.* https://www.politico.eu/article/poland-challenge-eu-climate-laws-fit-for-55-before-european-union-court-justice-minister-anna-moskwa/